SPACE TELESCOPE SCIENCE INSTITUTE

SYMPOSIUM SERIES: 18

Series Editor S. Michael Fall, Space Telescope Science Institute

PLANETS TO COSMOLOGY:
ESSENTIAL SCIENCE IN THE FINAL YEARS OF THE *HUBBLE*
SPACE TELESCOPE

This volume is based on a meeting held at the Space Telescope Science Institute on 3–6 May 2004.

With some uncertainty concerning *Hubble*'s next Servicing Mission still hanging, identifying the most crucial science to be performed by this superb telescope has become of paramount importance. With this goal in mind, the symposium examined a wide range of topics at the forefront of astronomy and astrophysics.

This book represents a collection of review papers, written by world experts, with a special emphasis on future research.

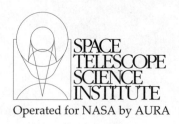

SPACE
TELESCOPE
SCIENCE
INSTITUTE
Operated for NASA by AURA

Other titles in the Space Telescope Science Institute Series.

Planets to Cosmology: Essential Science in the Final Years of the *Hubble Space Telescope*

Proceedings of the
Space Telescope Science Institute Symposium,
held in Baltimore, Maryland
May 3–6, 2004

Edited by
MARIO LIVIO
Space Telescope Science Institute, Baltimore, MD 21218, USA

STEFANO CASERTANO
Space Telescope Science Institute, Baltimore, MD 21218, USA

Published for the Space Telescope Science Institute

SPACE
TELESCOPE
SCIENCE
INSTITUTE
Operated for NASA by AURA

CAMBRIDGE
UNIVERSITY PRESS

CAMBRIDGE UNIVERSITY PRESS
Cambridge, New York, Melbourne, Madrid, Cape Town, Singapore, São Paulo

Cambridge University Press
The Edinburgh Building, Cambridge CB2 2RU, UK

Published in the United States of America by Cambridge University Press, New York

www.cambridge.org
Information on this title: www.cambridge.org/9780521847582

First published 2006

Printed in the United Kingdom at the University Press, Cambridge

A catalog record for this publication is available from the British Library

ISBN-13   978-0-521-84758-2 hardback
ISBN-10   0-521-84758-3 hardback

Contents

Participants

Afanas'ev, Sergei	Ioffe Physico-Technical Institute
A'Hearn, Mike	University of Maryland
Albrecht, Rudolf	European Space Agency
Arribas, Santiago	Space Telescope Science Institute
Bautista, Manuel	Instituto Venezolano de Investigaciones Científicas
Beckwith, Steve	Space Telescope Science Institute
Bell, Eric	Max Planck Institut für Astronomie
Blair, William	The Johns Hopkins University
Blandford, Roger	Stanford Linear Accelerator Center
Bromm, Volker	Space Telescope Science Institute
Bunker, Andrew	University of Exeter
Caldwell, John	Space Telescope Science Institute
Calvani, Humberto	The Johns Hopkins University
Calzetti, Daniela	Space Telescope Science Institute
Carpenter, Kenneth	NASA Goddard Space Flight Center
Casertano, Stefano	Space Telescope Science Institute
Chakrabarty, Dalia	Rutgers University
Challis, Peter	Harvard University
Chanial, Pierre	NASA Goddard Space Flight Center
Charbonneau, David	California Institute of Technology
Chiaberge, Marco	Istituto di Radioastronomia–CNR
Clampin, Mark	Space Telescope Science Institute
Clarke, Cathie	Institute of Astronomy–University of Cambridge
Crone, Mary	Skidmore College
Dalla Bontà, Elena	Padua University
de Mello, Duilia	NASA Goddard Space Flight Center
de Zeeuw, Timothy	Sterrewacht Leiden
Debes, John	Pennsylvania State University
Dijkstra, Mark	Columbia University
Doxsey, Rodger	Space Telescope Science Institute
Ebbets, Dennis	Ball Aerospace
Feldman, Paul	The Johns Hopkins University
Ferguson, Harry	Space Telescope Science Institute
Ferrarese, Laura	Rutgers University
Festou, Michel	Observatoire Midi Pyrénées
Floyd, David	Space Telescope Science Institute
Franx, Marijn	Leiden Observatory
French, Richard	Wellesley College
Freudling, Wolfram	Space Telescope–European Coordinating Facility
Gallagher, Jay	University of Wisconsin–Madison
Garcia, Javier	Instituto Venezolano de Investigaciones Científicas
García-Marín, Macarena	Instituto de Estructura de la Materia–CSIC
Giacconi, Riccardo	Associated Universities Inc.
Gilliland, Ronald	Space Telescope Science Institute
Godon, Patrick	Space Telescope Science Institute

Grebel, Eva	University of Basel
Gull, Theodore	NASA Goddard Space Flight Center
Haiman, Zoltan	Columbia University
Hartnett, Kevin	NASA Goddard Space Flight Center
Hasan, Hashima	NASA Headquarters
Hauser, Michael	Space Telescope Science Institute
Heap, Sara	NASA Goddard Space Flight Center
Huchra, John	Harvard-Smithsonian Center for Astrophysics
Huffman, Deborah	Fernbank Science Center
Infante, Leopoldo	P. Universidad Catolica de Chile
Jain, Bhuvnesh	University of Pennsylvania
Jaret, Steven	Fernbank Science Center
Jeletic, Jim	NASA Goddard Space Flight Center
Kamp, Inga	Space Telescope Science Institute
Kirshner, Robert	Harvard-Smithsonian Center for Astrophysics
Larsen, Søren	European Southern Observatory
Leckrone, David	NASA Goddard Space Flight Center
Livio, Mario	Space Telescope Science Institute
Macchetto, Duccio	Space Telescope Science Institute
Madrid, Juan	Space Telescope Science Institute
Maíz Apellániz, Jesús	Space Telescope Science Institute
Malhotra, Sangeeta	Space Telescope Science Institute
Margon, Bruce	Space Telescope Science Institute
Mathews, Grant	University of Notre Dame
Matters, Bonnie	NASA Goddard Space Flight Center
Mazzuca, Lisa	NASA Goddard Space Flight Center
McLean, Brian	Space Telescope Science Institute
McLeod, Kim	Whitin Observatory, Wellesley College
Meixner, Margaret	Space Telescope Science Institute
Meléndez, Marcio	Instituto Venezolano de Investigaciones Científicas
Meylan, Georges	Space Telescope Science Institute
Miller, Eric	University of Michigan
Miller, Lance	Oxford University
Nota, Antonella	Space Telescope Science Institute
O'Dowd, Matt	University of Melbourne
Papovich, Casey	Steward Observatory
Peterson, Bradley	Ohio State University
Puzia, Thomas	Space Telescope Science Institute
Reid, Iain	Space Telescope Science Institute
Richstone, Douglas	University of Michigan
Riess, Adam	Space Telescope Science Institute
Robberto, Massimo	Space Telescope Science Institute
Savage, Blair	University of Wisconsin
Schreier, Ethan	Associated Universities Inc.
Scowen, Paul	Arizona State University
Sembach, Kenneth	Space Telescope Science Institute
Shapley, Alice	University of California–Berkeley
Silverberg, Robert	NASA Goddard Space Flight Center
Somerville, Rachel	Space Telescope Science Institute
Sonneborn, George	NASA Goddard Space Flight Center

Sparks, William	Space Telescope Science Institute
Stanway, Elizabeth	Institute of Astronomy
Stiavelli, Massimo	Space Telescope Science Institute
Stocke, John	CASA, University of Colorado
Teplitz, Harry	Spitzer Science Center
Thien, Hilda	NASA Goddard Space Flight Center
Thompson, Rodger	University of Arizona/Steward Observatory
Tonry, John	University of Hawaii
Truran, James	University of Chicago
Tsvetanov, Zlatan	NASA Headquarters
Villaver, Eva	Space Telescope Science Institute
Weinberg, David	Ohio State University
Weymann, Ray	Carnegie Observatories
Wiseman, Jennifer	NASA Headquarters

Preface

The Space Telescope Science Institute Symposium on *Planets to Cosmology: Essential Science in the Final Years of the Hubble Space Telescope* took place during 3–6 May 2004.

These proceedings represent only a part of the invited talks that were presented at the symposium. We thank the contributing authors for preparing their manuscripts.

With some uncertainty concerning *Hubble*'s next Servicing Mission still hanging, identifying the most crucial science to be performed by this superb telescope has become of paramount importance. With this goal in mind, the symposium examined a wide range of topics at the forefront of astronomy and astrophysics. The result is a magnificent collection of results, with a special emphasis on future research.

We thank Sharon Toolan of ST ScI for her help in preparing this volume for publication.

Mario Livio
Stefano Casertano
Space Telescope Science Institute
Baltimore, Maryland

Hubble's view of transiting planets

By DAVID CHARBONNEAU

Harvard-Smithsonian Center for Astrophysics, 60 Garden Street, Cambridge, MA 02138, USA

The *Hubble Space Telescope* is uniquely able to study planets that are observed to transit their parent stars. The extremely stable platform afforded by an orbiting spacecraft, free from the contaminating effects of the Earth's atmosphere, enables *HST* to conduct ultra-high precision photometry and spectroscopy of known transiting extrasolar planet systems. Among *HST*'s list of successful observations of the first such system, HD 209458, are (1) the first detection of the atmosphere of an extrasolar planet, (2) the determination that gas is escaping from the planet, and (3) a search for Earth-sized satellites and circumplanetary rings. Numerous wide-field, ground-based transit surveys are poised to uncover a gaggle of new worlds for which *HST* may undertake similar studies, such as the newly-discovered planet TrES-1. With regard to the future of *Hubble*, it must be noted that it is the only observatory in existence capable of confirming transits of Earth-like planets that may be detected by NASA's *Kepler* mission. *Kepler* could reveal Earth-like transits by the year 2010, but without a servicing mission it is very unlikely that *HST* would still be in operation.

1. Introduction

When both the photometric transits and the radial velocity variations due to an extrasolar planet are observed, we are granted access to key quantities of the object that Doppler monitoring alone cannot provide. In particular, precise measurements of the planetary mass and radius allow us to calculate the average density and infer a composition. Such estimates enable a meaningful evaluation of structural models of these objects, including whether or not they possess a core of rocky material. These inferences, in turn, enable direct tests of competing scenarios of planet formation and evolution. Moreover, the transiting configuration permits numerous interesting follow-up studies, such as searches for planetary satellites and circumplanetary rings and studies of the planetary atmosphere.

In this review, I discuss the status of work in the field with a focus on the key contributions, both past and near-future, enabled by the *Hubble Space Telescope*. I begin by reviewing the properties of the transiting planets discovered to date, as well as the numerous ground-based efforts to detect more of these objects. I then consider the various follow-up studies of these gas-giant planets that are enabled by *HST*, as well as *HST*'s potentially unique role in following-up Earth-sized objects detected by NASA's *Kepler* mission. I finish by discussing two *HST*-based searches for transiting gas-giant planets. Throughout this contribution, I restrict my attention to studies of *transiting* planets; for an introduction to the broader range of *HST*-based studies of extrasolar planets, see Charbonneau (2004).

2. The current sample of transiting planets

At the time of writing, there are six extrasolar planets known to transit the disks of their parent stars: HD 209458b (Charbonneau et al. 2000, Henry et al. 2000) was initially identified through Doppler monitoring, TrES-1 (Alonso et al. 2004) was discovered by the small-aperture, wide-field Trans-Atlantic Exoplanet Survey (TrES) network, and the four others were found by Doppler follow-up of a list of more than 100 candidates identified by the OGLE team (Udalski 2002a, 2002b, 2002c, 2003): OGLE-TR-56b (Konacki et al.

2003), OGLE-TR-111b (Pont et al. 2004), OGLE-TR-113b (Bouchy et al. 2004; Konacki et al. 2004), and OGLE-TR-132b (Bouchy et al. 2004).

As a result of the recent flurry of discoveries, we can construct a mass-radius plot for transiting extrasolar planets, shown in Figure 1. From this plot, it is clear that all but one of the planets have radii that are consistent with a value that is only modestly inflated (typically <15%) over the Jupiter radius. This is in keeping with the predictions of models which include the effects of stellar insolation (Guillot et al. 1996; Burrows et al. 2000; Baraffe et al. 2003; Bodenheimer et al. 2003; Chabrier et al. 2004). There is a growing consensus among theorists that the large radius of HD 209458b, 1.35 ± 0.06 R_{Jup}, cannot be explained by stellar insolation effects alone, and indeed Figure 1 points to its anomalous nature. Guillot & Showman (2002) and Showman & Guillot (2002) considered a model in which a modest fraction of the incident stellar radiation in converted into mechanical energy (winds), which could fill the energy decrement and hence explain the large radius. It is not clear, however, why this mechanism would not work as efficiently for the other objects. An alternative hypothesis is that the orbital eccentricity of HD 209458b is continually pumped by the presence of a more distant and, as of yet, undetected planet; the damping of the eccentricity in turn provides the energy source needed to explain the large radius (Bodenheimer et al. 2003). Fortunately, it should be possible to test this model observationally, initially through the offset in the timing of the secondary eclipse (indicating a non-zero orbital eccentricity; Charbonneau 2003b), and subsequently through the detection of the second planet via precise radial velocity measurements. Solving the mystery of HD 209458b will be an engaging challenge with a resolution likely in the next couple years.

HD 209458b and TrES-1 are very similar in mass and equilibrium temperature, yet have dramatically different radii. As a result, we must abandon the simple expectation that the radius of an extrasolar gas-giant planet will be determined solely by its mass and degree of stellar insolation. Rather, we now know that the radii can be altered dramatically by additional processes. Identifying these mechanisms and understanding their implications for the structure of these planets will be a rewarding task for the near future.

As discussed in Section 4, only relatively nearby systems will be bright enough for the majority of follow-up observations with *HST* that have been completed for HD 209458b. Indeed, of the six known systems, only HD 209458b and TrES-1 satisfy the brightness criterion. Since we would like to increase the number of similarly accessible planets, I consider next the current ground-based surveys for these objects.

3. Ground-based surveys for new targets

Of the roughly 20 ground-based transit surveys (Horne 2003; Charbonneau 2003a)† that are either in operation or planned for the near future, only those that survey bright ($V < 13$) stars will find planets that may be pursued with the *HST*-based techniques described in the following section. Among these surveys are SuperWASP (Street et al. 2003), PASS (Deeg et al. 2004), HATnet (Bakos et al. 2004), Vulcan (Borucki et al. 2001), KELT (Pepper et al. 2003), and TrES (e.g., Alonso et al. 2004). All of these surveys use fast optics (typically consisting of a commercially available camera lens) and a correspondingly coarse pixel scale (typically 10–20 arcsec/pixel) to monitor thousands (and sometimes tens of thousands) of field stars. The dominant challenge facing such searches is no longer that of obtaining the requisite photometric precision (better than 1%) and

† http://star-www.st-and.ac.uk/~kdh1/transits/table.html

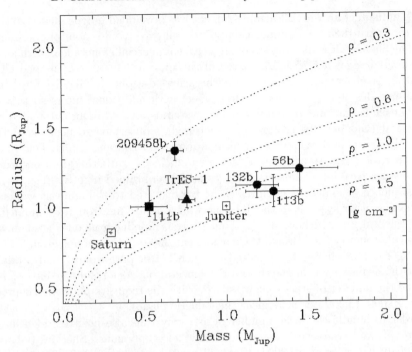

FIGURE 1. Radius versus mass for all known transiting planets. The values and uncertainties are taken from Brown et al. (2001) for HD 209458b, Sozzetti et al. (2004) for TrES-1, Torres et al. (2004a) for OGLE-TR-56b, Pont et al. (2004) for OGLE-TR-111b, Bouchy et al. (2004) for OGLE-TR-113b, and Moutou et al. (2004) for OGLE-TR-132b. Lines of constant density are also indicated. *Reprinted with permission from Sozzetti et al. (2004).*

sufficient phase coverage (typically 300 hours) on a sufficient number of stars (>5000), but rather that of identifying the astrophysical false positives (Latham 2003; Charbonneau et al. 2004b; Mandushev et al. 2004; Torres et al. 2004b), which are predicted to occur with a frequency of roughly 10 times that of true planetary systems (Brown 2003). These impostors each consist of a stellar system containing an eclipsing binary that precisely mimics the single-band light curve of a gas-giant planet transiting a Sun-like star. Fortunately, the stars surveyed by these projects are bright, which has several ramifications. First, the targets typically have well-determined 2MASS‡ colors and USNO-B proper motions (Monet et al. 2003)¶, so that the likelihood that a given target is a nearby late-type dwarf can be readily evaluated. Second, the stars are easily accessible to spectrographs mounted on modest-aperture telescopes, where observing time is often easily available: Initial spectroscopic monitoring with a lower Doppler precision (∼1 km s⁻¹; Latham 2003) is one effective way to reject the majority of such contaminants before gathering precise (∼10 m s⁻¹) radial velocity observations at much larger observatories where observing time is often heavily over-subscribed.

3.1. *The TrES Network*

The Trans-Atlantic Exoplanet Survey (TrES) consists of three telescopes: Sleuth† (Charbonneau 2003a; O'Donovan et al. 2004), STARE‡ (Brown & Charbonneau 2000), and

‡ http://www.ipac.caltech.edu/2mass/releases/allsky/
¶ http://www.nofs.navy.mil/projects/pmm/catalogs.html
† http://astro.caltech.edu/~ftod/tres/sleuth.html
‡ http://www.hao.ucar.edu/public/research/stare/stare.html

PSST (Dunham et al. 2004). Each instrument was developed independently, but with the intent that with their very similar capabilities, all three could monitor the same field of view. The Sleuth telescope consists of an $f/2.8$ commercial camera lens with an aperture of 10 cm. It images a $5.7° \times 5.7°$ patch of sky onto a $2 k \times 2 k$ thinned CCD, with a resulting pixel scale of 9.9 arcsec. Sleuth gathers data in the SDSS r filter (the other systems use Bessell R), but has a filter wheel with additional filters so that accurate colors of each target may be obtained. The system is located in an automated clamshell enclosure at Palomar Observatory, and each night's observing sequence is handled in a completely automated fashion by a local workstation running Linux. Due to its automated operation, Sleuth achieves a very high efficiency, gathering data on nearly every night weather permits. Calibration images are also obtained nightly. Each morning the entire sequence of images is compressed and transferred to the analysis workstation at Caltech. The other systems in the TrES network operate in a roughly similar fashion.

Over the past year, we have switched our analysis pipeline from one based on weighted-aperture photometry (i.e., Brown & Charbonneau 2000) to that of image subtraction (i.e., Allard 2000). This change has resulted in much better performance and produces time series with an rms very near the theoretical expectation. As a result, we have significantly increased the number of stars that we survey with the requisite photometric precision to roughly 10,000–15,000 per field (near the galactic plane).

In order to handle the large number of astrophysical false positives identified by the TrES survey, we have recently assembled a new telescope named Sherlock (Charbonneau et al. 2004b; Kotredes et al. 2004) that will be dedicated to such follow-up work. Sherlock is located in the same enclosure as Sleuth at Palomar Observatory. It is based on a commercially available $f/6.3$ Schmidt-Cassegrain telescope with an aperture of 25 cm. It images a $0.5° \times 0.5°$ field-of-view onto a $1 k \times 1 k$ thinned CCD, and is equipped with a filter wheel housing a selection of relatively narrow-band interference filters. Each night, we identify any active candidates that are predicted to present eclipses, and perform high-cadence multi-color photometry of the highest-priority target. Sherlock is able to distinguish several forms of false positives for two reasons: First, the increased angular resolution (1.7 arcsec/pixel) compared to the TrES instruments (10 arcsec/pixel) resolves most stellar blends that occur due to projection along the line-of-sight; in such a scenario, the light curve of the isolated source as seen by Sherlock typically presents a significantly deeper eclipse, which is no longer consistent with a transiting planet. Second, stellar blends (physically associated or not) in which the blending and occulted stars are of significantly different temperature will present eclipse depths that vary in color in excess of the small effects due to limb-darkening. The follow-up photometry provided by Sherlock is complementary to spectroscopic work similarly aimed at identifying the impostors (i.e., Latham 2003). Notably, we plan to operate Sherlock in a fully-automated manner, and the small amount of labor, compared to the spectroscopic follow-up, makes this method an appealing one by which to reject the bulk of such astrophysical false positives.

3.2. *The New World of TrES-1*

Alonso et al. (2004) present the detection of the planet TrES-1, a transiting hot Jupiter with an orbital period of 3.03 days. This work, combined with a detailed analysis of the stellar spectrum by Sozzetti et al. (2004), permits a precise estimation of the planetary mass $M_p = 0.76 \pm 0.05 \, M_{\mathrm{Jup}}$ and radius $R_p = 1.04^{+0.08}_{-0.05} \, R_{\mathrm{Jup}}$. As noted above, the 3-$\sigma$ discrepancy between the value of the radius of TrES-1 and that of HD 209458b, $R_p = 1.35 \pm 0.06 \, R_{\mathrm{Jup}}$, despite the similarity in mass and degree of stellar insolation, is an interesting puzzle with several possible resolutions. As discussed in the following section, *HST* may play a central role in identifying the correct answer.

It is critical to note that TrES-1 is only the second transiting planet for which the bulk of *HST*-based studies (described in the next section) may be pursued; planets identified by the OGLE survey orbit stars that are simply too faint. In the following section, we review these *HST*-based investigations.

4. *HST* studies of transiting extrasolar gas giants

Nearly the entirety of *HST*-based studies of transiting exoplanets to date has focused on HD 209458 (one exception is the ACS-HRC campaign to observe two transits of OGLE-TR-56)†. I summarize these efforts below; for a more detailed description, see Charbonneau (2003b; 2004).

4.1. *Improved estimates of planetary parameters*

Ground-based efforts to obtain precise time series photometry of HD 209458 are limited to a precision of $\sim$2 mmag and a cadence of $\sim$10 minutes (Charbonneau et al. 2000; Henry et al. 2000; Jha et al. 2000; Deeg, Garrido, & Claret 2001). With data of this quality, estimations of the planetary radius (and other parameters) are confounded by a significant degeneracy between the planetary and stellar radii and the orbital inclination. In short, by increasing the planetary and stellar radii in tandem to preserve their ratio, the same transit depth can be produced, and reducing the orbital inclination preserves the chord length across the star to match the observed transit duration. Since independent estimates of the stellar radius from stellar models (e.g., Cody & Sasselov 2002) were typically limited to a precision of $\sim$10%, estimates of the planetary radius retained this significant uncertainty.

The ultra-precise light curve by Brown et al. (2001) was produced by using STIS to disperse the large number of photons over as many pixels as possible (to retain a high observing efficiency and mitigate flat-fielding effects), and subsequently summing the recorded counts to produce a photometric index with a typical precision of 1.1×10^{-4} and a cadence of 80 s. The quality of these data breaks the former degeneracy (principally as a result of the precise measurement of the slope and duration of ingress and egress). As a result, the authors derived a precise estimate of the planetary radius, $R_p = 1.35 \pm 0.06 \ R_{\rm Jup}$, as well as that of the star, $R_s = 1.15 \pm 0.05 \ R_{\rm Sun}$ (consistent with, but more precise than the *Hipparcos* value). This estimate is still subject to the uncertainty in the independent estimate of the stellar mass, M_s, but the effect is small since the uncertainty in the radius is only weakly dependent upon that of the mass, i.e., $\Delta R_p / R_p \simeq 0.3 \, \Delta M_s / M_s$.

More recently, Charbonneau et al. (2004a) gathered new STIS data at lower resolution, but spanning a large wavelength range, 290–1080 nm. The data can be subdivided into various effective bandpasses prior to forming the photometric time series, and the resulting light curves clearly show the predicted color-dependence due to limb-darkening. As a result, it should be possible to assume limb-darkening coefficients based on a stellar model, and derive a very precise estimate of the planetary radius.

4.2. *Searches for planetary satellites, circumplanetary rings, and reflected light*

Brown et al. (2001) also noted that the STIS data described above was sufficiently precise that terrestrial-sized objects present in the HD 209458 system could be revealed. This sensitivity was unprecedented: although ground-based searches for Earth-sized objects are possible for small stars such as the M-dwarf binary CM Dra (e.g., Doyle et al. 2000;

---

† http://www.stsci.edu/cgi-bin/get-proposal-info?9805

Deeg et al. 2000), no instrument had previously demonstrated the ability to see such small objects in transit across a star the size of the Sun. Brown et al. (2001) concluded that they could exclude the presence of a planetary satellite with radius larger than 1.2 R_{Earth} (although not for all possible values of the putative satellite's orbital period and phase). Similarly, they excluded the presence of opaque circumplanetary rings with a radius greater than 1.8 R_p, since such a ring system would have resulted in large deviations during ingress and egress, which were not seen.

Brown et al. (2001) also examined their data for offsets in the times of the center of the planetary eclipses; such offsets would result from a massive planetary satellite. Since the centroid of each eclipse could be measured with a precision of 6 s, they were able to exclude satellites more massive than 3 M_{Earth}. More recently, Schultz et al. (2003) used *HST*'s Fine Guidance Sensors (FGS) to obtain photometry with an exceptional cadence of 0.025 s and a SNR of 80. By targeting times of ingress and egress, they will place stringent constraints on any timing variations in this system.

Exquisite photometry of a hot-Jupiter system might also enable the detection of the light reflected from the planet. This effect has been sought using a spectroscopic technique from the ground for several hot-Jupiter systems (Charbonneau et al. 1999; Collier Cameron et al. 2002; Leigh et al. 2003), but as of yet, only upper limits have been obtained. The eclipsing geometry of the HD 209458 system presents an attractive opportunity; precise photometry during times of secondary eclipse could detect the decrement in light as the planet passes behind the star. Not only would such data enable an evaluation of the wavelength-dependent albedo, but these data would also constrain the orbital eccentricity through the timing of the secondary eclipse (Charbonneau 2003b). This project is also being pursued by the *MOST* satellite (Walker et al. 2003; Matthews et al. 2004)†, which has achieved unprecedented precision despite its small aperture. However, *HST* retains the spectroscopic information (either through STIS, or grism modes of ACS), whereas *MOST* has a single fixed bandpass.

4.3. *Atmospheric transmission spectroscopy*

Based on theoretical predictions (Seager & Sasselov 2000; Brown 2001; Hubbard et al. 2001) that hot Jupiters such as HD 209458b should present strong alkali metal features in their transmission spectra, Charbonneau et al. (2002) pursued this effect with STIS. They detected an increase in the transit depth of $(2.32 \pm 0.57) \times 10^{-4}$ in a 1.2 nm bandpass centered on the Na D lines at 589.3 nm. After ruling out alternate explanations for this effect, they concluded that this decrement was indeed due to absorption from atomic sodium in the planetary atmosphere. Notably, the observed signal was only 1/3 that predicted from their fiducial model, a cloudless atmosphere with a solar abundance of sodium in atomic form. One explanation for this decrease may be the presence of clouds high in the atmosphere, effectively reducing the size of the atmosphere viewed in transmission. Such clouds would likely reduce other transmission spectral features in a similar fashion, and indeed recent ground-based work in the infrared seems to confirm this interpretation (Deming et al. 2004).

Using STIS in the UV, Vidal-Madjar et al. (2003) detected a 15% decrement in the flux at Lyα during times of eclipse, which they interpret as evidence for ongoing atmospheric escape of a significant quantity of hydrogen (Lecavelier des Etangs et al. 2004). In a more recent paper, Vidal-Madjar et al. (2004) also present evidence of a corresponding decrement in lines of carbon and oxygen, indicating their presence in the cloud of material surrounding the planet; however, the effect for these features is detected with significantly less statistical significance than that of Lyα. The precise mechanism of escape, and the connection (if any) of this extended material to the inflated value of the radius remain open questions that demand further study.

5. Connection to the *Kepler* mission

The NASA *Kepler*† mission (Borucki et al. 2003a; 2003b) is scheduled for launch in 2007, and may make the first detection of an extrasolar Earth-like planet. Indeed, if most Sun-like stars have such planets, dozens should be detected. *Kepler* will therefore set the scale for future efforts to directly characterize such extrasolar Earths, since it will tell us how common these objects are, and, by inference, how far we might expect to look to locate the closest example.

As with the ground-based searches, *Kepler* will need robust techniques by which to discriminate true planetary transits from astrophysical false positives, notably stellar blends. It should be noted, therefore, that *HST*/ACS is likely the only instrument in existence that will be capable of confirming the photometric signals detected by the *Kepler* Mission. More importantly, *HST*/ACS would be able to provide photometry over several bandpasses that are distinct from the single, fixed bandpass used by *Kepler*. As discussed in Section 3, such photometry during times of eclipse can be used to search for a color-dependence to the transit depth, which would indicate that such a candidate is an impostor.

The following SNR calculation is adapted from one for upcoming DD observations of TrES-1 (T. M. Brown, R. L. Gilliland, et al., personal communication). Using ACS/HRC with the G800L grism to disperse the light from a $V = 12.0$ star over roughly 200 pixels allows collection of 5×10^7 e- per exposure while staying 30% under saturation. Given the much higher efficiency of the G800L on ACS than the similar capabilities on STIS, and the significantly fainter typical magnitude for the *Kepler* candidate assumed here ($V = 12.0$) than HD 209458 ($V = 7.6$), it makes sense to use ACS (as opposed to STIS), regardless of STIS loss. Adopting an exposure time of 115 s, and using the 512 square subarray nearest the readout amp for HRC (which requires a readout overhead of 35 s), results in a net cadence of 150 s and an observing efficiency of 77%. The SNR per exposure will be 7150. Over the 13 hour duration of the eclipse (and accounting for the roughly 55 minute visibility per *HST*-orbit for the *Kepler* field-of-view), roughly 179 exposures could be obtained, for a photon-noise-limited precision of 1.05×10^{-5}. Since the determination of the eclipse depth is also affected by the precision of the out-of-eclipse data, this reduces to 1.48×10^{-5} (for a comparable duration out of eclipse). This represents a 5.7-σ detection of a Earth-sized object across a star with the solar radius. Of course, this calculation has not considered the effect of systematic errors, which may limit the precision to a level significantly worse than the theoretical photon-noise limit. Furthermore, *Kepler* candidates that are fainter than $V = 12.0$ would be detected with less confidence, unless the ratio of planet and stellar radii is increased.

Short-period Earth-sized planets may be detected by *Kepler* early in the mission, and *HST*/ACS may still be operational at this point. However, true Earth analogs will be identified only as early as 2010, since these present transits only once a year, and two such events are required for a period estimate (*Kepler* requires three events for a reliable detection). Without a servicing mission, it is very unlikely that *HST* would still be in operation at that time. Given that the search for Earth-like planets is so central to future

† http://www.kepler.arc.nasa.gov/

NASA plans, the impact of the loss of *HST* to these plans should be considered in close detail.

6. *Hubble* as a survey instrument

The discussion above has focused on the use of *HST* to conduct follow-up studies of transiting extrasolar planets identified by other telescopes. However, *HST* has also proven to be a powerful transit survey instrument in its own right. The small field-of-view precludes the simultaneous survey of a sufficiently large number of bright stars, as is done for many of the ground based surveys. However, for certain fields-of-view, *HST*'s sharp point-spread function enables precise photometry of thousands of stars for which ground-based work cannot proceed. To date, *HST* has surveyed two such fields for transiting planets: the globular cluster 47 Tuc, and the Galactic Bulge.

6.1. *A transit search in the globular cluster 47 Tuc*

The use of *HST* as a survey instrument for transiting planets was pioneered by Gilliland et al. (2000). In July 1999, they monitored 34,000 main-sequence stars in the globular cluster 47 Tuc for 8.3 days. Star clusters make attractive hunting grounds for transiting planets for several reasons. First, the uniformity of age and metallicity of the cluster members provides the ideal laboratory setting within which to study the dependence of the planet population on these quantities. Second, since the stellar radii may be inferred from measurements of the apparent brightness and an evolutionary models for the cluster, the radius of the transiting secondary can also be reliably inferred. A key element in the design of the Gilliland et al. (2000) experiment was that a large range of stellar radii could be monitored for transiting planets, since the poorer photometric precision obtained for much smaller (and less massive stars) was counterbalanced by a corresponding increase in the depth of the transit. In particular, the stars surveyed ranged in size from roughly 1.5 R_{Sun} (corresponding to slightly evolved stars with masses of 0.87 M_{Sun}, at visual magnitudes of $V = 17.4$) to main-sequence stars with radii of 0.51 R_{Sun} (corresponding to masses of 0.55 M_{Sun} and a visual magnitudes of $V = 21.9$).

The survey found no stars presenting planetary transits, which was very much at odds with the findings of the Doppler surveys: in the local solar neighborhood, roughly 1% of Sun-like stars have hot Jupiters; folding in the recovery rates and efficiencies specific to the *HST* 47 Tuc survey, approximately 17 such objects would have been expected. The core result of the survey was thus that the population of hot Jupiters in 47 Tuc was at least an order of magnitude below that of the local solar neighborhood. This disparity may be due in part to the decreased metallicity of 47 Tuc ([Fe/H] $= -0.7$), as the lower-metallicity environment may impede planet formation and/or migration. A likely additional effect is that of crowding: at the typical location of the Gilliland et al. (2000) observations (one arcminute from the core), the stellar density is roughly 10^3 M_{Sun} pc^{-3}; gas giant planets at several AU from the star could be disrupted in the typical planetary system if it suffers a close dynamical encounter prior to the inward migration of the planet (at which point it is sufficiently bound to withstand such disruption).

Due to the uniformity of the stellar population and the data set, and the extensive set of detection tests performed, the Gilliland et al. (2000) result remains one of the few well-characterized (and hence astrophysically useful) null results for a survey of transiting extrasolar planets. Wide-field surveys such as TrES are of great interest because they may deliver bright, transiting planets for which the parameters can be accurately measured, and which are amenable to follow-up studies. However, these wide-field surveys are unlikely to yield meaningful constraints of the rate-of-occurrence of hot Jupiters as a function of the properties and environment of the central star, due to the very diverse nature of the targets surveyed in their magnitude-limited field samples.

6.2. *A transit search in a galactic bulge field*

K. Sahu and collaborators (program GO-9750†) have recently undertaken an ambitious *HST*-based transit search that seeks to capitalize on the increased (relative to WFPC2) field-of-view and sensitivity allowed by ACS/WFC. In February 2004, they observed a field in the galactic bulge for seven days with ACS/WFC. In the 202×202-arcsecond field-of-view, they expect to monitor roughly 167,000 F, G, and K dwarfs brighter than $V = 23$, for which they will obtain a photometric precision sufficient to detect the transit of a Jupiter-sized planet. If the rate of occurrence of hot Jupiters for these stars is the same as that in the solar neighborhood (roughly 1%), they may detect more than 100 such planets. The number of disk and bulge stars are approximately equal, and the membership of a given star to a population will be determined by proper motions from data obtained at a different epoch (these data are also in hand). Furthermore, the metallicity of stars in the field-of-view is expected to vary by more than 1.5 order of magnitude. As a result, this dataset might permit the detection of a sufficiently large and diverse group of hot Jupiters so that the dependence of rate-of-occurrence and planetary radius upon several properties of the primary (notably stellar type, disk vs. bulge membership, and metallicity) could be disentangled.

A note of caution must be sounded, however. As discussed earlier, the primary challenge facing transit surveys is no longer that of photometric precision and phase coverage, but rather that of rejecting the astrophysical false positives, whose photometric light curves precisely mimic that of a planetary transit (Charbonneau et al. 2004b; Latham 2003). For the brightest targets, VLT radial velocity monitoring may reveal the spectroscopic orbit. More importantly, a detailed spectroscopic analysis for blends of eclipsing binaries could be performed. However, even for the much brighter stars that have been surveyed for transits, it has proven extremely difficult to identify hierarchical triples, in which the light from a third, bright star dilutes the photometric and spectroscopic variability of the eclipsing binary. Two examples of the degree to which such systems may confound researchers are given by Torres et al. 2004b (for a system identified by the OGLE survey, Udalski et al. 2002a, 2002b, 2003) and Mandushev et al. 2004 (for a candidate found by the TrES network). For the bulk of candidates, however, no spectrograph in existence will be able to recover the Doppler orbit (for a detailed presentation of the signal-to-noise calculation, see Charbonneau 2003a). The investigators will then need to rely upon more indirect considerations (such as colors and proper motions) to argue in favor of a planetary interpretation. Whether this argument can be made convincingly for an object that displays a transit light curve but no measurable Doppler orbit remains to be seen. Moreover, since the planetary radius is approximately degenerate with mass across two orders of magnitude (0.5–80 $M_{\rm Jup}$; Burrows et al. 2001), the value of each individual object will be diminished relative to those identified for brighter stars. These concerns aside, however, the prospect of perhaps doubling the number of known extrasolar planetary systems is a fascinating one indeed, and the results from this survey are eagerly awaited.

† http://www.stsci.edu/cgi-bin/get-proposal-info?9750

7. Final note regarding STIS

As I was finalizing this contribution, the STIS spectrograph went offline due to the failure of an internal 5V power supply‡. As described above, STIS was the most productive *HST* instrument with regards to follow-up studies of hot Jupiters. Fortunately, many investigations requiring high photometric and/or spectroscopic stability can be accomplished with ACS (using a grism element to handle to the high photon rates), as well as NICMOS and FGS. Notably, the typical brightness of the transiting hot Jupiters that will be identified over the next couple years by the numerous ongoing wide-field surveys is optimally suited for ACS rather than STIS, regardless of STIS loss.

REFERENCES

ALARD, C. 2000 *A&AS* **144**, 363.

ALONSO, R., ET AL. 2004 *ApJ* **613**, L153.

BAKOS, G., NOYES, R. W., KOVÁCS, G., STANEK, K. Z., SASSELOV, D. D., & DOMSA, I. 2004 *PASP* **116**, 266.

BARAFFE, I., CHABRIER, G., BARMAN, T. S., ALLARD, F., & HAUSCHILDT, P. H. 2003 *A&A* **402**, 701.

BODENHEIMER, P., LAUGHLIN, G., & LIN, D. N. C. 2003 *ApJ* **592**, 555.

BODENHEIMER, P., LIN, D. N. C., & MARDLING, R. A. 2001 *ApJ* **548**, 466.

BORUCKI, W. J., CALDWELL, D., KOCH, D. G., WEBSTER, L. D., JENKINS, J. M., NINKOV, Z., & SHOWEN, R. 2001 *PASP* **113**, 439.

BORUCKI, W. J., ET AL. 2003a. In *Scientific Frontiers in Research on Extrasolar Planets* (eds. Drake Deming & Sara Seager). ASP Conf. Ser. 294, p. 427. ASP.

BORUCKI, W. J., ET AL. 2003b *Proc. SPIE* **4854**, 129.

BOUCHY, F., PONT, F., SANTOS, N. C., MELO, C., MAYOR, M., QUELOZ, D., & UDRY, S. 2004 *A&A* **421**, L13.

BROWN, T. M. 2001 *ApJ* **553**, 1006.

BROWN, T. M. 2003 *ApJ* **593**, L125.

BROWN, T. M. & CHARBONNEAU, D. 2000. In *Disks, Planetesimals, and Planets* (eds. F. Garzón, C. Eiroa, D. de Winter, & T. J. Mahoney). ASP Conf. Ser. 219, p. 584. ASP.

BROWN, T. M., CHARBONNEAU, D., GILLILAND, R. L., NOYES, R. W., & BURROWS, A. 2001 *ApJ* **552**, 699.

BURROWS, A., GUILLOT, T., HUBBARD, W. B., MARLEY, M. S., SAUMON, D., LUNINE, J. I., & SUDARSKY, D. 2000 *ApJ* **534**, L97.

BURROWS, A., HUBBARD, W. B., LUNINE, J. I., & LIEBERT, J. 2001 *Reviews of Modern Physics* **73**, 719.

CHABRIER, G., BARMAN, T., BARAFFE, I., ALLARD, F., & HAUSCHILDT, P. H. 2004 *ApJ* **603**, L53.

CHARBONNEAU, D. 2003a. astro-ph/0302216.

CHARBONNEAU, D. 2003b. In *Scientific Frontiers in Research on Extrasolar Planets* (eds. Drake Deming and Sara Seager). ASP Conf. Ser. 294, p. 449. ASP.

CHARBONNEAU, D. 2004. In *Stars as Suns: Activity, Evolution, and Planets* (eds. A. K. Dupree & A. O. Benz), IAU Symposium 219. p. 367. ASP.

CHARBONNEAU, D., BROWN, T. M., DUNHAM, E. W., LATHAM, D. W., LOOPER, D. L., & MANDUSHEV, G. 2004a. In *The Search for Other Worlds: Fourteenth Astrophysics Conference* (eds. Stephen S. Holt & Drake Deming). AIP Conference Proceedings, Volume 713, p. 151. American Institute of Physics.

CHARBONNEAU, D., BROWN, T. M., GILLILAND, R. L., & NOYES, R. W. 2004b. In *Stars as Suns: Activity, Evolution, and Planets* (eds. A. K. Dupree & A. O. Benz). IAU Symposium 219, p. 367. ASP.

CHARBONNEAU, D., BROWN, T. M., LATHAM, D. W., & MAYOR, M. 2000 *ApJ* **529**, L45.

‡ http://www.stsci.edu/hst/stis/

CHARBONNEAU, D., NOYES, R. W., KORZENNIK, S. G., NISENSON, P., JHA, S., VOGT, S. S., & KIBRICK, R. I. 1999 *ApJ* **522**, L145.

CODY, A. M. & SASSELOV, D. D. 2002 *ApJ* **569**, 451.

COLLIER CAMERON, A., HORNE, K., PENNY, A., & LEIGH, C. 2002 *MNRAS* **330**, 187.

DEEG, H. J., ALONSO, R., BELOMONTE, J. A., ALSUBAI, K., HORNE, K., & DOYLE, L. R. 2004 *PASP* **116**, 985.

DEEG, H. J., DOYLE, L. R., KOZHEVNIKOV, V. P., BLUE, J. E., MARTÍN, E. L., & SCHNEIDER, J. 2000 *A&A* **358**, L5.

DEEG, H. J., GARRIDO, R., & CLARET, A. 2001 *New Astronomy* **6**, 51.

DEMING, D., BROWN, T. M., CHARBONNEAU, D., HARRINGTON, J., & RICHARDSON, L. J. 2005 *ApJ*, submitted.

DOYLE, L. R., ET AL. 2000 *ApJ* **535**, 338.

DUNHAM, E. W., MANDUSHEV, G. I., TAYLOR, B., & OETIKER, B. 2004 *PASP* **116**, 1072.

GILLILAND, R. L., ET AL. 2000 *ApJ* **545**, L47.

GUILLOT, T., BURROWS, A., HUBBARD, W. B., LUNINE, J. I., & SAUMON, D. 1996 *ApJ* **459**, L35.

GUILLOT, T. & SHOWMAN, A. P. 2002 *A&A* **385**, 156.

HENRY, G. W., MARCY, G. W., BUTLER, R. P., & VOGT, S. S. 2000 *ApJ* **529**, L41.

HORNE, K. 2003. In *Scientific Frontiers in Research on Extrasolar Planets* (eds. Drake Deming & Sara Seager). ASP Conf. Ser. 294, p. 361. ASP.

HUBBARD, W. B., FORTNEY, J. J., LUNINE, J. I., BURROWS, A., SUDARSKY, D., & PINTO, P. 2001 *ApJ* **560**, 413.

JHA, S., CHARBONNEAU, D., GARNAVICH, P. M., SULLIVAN, D. J., SULLIVAN, T., BROWN, T. M., & TONRY, J. L. 2000 *ApJ* **540**, L45.

KONACKI, M., ET AL. 2004 *ApJ* **609**, L37.

KONACKI, M., TORRES, G., JHA, S., & SASSELOV, D. D. 2003 *Nature* **421**, 507.

KOTREDES, L., CHARBONNEAU, D., LOOPER, D. L., & O'DONOVAN, F. T. 2004. In *The Search for other Worlds: Fourteenth Astrophysics Conference* (eds. Stephen S. Holt & Drake Deming). AIP Conference Proceedings, Volume 713, p. 173. American Institute of Physics.

LATHAM, D. W. 2004. In *Scientific Frontiers and Research in Extrasolar Planets* (eds. D. Deming and S. Seager). ASP Conf. Ser. 294, p. 409. ASP.

LECAVELIER DES ETANGS, A., VIDAL-MADJAR, A., MCCONNELL, J. C., & HÉBRARD, G. 2004 *A&A* **418**, L1.

LEIGH, C., COLLIER CAMERON, A., UDRY, S., DONATI, J., HORNE, K., JAMES, D., & PENNY, A. 2003 *MNRAS* **346**, L16.

MANDUSHEV, G., TORRES, G., LATHAM, D. W., CHARBONNEAU, D., ALONSO, R., DUNHAM, E. W., WHITE, R. J., BROWN, T. M., & O'DONOVAN, F. T. 2004 *ApJ*, submitted.

MATTHEWS, J. M., KUSCHING, R., GUENTHER, D. B., WALKER, G. A. H., MOFFAT, A. F. J., RUCINSKI, S. M., SASSELOV, D., & WEISS, W. W. 2004 *Nature* **430**, 51.

MONET, D. G., ET AL. 2003 *AJ* **125**, 984.

MOUTOU, C., PONT, F., BOUCHY, F., & MAYOR, M. 2004 *A&A* **424**, L31.

O'DONOVAN, F. T., CHARBONNEAU, D., & KOTREDES, L. 2004. In *The Search for other Worlds: Fourteenth Astrophysics Conference* (eds. Stephen S. Holt & Drake Deming). AIP Conference Proceedings, Volume 713, p. 169. American Institute of Physics.

PEPPER, J., GOULD, A., & DEPOY, D. L. 2003 *Acta Astronomica* **53**, 213.

PONT, F., BOUCHY, F., QUELOZ, D., SANTOS, N. C., MELO, C., MAYOR, M., & UDRY, S. 2004 *A&A* **426**, L15.

SCHULTZ, A. B., ET AL. 2003. In *Scientific Frontiers in Research on Extrasolar Planets* (eds. Drake Deming & Sara Seager). ASP Conf. Ser. 294, p. 479. ASP.

SEAGER, S. & SASSELOV, D. D. 2000 *ApJ* **537**, 916.

SHOWMAN, A. P. & GUILLOT, T. 2002 *A&A* **385**, 166.

SOZZETTI, A., ET AL. 2004 *ApJ* **616**, L167.

STREET, R. A., ET AL. 2003. In *Scientific Frontiers in Research on Extrasolar Planets* (eds. Drake Deming & Sara Seager). ASP Conf. Ser. 294, p. 405. ASP.

TORRES, G., KONACKI, M., SASSELOV, D. D., & JHA, S. 2004a *ApJ* **609**, 1071.

TORRES, G., KONACKI, M., SASSEOLOV, D. D., & JHA, S. 2004b *ApJ* **614**, 979.

UDALSKI, A., PIETRZYNSKI, G., SZYMANSKI, M., KUBIAK, M., ZEBRUN, K., SOSZYNSKI, I., SZEWCZYK, O., & WYRZYKOWSKI, L. 2003 Acta Astronomica 53, 133.

UDALSKI, A., SZEWCZYK, O., ZEBRUN, K., PIETRZYNSKI, G., SZYMANSKI, M., KUBIAK, M., SOSZYNSKI, I., & WYRZYKOWSKI, L. 2002c Acta Astronomica 52, 317.

UDALSKI, A., ZEBRUN, K., SZYMANSKI, M., KUBIAK, M., SOSZYNSKI, I., SZEWCZYK, O., WYRZYKOWSKI, L., & PIETRZYNSKI, G. 2002b Acta Astronomica 52, 115.

UDALSKI, A., ET AL. 2002a Acta Astronomica 52, 1.

VIDAL-MADJAR, A., LECAVELIER DES ETANGS, A., DÉSERT, J.-M., BALLESTER, G. E., FERLET, R., HÉBRARD, G., & MAYOR, M. 2003 Nature 422, 143.

VIDAL-MADJAR, A., ET AL. 2004 ApJ 604, L69.

WALKER, G., ET AL. 2003 PASP 115, 1023.

Unsolved problems in star formation

By C. J. CLARKE

Institute of Astronomy, Madingley Road, Cambridge, CB3 OHA, UK

The study of star formation is currently benefiting from a wealth of new observational data, exploiting the high-sensitivity, wide-field, high-resolution capabilities of a diverse range of space and ground-based instrumentation. In parallel with this, high performance computing is enabling theorists to tackle key problems which—due to their complex geometry and non-linear nature—had long been recognized to be beyond the reach of analytical theory. In this review, rather than reporting progress in each of these areas, I will instead set out some scientific questions that one would expect to be answered before one would regard star formation as a topic that was largely solved. I have accordingly selected three areas: 1) molecular clouds and their relationship to the stars they form and to the wider galactic disk, 2) the question of the determinants of stellar mass (i.e., the IMF), and 3) the issue of protostellar disk dispersal and its relation to planet formation. For each topic, I outline areas of consensus, recent results, and discuss the key problems that can plausibly be addressed in the next five years.

---

1. Introduction

In this contribution I have selected three main issues in contemporary star-formation studies. I have chosen these themes because 1) they represent important areas of uncertainty in our current understanding, 2) they involve a synergy between theory and observation, and 3) they span the range of length scales—from planetary to galactic scales—that are involved in different aspects of the star-formation process. Unlike most of the other contributions in this volume, these topics have not been selected according to *HST*'s contribution in these areas. Nevertheless, the role played by *HST* in key areas—such as the study of pre-main sequence stars beyond the Galaxy and the imaging of photoevaporative flows from disks around young stars—will be duly emphasized.

I start with two longstanding problems concerning the conversion of gas into stars. First, why does a $\sim 10^5$ $M_\odot$ Giant Molecular Cloud fragment into a large number of stellar-mass objects, rather than collapsing monolithically into a supermassive object? Secondly, why is star formation *inefficient* (in the sense that only a small fraction of the mass of a Giant Molecular Cloud can turn into stars on a cloud-internal free-fall timescale in order that the observed level of star formation in the galaxy is not exceeded)? Turbulent fragmentation models—which model clouds with the supersonic internal motions and non-linear density structures seen in real Giant Molecular Clouds—provide an attractive solution to the first problem and, it has been argued, may also contribute to the solution of the second, although feedback from recently formed stars must also play a role in this. I briefly review the successes of these models and highlight areas for future work: what are the most critical observations that can be used to confront these models and, equally fundamentally, what is the origin of this turbulence observed in molecular clouds?

I then turn to the IMF, an understanding of whose origin is regarded, by many, as being the most important contribution that star-formation studies can make to wider astrophysics. The literature abounds with 'successful' IMF theories—at least those that can reproduce the form of the IMF for masses within an order of magnitude or so of a solar mass. Given this degeneracy, it is perhaps most profitable to concentrate on the relatively uncharted waters of the extremes of the IMF, where there is some preliminary evidence for an environmental dependence of the mass function. I emphasize the value of ongoing observational studies that are providing the required calibration of the mass-luminosity relationship at the low mass end and also review some recent work on the demography of massive stars which can in principle throw some light on the unsolved problem of the mechanism for massive star formation.

Finally, I consider the interface between star and planet-formation studies, in particular the open question of what disperses the disks around young stars. At present, the study of such disks is as close as one can come to an observational study of the planet-formation process and it is important to understand not only the timescale but the mechanism by which disks—the reservoirs of planet-forming material—are dispersed during the pre-main sequence lifetime of a star. I will describe the first observational evidence that disks are cleared from the inside out (which is a rational expectation, given the shorter dynamical timescales at small radii), and how forthcoming observations may be used to determine whether it is planet formation that is emptying out the inner regions of such discs.

2. Turbulent fragmentation models—where next?

2.1. *Turbulent fragmentation models: The story so far*

The last decade has seen the proliferation of simulations of the internal dynamics of molecular clouds. All such models inject the clouds with supersonic internal bulk motions (loosely termed 'turbulence'), an elementary requirement given the observed superthermal line widths within molecular clouds (Larson 1981; Falgarone, Puget, & Perault 1992; Caselli & Myers 1995), and many also incorporate magnetic fields (again an observational requirement: see the review by Crutcher 1999). The initial motivation of these models was to study the decay of the turbulence and thus to test over what timescales could bulk motions support the cloud against collapse, a question of obvious relevance to the star formation efficiency issue mentioned above. Although it was at this time commonly hypothesized (e.g., Arons & Max 1975) that the magnetic field could cushion the dissipation of turbulent energy—and thus extend cloud lifetimes and reduce the star formation rate per free-fall time—simulations demonstrated that undriven magnetohydrodynamical turbulence in fact decays on a cloud-crossing timescale (Stone, Ostriker, & Gammie 1998; Mac Low et al. 1998). Since then, calculations have divided into those that support clouds over longer timescales by the continual replenishment of turbulent energy (e.g., Padoan & Nordlund 1999; Klessen, Heitsch, & Mac Low 2000; Heitsch, Mac Low, & Klessen 2001; Vazquez-Semadeni, Ballesteros-Paredes, & Klessen 2003) and those that instead introduce a one-off shot of kinetic energy which is then allowed to decay through shock dissipation (e.g., Ostriker, Gammie, & Stone 1999; Bate, Bonnell, & Bromm 2002a,b, 2003). In the latter case, it is naturally necessary to invoke feedback effects (e.g., photoionization or stellar winds) to prevent clouds turning into stars with 100% efficiency on a free-fall timescale.

Following on from this initial interest in cloud lifetimes, subsequent simulations have attempted to reproduce various statistical descriptors for the internal structure and kinematics of molecular clouds (see Padoan et al. 1998; Padoan et al. 1999; Ossenkopf, Klessen, & Heitsch 2001; Gammie et al. 2003). A more recent development has been the attainment in these models of a numerical resolution which also permits the study of *star formation* in such clouds. Currently, such resolution in large scale cloud simulations has only been attained using high resolution SPH calculations (where the resolution limit—around a Jupiter mass—is below the opacity limit for fragmentation; Bate, Bonnell, & Bromm 2002a,b, 2003), although Eulerian (AMR) codes are currently able to model star formation within molecular cloud *cores* at a similar resolution (see Klein et al. 2003).

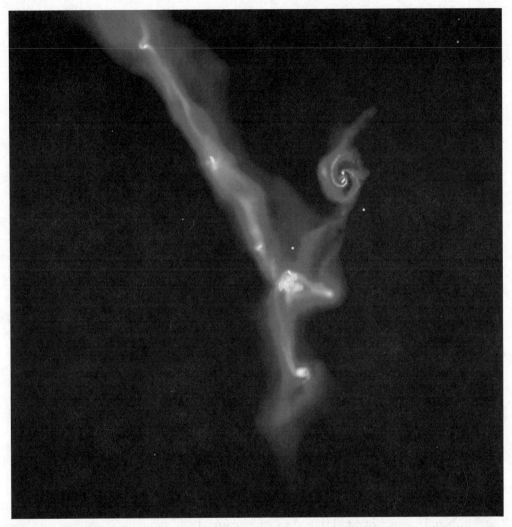

FIGURE 1. Snapshot from the simulations of Bate, Bonnell, & Bromm 2003, demonstrating the filamentary structure in the gas in turbulent fragmentation calculations and the formation of a number of sites of multiple star formation. The site in the upper right demonstrates the formation of a multiple system through disk fragmentation, and a low mass escaper can be seen in its vicinity. The frame is of width 0.025 pc and depicts a moment 2×10^5 years after the onset of cloud collapse. Such stellar systems would still be deeply embedded in their natal gas.

Such turbulent fragmentation calculations are characterized by a number of generic features (Bate, Bonnell, & Bromm 2003). First, the initial injection of supersonic turbulence induces strong shocks, with gravitational fragmentation of shocked layers giving rise to a network of filamentary structures. Thereupon further fragmentation within filaments gives rise to the production of several star formation sites, each typically comprising around 10 objects (see Fig. 1). In the calculations, around half of the objects result from primary fragmentation of the filaments and around two-thirds derive from fragmentation of circumstellar disks, the former predominantly (but not exclusively) ending up as stars and the latter predominantly (but not exclusively) as brown dwarfs. The highly unstable nature of non-hierarchical few body ensembles means that these small clusters are very short-lived: the systems achieve more or less hierarchical configurations over a few internal crossing timescales through the formation and hardening of binaries and the ejection of cluster members. Although this process has been well studied as a purely Nbody problem (e.g., van Albada 1968; Sterzik & Durisen 1995, 1998), the present situation is considerably more complex, since such dynamical interactions occur in the presence of significant quantities of gas (both in the form of circumstellar disks and in continued inflow down the filaments). Consequently, the stars continue to accrete during the break-up of the cluster, this accretion phase being terminated for each object as it is ejected from its natal gas reservoir. The interplay between chaotic few body dynamics and the accretion process results in a highly inequitable distribution of final stellar masses (Bonnell et al. 1997; Delgado-Donate, Clarke & Bate 2003a): objects that acquire high masses early on are relatively hard to dislodge from the deepest part of the mini-cluster potential and therefore continue to have access to a copious reservoir of accretable gas. Objects whose orbital trajectories result in a slower initial gas acquisition are instead more susceptible to slingshot ejection and are thereby deprived of a reservoir of accretable material, a fact invoked by Reipurth & Clarke (2001) to explain the formation of brown dwarfs. Such a scenario thus builds up an IMF that is—within the finite sample statistics of the simulations—broadly compatible with the observed IMF (Bate, Bonnell, & Bromm 2003). Although, as noted above, the observed IMF is also reproducible by a wide range of scenarios involving quite different physics, this is the first time that an IMF has 'emerged' from a hydrodynamical simulation of clustered star formation, and it is obviously encouraging that the products of such simulations are in this respect perfectly compatible with observations.

In other respects, also, these turbulent fragmentation calculations are successful in reproducing a variety of observational diagnostics. For example, they produce a suitably high fraction of binary and higher order multiple systems and a velocity dispersion of stars that—at a few km s^{-1}—is compatible with the observed kinematics of young stars. Although Reipurth & Clarke (2001) hypothesized that low-mass objects (brown dwarfs) might be ejected from these multiples with significantly higher velocities than their heavier (stellar) counterparts, the ejection velocities in the simulations are in fact quite insensitive to mass, a result that is necessary in order to explain the similar brown-dwarf fractions found in open clusters compared with the field (Moraux & Clarke 2004). A further appealing aspect of the simulations is that accretion disks are truncated by gravitational interactions and thus stars emerge from their natal mini-clusters with a wide range of initial disk masses and radii (see below, however, for a discussion of the predicted disk sizes in brown dwarfs). Such an initial spread is necessary in order to explain the fact that the 'clock' for disc dispersal is observed to vary widely from star to star (Armitage, Clarke, & Palla 2003). Finally, the best evidence in favor of an origin of stars in compact mini-clusters is the direct evidence of small scale clustering amongst the youngest stars. For example, Reipurth (2000) found a very high fraction of multiple sources ($N \geqslant 3$) in a survey of young (10^5 years) deeply embedded outflow sources. Likewise, high resolution imaging of the environment of the nearby Herbig Ae star, HD 104237, using STIS coronagraphy and complementary ground-based imaging, revealed that the system is at least a (non-hierarchical) sextuplet (Grady et al. 2004; Feigelson, Lawson, & Garmire 2003). Most recently, Lada et al. (2004), found that although optically bright stars in the Orion Nebula Cluster are rather smoothly distributed (see also Bate, Clarke, & McCaughrean 1998), deep near-infrared imaging reveals considerable sub-structure in the stellar groupings in the background molecular cloud, consistent with the notion that stellar clusters are hierarchically assembled from small N groupings (Scally & Clarke 2002; Bonnell, Bate, & Vine 2003).

These results paint an apparently rosy picture for the success of turbulent fragmentation models, almost suggesting that the subject is unsuitable for inclusion in a review of unsolved problems! However, there clearly *are* unsolved problems with this scenario that fall into several categories.

2.2. *Unsolved problems*

2.2.1. *Missing physics*

The first unsolved problem is that, despite their undoubted empirical success in many areas, these simulations undoubtedly neglect or over-simplify important physics. Most obviously, they omit magnetic fields for the technical reason that magnetic fields have not thus far been incorporated successfully in SPH calculations (although see Price & Monaghan 2004). Proponents of these models would argue that they apply to regions of the cloud which have—by whatever means—solved their magnetic support problem so that they have achieved supercritical mass-to-flux ratios, and would furthermore argue that the empirical successes described above lend some credence to this view. Nevertheless, such arguments can hardly be regarded as a substitute for detailed modeling. Until recently, simulations of MHD turbulence have omitted physical diffusion processes and hence imposed a fixed mass to flux ratio. The simulations of Li & Nakamura 2004, which include the effects of ambipolar diffusion, provide a first step towards the study of MHD turbulence in the absence of flux freezing, albeit in sheet geometry.

Another oversimplification in the current generation of turbulent fragmentation calculations is the use of a simple piecewise polytropic equation of state, motivated by the results of radiative transfer calculations in spherical symmetry (Larson 1969; Masunaga & Inutsuka 1999). It is not at all clear that such an effective equation of state is necessarily valid for structures collapsing with filamentary/disc geometry. Since the fragmentation of circumstellar discs (which plays such a prominent role in the turbulent fragmentation calculations) is sensitive to the ratio of cooling time to the local orbital timescale (Gammie 2001; Rice et al. 2003a), it is vitally important to apply the correct thermodynamic description to the gas in such disks. The recent work of Whitehouse & Bate (2004), which incorporates radiative diffusion in SPH, is thus of particular importance to future modeling of this problem.

2.2.2. *Observational tests*

The second category of unsolved problems associated with these calculations is observational, notwithstanding the wide variety of observational areas where the models have been a striking success. One shortcoming—currently shared by *all* star formation calculations—is the inability of such models to produce enough extreme mass ratio binaries. Although the relative lack of such systems observed at small separations is well known (the 'brown dwarf desert,' Marcy & Butler 1998; Zucker & Mazeh 2001), the binary population as a whole (i.e., including all separations) shows a marked preference for unequal mass pairs (for example, in the classic compilation of Duquennoy & Mayor (1991), the companion mass ratio distribution for solar mass stars is rising towards the incompleteness limit at $q \sim 0.2$). The reason that *all* models fail to produce such unequal mass pairs is that—regardless of the mechanism and geometry of the initial binary fragmentation—the continued accretion of gas from the parent core tends to equalize the binary component masses (Bate 2000). [Note that although calculations may often report the production of pairs that differ greatly in mass, this result does *not* persist in calculations that are run to the point that the majority of gas is accreted; see Delgado-Donate et al. (2004).] In the turbulent fragmentation calculations, long lived low-mass companions *only* result from the incomplete dynamical decay of complex multiples: in some cases, low mass objects (brown dwarfs) are deposited in wide (hierarchical) orbits instead of being completely ejected from the system (Delgado-Donate et al. 2004). An obvious observational test of this scenario would be to examine the wide star-brown dwarf pairs discovered by 2MASS (Gizis et al. 2001), to ascertain whether the 'primaries' of these systems are in fact themselves multiple systems when observed at high resolution.

Notwithstanding the result of this test, it would seem clear that, overall, the incidence of low-mass companions in the simulations is insufficient to match the Duquennoy & Mayor statistics for binaries with solar-mass primaries. It should be noted, however, that only the closer pairs in the Duquennoy & Mayor (1991) sample resulted from their homogeneous radial velocity survey and that the results on wider pairs—which contribute most of the low-mass companions—are culled heterogeneously from the literature. Given the theoretical difficulty of producing extreme mass ratio pairs, it would perhaps be timely to revisit this problem observationally, so that the magnitude of the discrepancy is clear.

A better known problem with the turbulent fragmentation models concerns their predictions for the sizes of disks around brown dwarfs. As stressed by Bate, Bonnell, & Bromm (2003), the reason that disks around brown dwarfs are smaller in the simulations than their counterparts around stars is mainly due to the fact that, whereas brown dwarfs are ejected from their natal clusters following close interactions that prune their disks, stars tend to remain in gas-rich regions after interactions and hence have the opportunity to re-grow their disks. Consequently, the incidence of disks of size >10 AU is an order of magnitude higher in the simulations for stars than for brown dwarfs. Unfortunately, the resolution limit of the simulations means that it is impossible to make detailed predictions about the incidence of disks around brown dwarfs on a scale of <10 AU, this being the scale on which there is now ample observational evidence (based both on L-band excess and spectroscopic accretion diagnostics) for disks and/or accretion in brown dwarfs (Muench et al. 2001; Natta et al. 2002; Liu, Najita, & Tokunaga 2003; Jayawardhana, Mohanty, & Basri 2003; Jayawardhana et al. 2004). However, although we currently have no evidence that the majority of brown dwarfs have disks larger than 10 AU (which would contradict current turbulent fragmentation models), the data hint at a possible conflict: if brown dwarf discs are smaller than stellar discs, then one might reasonably expect them to fade more quickly (Armitage & Clarke 1997), and yet there is no evidence for this from the relative statistics of disks around stars and brown dwarfs in objects of similar ages (Jayawardhana et al. 2004). Pending a fuller understanding of the processes driving disk evolution (see Section 4.2), it is hard to make a more definitive statement than this at the present time. Evidently the issue will only be settled by measurements that are sensitive to dust at larger radii, so that the *Spitzer Space Telescope* and ALMA will play a vital role, particularly since the latter holds the prospect for imaging extended dust disks around brown dwarfs.

In other observational areas, these turbulent fragmentation models remain untested at the present time. Whereas to date most emphasis has been placed on comparing the properties of the resulting *stars* (and brown dwarfs) with observations (see Bate, Bonnell, & Bromm 2002a,b 2003; Goodwin, Whitworth, & Ward-Thompson 2004a; Delgado-Donate et al. 2004), relatively little attention has been paid to the issue of whether the gaseous structures produced are compatible with the observed properties of molecular clouds. This question has been extensively examined in the case of calculations that do not follow the formation of individual stars (see Padoan et al. 1998; Padoan et al. 1999; Ossenkopf, Klessen, & Heitsch 2001; Gammie et al. 2003; Ballesteros-Paredes, Klessen, & Vazquez-Semadeni 2003), but it is obviously impossible to assess the relationship between the stellar and gas properties in these simulations. Probably the simplest discriminant between these turbulent fragmentation models (where young stars arise in few body clusters) and conventional models where one or two stars form per molecular cloud core, lies not in the overall velocity dispersion of the stars but in the *relative* velocity between the stars and the gas against which they are projected. To date, such an exploration has been hampered by the relative paucity of stellar radial velocity data, especially in the most deeply embedded regions. The recent demonstration (Doppmann, Jaffe, & White 2003) that radial velocities can be measured even in deeply embedded environments such as Ophiuchus, using photospheric features in the infrared, is thus particularly to be welcomed.

2.2.3. *Whence the turbulence?*

The third category of unsolved problems concerns the initial conditions of the turbulent fragmentation simulations. The turbulence injected at the outset of the simulations plays a vital role in the subsequent evolution of the system, since it generates the non-linear density structure that sets the stage for the fragmentation of the gas into multiple small N stellar clusters. In the absence of such turbulence, the cloud would instead have collapsed monolithically into a single object. Moreover, the turbulence also endows the cloud with local vorticity, and hence provides the angular momentum required to produce disks and binary systems. In these respects, the turbulence is of fundamental importance. However, one should note that, although in the case of simulations starting with individual cores the details of the kinetic energy input into the cores (for example, the power spectrum employed or the level of turbulent energy) do affect the properties of the resulting binary stars (Goodwin, Whitworth, & Ward-Thompson 2004b) and the IMF in the substellar regime (Delgado-Donate, Clarke, & Bate 2003b), the use of different initial turbulent power spectra applied to the whole cloud has very little effect on the resulting stellar population (Bate, private communication). This result can be understood inasmuch as the distribution of stellar masses produced is a consequence of competitive accretion in the context of chaotic few body dynamics (Bonnell et al. 1997; Delgado-Donate, Clarke, & Bate 2003a); although such few body ensembles are, in effect, set up by the gravitational amplification of structures originally endowed by the turbulence, detailed memory of these progenitor structures appears to be erased during the competitive accretion process.

To some extent, then, this result relieves the turbulence from having to have a very specific set of properties (apart from being supersonic, and thus producing non-linear density fluctuations in shocks) in order to give rise to a robust set of stellar parameters. Nevertheless, the fact that such supersonic motions are observed within molecular clouds, and that they have such profound implications for the star formation in the cloud, makes the origin of such motions of some considerable interest. Ideally one would like a theory for their origin that could both reproduce those properties that can be inferred from observations [specifically the observed size-linewidth relation (Larson 1981), which constrains the power spectrum of the velocity field] and could also make predictions for those properties that *cannot* be deduced directly from observations.

Of these latter, the most important question is simply whether the turbulence represents the one-off injection of kinetic energy during the process that assembles a giant molecular cloud, or whether this kinetic energy is continually replenished, so as to at least partially offset the dissipation of kinetic energy in shocks. In the former case, cloud lifetimes are short: gas is converted into stars within a few crossing times (typically a few Myr) and an excessively high star formation efficiency can be avoided only through invoking some form of destructive feedback process that can disperse the cloud once only

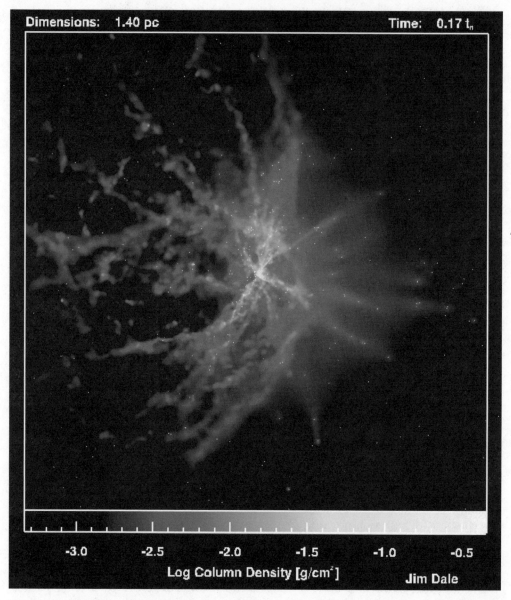

FIGURE 2. Photoionized cloud structure from Dale et al. 2004. Photoionizing radiation exaggerates the inhomogeneous nature of the initial conditions by the preferential acceleration of material in low density channels between filaments.

a few per cent of its mass is converted into stars (see Elmegreen 2000a for arguments in favor of such a scenario). Stellar winds and photoionization are the most likely feedback agents in that respect (Whitworth 1979; Franco et al. 1996; Clarke & Oey 2002), although it is only very recently that simulations have developed the capability to model such feedback in cloud structures that are realistically inhomogeneous (Dale et al. 2004; see Fig. 2). If, instead, the turbulent energy is re-supplied, then clouds can be globally supported against collapse over many free-fall timescales. Simulations that employ an ad hoc forcing of the velocity field at the large scale (Vazquez-Semadeni, Ballesteros-Paredes, & Klessen 2003) develop a characteristic star formation history: an initial burst

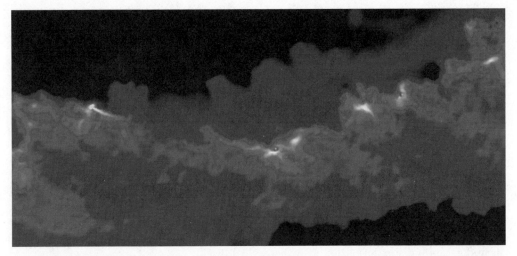

FIGURE 3. Simulation of a clumpy ISM passing through spiral shocks from Bonnell et al. 2004 (in preparation). The resulting clouds bear a visual resemblance to observed molecular clouds and, more significantly, reproduce the observed size-linewidth relation in molecular clouds

of star formation corresponding to the collapse of denser regions and then an extended period of quiescence, in which the residual cloud is supported by the continued resupply of turbulent support. Such models can therefore produce star formation efficiencies that are appropriately low (globally) without demanding an extraordinarily efficient feedback mechanism to quench star formation. On the other hand, the lack of observed molecular clouds that are 'quiescent' (in the sense of lacking any evidence of recent star formation—such as embedded protostars) argues against this quiescent period being extensive, and so would still require some agent to disperse the cloud after the onset of star formation.

Currently, it is generally believed that the source of the turbulence is external to the cloud, rather than deriving from star formation within the cloud. The best evidence for this is simply that the observational indicators of turbulent motions—superthermal line widths of molecular gas—are observed to be present in the Maddalena Cloud, which has not undergone any significant star formation to date (Williams & Blitz 1998). This suggests that molecular clouds are born (assembled) with their supersonic internal motions already in place and that to understand their origin, one has to understand the process of molecular cloud formation and, in particular, the relationship between molecular clouds and the wider environment of the galactic disk. Such studies are in their infancy, because of the extreme computational requirements of performing simulations that both model the global environment and yet have sufficient resolution to model the internal dynamics of molecular clouds. Recent works that have explored molecular cloud formation in the context of global disk simulations include Kim, Ostriker, & Stone (2003)—which follows the evolution of the magnetorotational instability (MRI) in galactic disks—and Wada, Meurer, & Norman (2002), who instead rely on large scale gravitational instabilities. In both cases, however, the molecular clouds thus formed are close to the resolution limit of the calculations and thus it is not possible to examine the kinematic state of the clouds produced—nor whether these clouds are subject to continued driving of their internal motions from the larger scale environment once they are formed. Bonnell et al. (2004), by contrast, have considered considerably simpler physics, namely the effect of collisions in a clumpy medium subject to a prescribed spiral potential, and, through such simplification, have attained the capability of resolving both the large scale parent medium and the internal structure and kinematics of the clouds thus produced. Pilot simulations exhibit an encouraging resemblance to observed molecular clouds (see Fig. 3).

In summary, then, turbulent fragmentation calculations present an appealing scenario that is compatible with a range of fundamental stellar data, such as the IMF, the ubiquity of binary systems and the level of stellar velocities in star-forming regions. Such models, however, omit important physics (magnetic fields and realistic treatment of radiative transfer), have not yet been properly tested with regard to the predicted properties of the *gas* and, in common with all star-formation scenarios, have problems generating extreme mass-ratio binaries. Perhaps most importantly, they employ ad hoc initial conditions and cannot themselves answer the question of the origin of such conditions (specifically the presence of a supersonic internal velocity field). The origin of these motions must be sought in simulations that instead model the larger scale ISM, but with sufficient dynamic range of resolution so as to be able to study the internal properties of the star forming clouds produced. Such simulations are in their infancy, due to the formidable computational challenge they present, but must surely represent the next big challenge in understanding star formation at a theoretical level.

3. IMF issues

3.1. *Variations in the IMF at the low mass end*

In recent years, several groups have obtained the first (arguably) complete censuses of stellar and substellar objects in nearby star forming regions (see, for example, Muench et al. 2002; Luhmann et al. 2003a,b. Whereas the IMF in the solar neighborhood contains contributions from stars that potentially originated in a range of environments and at a range of epochs, the obvious advantage of studying the mass function in star-forming regions is that one can, in principle, study the influence of environment on the IMF.

Before describing the results of these surveys, however, it is necessary to issue an important caveat. Mass functions are generally constructed either through placing stars on a Hertzsprung-Russell diagram and deriving masses (and ages) through the use of theoretical pre-main sequence tracks, or else through performing Monte Carlo experiments in order to constrain the combinations of star-formation history and IMF that would reproduce the observed luminosity function in some waveband. In either case, therefore, the derived IMF is sensitive both to the reddening and to the theoretical pre-main sequence evolutionary tracks employed. Although there has been considerable progress in recent years in extending such tracks to low masses (due largely to the more realistic treatment of the atmospheres of these objects, which deviate strongly from black body emitters: see Baraffe et al. 1998; Chabrier et al. 2000; Baraffe et al. 2002), it is strongly desirable that such tracks be empirically calibrated, using dynamical masses of objects in pre-main sequence binaries. Hillenbrand & White (2004) have recently reported on the current status of such attempts, demonstrating that a good convergence between masses predicted by tracks and dynamically obtained masses is found only for stars more massive than $1.2\ M_\odot$. Moreover, there is a serious lack of dynamical calibrators at the low-mass end, the minimum dynamical mass of a pre-main sequence star being $0.3\ M_\odot$. This means that all conclusions about possible variations in the substellar IMF in star-forming regions (see below) have to be interpreted in this light. This uncertainty makes the current enterprise of using infrared spectroscopy to increase the census of dynamical masses for low-mass pre-main sequence stars (e.g., Prato et al. 2002) of particular timeliness.

Leaving aside these uncertainties, the current situation is that the IMF shows no significant variations between different star-forming regions, apart from the fact that in the Taurus dark cloud there is an apparent deficit of brown dwarfs compared with the other regions (Luhmann et al. 2003a). A number of explanations have been proposed for this (see, e.g., Kroupa & Bouvier 2003; Goodwin, Whitworth, & Ward-Thompson 2004b; Delgado-Donate, Clarke, & Bate 2003b), which relate to such environmental differences as the lower levels of turbulence and the absence of OB stars in Taurus. It has also been argued that the deficit of brown dwarfs in Taurus is due to the diffuse nature of the region and the fact that brown dwarf surveys were initially centered on the relatively small areas of the cloud where the stars are concentrated: a differential velocity at birth between stars and brown dwarfs could then account for this apparent difference, even if the IMF were normal in Taurus. Surveys with increasingly large areal coverage are, however, so far failing to unearth the missing population of brown dwarfs in Taurus (e.g., Briceno et al. 2002), and, in any case, the current generation of turbulent fragmentation models do *not* predict that brown dwarfs are born with a higher velocity dispersion than stars.

3.2. The characteristic mass for star formation—does it vary?

Apart from this possible evidence of variation of the IMF in the substellar regime, the apparent uniformity of the IMF makes it very hard to discriminate between contending models, since its rather featureless (log-normal/multi power law) form is readily replicated by models involving a diversity of physical processes (see, e.g., Adams & Fatuzzo 1996 for a discussion of this point). The most noteworthy feature of the IMF is that, in all well-studied regions where the mass function can be determined over a large dynamic range, there is a flattening of the IMF below around a solar mass (Scalo 1986; Kroupa, Tout, & Gilmore 1990; Kroupa 2002), which endows the stellar population with a mean, or characteristic, mass that is roughly this value. Although a variety of physical processes may be invoked to replicate the functional form of the IMF, the existence of an (arguably universal?) mass scale of around 10^{33} g presents a more concrete challenge to models. Although this mass scale can be interpreted as a Jeans mass for typical dense cores in molecular clouds (e.g, Larson 1998), this explanation only pushes back the explanation as to what sets this scaling in molecular clouds. Clarke & Bromm (2003) link this mass scaling to the temperature and mean internal pressure of molecular clouds, the latter of which is influenced by the potential well of the parent galaxy, and map out how the characteristic mass would be expected to vary with cosmic epoch and host galaxy.

Such speculations are, however, extraordinarily difficult to test observationally, since to detect any variation of the characteristic mass with environment and epoch, it is necessary to deduce information on the IMF on mass scales both larger than *and* less than the characteristic mass at which the IMF flattens. Unfortunately, the IMF is only well determined observationally on both sides of this mass scale in the nearby star forming regions discussed above, where one is necessarily not probing a very diverse array of environments. (Note however that *HST* has enabled the IMF to be probed down to roughly a solar mass in the 30 Doradus region of the LMC, and has shown *no* evidence for any variation in the characteristic mass scale, even in this rather exotic environment in the close vicinity of the R136 starburst cluster; Brandner et al. 2001). At more distant cosmic epochs, by contrast, one's information on the IMF is instead derived over limited dynamic ranges of mass in each environment (for example, globular clusters probe the IMF only up to ∼0.8 $M_\odot$, whereas population synthesis models in Lyman break galaxies constrain the IMF slope for stars only down to spectral type of early B; Pettini et al. 2003). Theoretical interest in the IMF is therefore somewhat stalled, but a convincing demonstration for a significantly different characteristic mass scale in some region would revitalize this field (see, for example, Smith & Gallagher 2001; d'Antona & Caloi 2004 for arguments for significantly different IMFs in populous clusters).

3.3. *The high-mass tail*

We may summarize the discussion in this section so far as showing that the IMF is probably somewhat variable at the low-mass end (but we don't know why) and that the characteristic mass is apparently rather invariant in the environments we can study (but again we don't know why). At the high-mass end, the IMF seems to follow the Salpeter slope in all regions studied (Salpeter 1955), with the possible exception of the field population of the Magellanic Clouds, where the slope is apparently much steeper (Massey 2002). The maximum mass of stars observed in populous clusters such as 30 Doradus implies that the upper end of the IMF is not limited purely by finite sampling of an underlying distribution that follows the Salpeter distribution to arbitrarily high mass: instead there is evidence for an upper limit to the mass function of $\sim$200 $M_\odot$ (Weidner & Kroupa 2004; see also Elmegreen 2000b).

However, such simplicity of the mass function at the high-mass end in no way reflects any consensus as to the mechanism by which massive stars are produced, and this issue certainly deserves to be flagged as a major unsolved problem in star formation. Over the years, there has been a persistent concern that massive star formation may not merely operate as a scaled-up analogue of the low-mass star formation process, due to the greater importance of radiation pressure on dust as a feedback agent opposing the accretion of gas onto the star. In an influential paper, Wolfire & Cassinelli (1987) argued that massive stars may only be formed by accretion if the dust in massive star formation regions is drastically depleted in small grains compared with the ISM. Subsequent calculations have shown that the disruptive role of radiation pressure is much less significant if the assumptions of spherical symmetry (Yorke & Sonnhalter 2002) and/or steady-state accretion (Edgar & Clarke 2003) are relaxed. In the meantime, however, Bonnell, Bate, & Zinnecker (1998) developed models in which massive stars are instead formed through stellar collisions, which occur through the hardening of binaries in the cores of clusters that are postulated to pass through a brief ultra-dense phase (stellar density around 10^8 stars pc^{-3}; see also Bonnell & Bate 2002; Bonnell, Vine, & Bate 2004). Evidently such densities exceed those found in observed regions of massive star formation by many orders of magnitude, and so it is hypothesized that these ultradense cluster cores are re-inflated by mass loss due to stellar feedback processes such as photoionization or stellar winds. Although many aspects of this process need considerably more exploration (for example, the effect of feedback within the collisional scenario; Edgar & Clarke 2004), there is currently no overwhelming theoretical counterargument to massive stars being formed through either accretion *or* collisions. Can observations of massive stars then place some constraints on the relative roles of these two processes?

Possible observational evidence comes in two flavors. The first concerns the existence of disks around young massive stars: there are now a number of examples of large scale ($\sim$10^4 AU) flattened gaseous structures in regions of massive star formation which have been interpreted as disks (Beltran et al. 2004; Chini et al. 2004). This interpretation is strengthened by the fact that such structures appear to be rotating about their minor axes, which are roughly aligned with the axes of observed outflows. It should be noted, however, that the observed velocity gradients are much too low for the flows to be centrifugally supported (given their masses as deduced from millimeter line and continuum observations), and thus these structures should be regarded, on these scales, as mildly flattened rotating inflows, rather than centrifugally supported disks. In simulations of the collision scenario, disks form around massive stars (typically high-mass binaries)

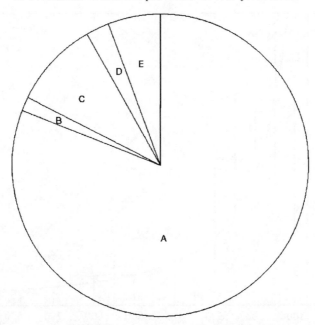

FIGURE 4. Pie chart describing the demography of O stars in the Milky Way. A denotes stars in clusters/associations, B indicates classical runaways, C marks stars that could arguably be runaways, D denotes isolated O stars with low-mass clusters detected, and E indicates isolated O stars with no low-mass clusters detected. Adapted from de Wit et al. (2004).

that have already formed through collisions (Bonnell, private communication) and the disks do not in this case provide the main conduit through which mass is acquired by the central stars. Such disks are thus expected to be less massive than the stars they contain and are moreover observed in the simulations at the stage that the cluster core is in the ultradense phase required for collisions to occur. It is not clear that such disks would survive the feedback processes that are invoked to re-inflate the core of the cluster to the substantially lower densities observed in regions of massive-star formation. Thus, the extended and very massive disks that have been detected recently may be hard to reconcile with the collision hypothesis, though more work is required in this area.

The other type of observational constraint on the origin of massive stars derives from their demography. The collision scenario *requires* that stars thus formed were at one point in the core of a massive cluster and that the density of the cluster core at this stage was extremely high. It is thus often argued that the tendency of massive stars to be found in clusters or associations (e.g., Mason et al. 1998; Testi, Palla, & Natta 1999; Maíz-Apellániz & Walborn 2003) and the fact that massive stars are usually located in the densest regions of young clusters (Bonnell & Davies 1998) are both qualitatively consistent with the notion that massive-star formation involves collective effects.

Recently, de Wit et al. (2004) have undertaken a thorough investigation of the demography of the brightest O stars in the Milky Way (see Fig. 4). Their results support a picture in which a high fraction of OB stars are either located in clusters or associations, or have good kinematical evidence for having been ejected from such regions. Nearly 10% of the sample are indeterminate in the sense that, although isolated, their kinematic data is arguably consistent with their having originated in clusters/associations. This however leaves 10% of the sample that appear to be genuinely isolated, with a low probability of origin in a cluster or association of OB stars. In around a quarter of these latter cases,

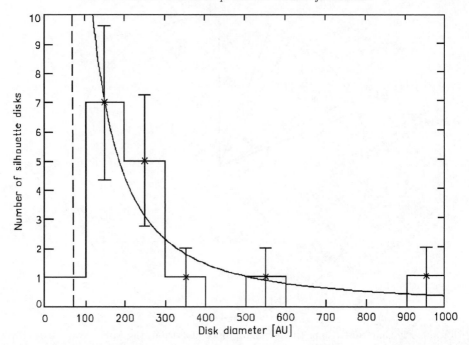

FIGURE 5. Histogram of disk sizes for the silhouette disks in the Orion Nebula Cluster, mainly derived from images from the HST Wide Field Planetary Camera. The dashed vertical line represents the resolution limit of 0.15 arcseconds. Courtesy of McCaughrean and Roddmann (in preparation).

deep imaging reveals a small cluster of low-mass stars around the OB star. These clusters are too low in mass to have plausibly produced the conditions for stellar collisions, but the survival of the cluster argues strongly against such stars being runaways.

These results provide no proof either way on how the *majority* of OB stars must have formed, since—although clustering in OB stars is suggestive of the collisional mode—it in no way excludes stars having formed through conventional accretion scenarios. However, the 10% or so of stars that are genuinely isolated from other OB stars, without possibility of being runaways, could *not* have formed through collisions. Therefore, at the very least, this study shows that some massive stars must form through accretion.

A final note on the controversy surrounding massive-star formation relates to the IMF. Perhaps the most puzzling aspect of the observed IMF is that it shows no features at the mass scale at which radiation pressure on dust becomes important ($\sim$10 $M_\odot$). Although recent simulations have shown that stars *can* form by accretion, despite the effect of radiation pressure on dust, this requires appropriate initial conditions: where these are not met, there is a tendency for stars to end up with masses close to this critical mass of $\sim$10 $M_\odot$ (Edgar & Clarke 2003, 2004). The fact that the observed IMF shows no such feature is strong evidence that—whatever the mechanism for massive star formation—it is more insensitive to radiative feedback on dust than many of the current models are.

4. Circumstellar disks and the link to planets

4.1. *The evidence for disks*

It has long been recognized that the formation of disks around young stars is a necessary consequence of angular momentum conservation during the collapse of rotating cloud cores and that, moreover, disk formation is a necessary precursor to planet formation (Lynden-Bell & Pringle 1974). Prior to *HST*, the existence of disks was mainly inferred from the spectral energy distribution of Classical T Tauri stars (Adams, Lada, & Shu 1988), from the presence of blueshifted wind emission (ascribed to occultation of the redshifted wind by an optically thick disk: Appenzeller et al. 1985; Edwards et al. 1987) and through direct imaging—at mm wavelengths—of a handful of the most extended disks (Sargent & Beckwith 1987; Koerner, Sargent, & Beckwith 1993). The high-resolution optical imaging capabilities of *HST* have transformed this situation, both through the detection of dust disks silhouetted against bright nebular emission in the Orion Nebula Cluster (McCaughrean & O'Dell 1996; Bally, O'Dell, & McCaughrean 2000) and through coronagraphic imaging of disks in scattered light (see Grady et al. 2003; Schneider et al. 2003). Nevertheless, it should be emphasized that it is still possible to image disks only on a rather large scale (see Fig. 5 for the size distribution of silhouette disks, noting the resolution limit at around 70 AU corresponding to ∼0.15 arcseconds at the distance of Orion; likewise coronagraphic imaging excludes regions within ∼0.5 arcsecond of the central star). Direct imaging of the innermost (potentially planet forming) regions of circumstellar disks is a challenge for the upcoming generation of optical interferometers and, at millimeter wavelengths, for ALMA.

4.2. *Disk dispersal—a link to planet formation?*

Since disks are believed to contain the raw material for planet formation, the study of such disks can in principle throw light on the mechanism by which planets form. For example, the observed lifetimes of disks (as evidenced by the decline in sources with L-band excesses in clusters older than a few Myr; Haisch, Lada, & Lada 2001) has been used to fuel the debate on whether giant gaseous planets form by core accretion or gravitational instability (Pollack et al. 1996; Kornet, Bodenheimer, & Rozycka 2002; Boss 1997, 2004). Other studies have instead examined the dispersal of disks through the age dependence of the *accretion* luminosity, a quantity that may be inferred from the U-band excess for stars of known spectral type (Gullbring et al. 1998). Such studies reveal a large spread in the accretion rates of stars of a given age in a given region, implying that disk evolution is controlled by disk clocks that vary from star to star (Hartmann et al. 1998; Armitage, Clarke, & Palla 2003). Most recently, there have been claims based on WFPC2 photometry (Robberto et al. 2004; Romaniello, Robberto, & Panagia 2004) for accretion rates at a given age being respectively much lower and much higher (compared with those measured in Taurus) for stars in the Orion Trapezium Cluster and in the Large Magellanic Cloud. It is currently unclear whether this is predominantly a function of the rather different stellar mass regimes of the three samples, or whether it demonstrates a systematic variation of disk clocks with environment (e.g., proximity of OB stars or metallicity). Similar studies in a range of environments (e.g., Sicilia-Aguilar et al. 2004), and over a range of stellar masses, will help to clarify this issue.

In order to make the link between disk dispersal and planet formation, however, it is important to understand the *mechanism* by which disks are dispersed. Are disks simply consumed (as is often assumed) by the process of planet formation? If this is the case, planet formation is a *self*-limiting process. Alternatively, is there another dispersive process (either internal or external to the disk) against which planet formation has to compete? In some ways, the latter possibility is more interesting, since it would in principle lead one to understand the factors that determine the incidence of planetary systems.

It is evident that in some star forming regions, the dominant disk dispersal agents are environmental. Although neither star-disk collisions (Scally & Clarke 2001) nor heating by diffuse X-rays (Alexander, Clarke, & Pringle 2004) are significant disk destruction agents even in the richest clusters, the Orion Nebula furnishes ample evidence that some of its disks are being photoevaporated by the ultraviolet flux of the cluster's central OB stars. This process is believed to be manifest as the proplyd phenomenon, whereby bright ionized envelopes are observed around the majority of stars in the core of the Orion Nebula Cluster (O'Dell, Wen, & Hu 1993; Bally, O'Dell, & McCaughrean 2000) and are well modeled as a flow of matter from the disk surface that is heated by the OB stars' far ultraviolet continuum (Johnstone, Hollenbach, & Bally 1998; Stoerzer & Hollenbach 1999). The predicted mass-loss rates in these flows, which have been confirmed by emission line modeling (Henney & O'Dell 1999), are very high. When combined with the disk masses in these systems, inferred from mm observations (Bally et al. 1998), the timescales for complete photoevaporation are very short (a few $\times 10^4$ years). This 'proplyd lifetime problem' has been suggested as evidence for the extreme youth of the main ionizing source in the Orion Nebula (θ^1C Ori), suggesting that we are observing Orion at a special moment in its evolution. However, the observation of similar proplyd-like structures in Carina (Smith, Bally, & Morse 2003) argues instead that in both cases the proplyd sources are those that have only very recently been exposed to the central ultraviolet source, and prior to that point they were screened by dense molecular gas.

Evidently, then, there is no difficulty in answering the question of what disperses circumstellar disks in the case of proplyd sources (note that although only the outer-most, most loosely bound, portions of the disks can be directly photoevaporated, Adams et al. (2004) have argued that the setting up of a Parker wind is also able to remove material from the inner disk). However, it would be misleading to conclude that this mechanism is applicable to the majority of disk systems, even in the Orion Nebula Cluster, since the efficiency of photoevaporation declines steeply for disks that are more than a few tenths of a parsec of the central OB stars. Scally & Clarke (2001) showed that for any plausible dynamical model of the Orion Nebula Cluster, the majority of the cluster's stars spend insufficient time in the central region for them to suffer significant mass loss due to photoevaporation. Since it is often postulated that the Orion Nebula Cluster represents an earlier evolutionary stage of open clusters such as the Pleiades (e.g., Kroupa, Aarseth, & Hurley 2001), this suggests that one would not expect open clusters to be deficient in planets due to the photoevaporation of their natal disks.

For the bulk of stars in the Galaxy, therefore, one can neglect disk dispersal due to environmental effects. For these stars, therefore, one has to seek an *intrinsic* mechanism to explain the fact that circumstellar diagnostics fade over a 1–10 Myr timescale. A possible clue as to the dispersal mechanism may be provided by the longstanding observation (Skrutskie et al. 1990; Hartigan et al. 1990) that stars apparently lose their disks (as evidenced by excess emission in the near infrared) over a timescale that is a small fraction (1–10%) of the period over which the star possessed a disk. Such 'two timescale' behavior (Clarke, Gendrin & Sotomayor 2001) is based on the relative lack of 'transition objects' (objects with infrared colors that are intermediate between that of an optically thick disk and a pure stellar photosphere; Kenyon & Hartmann 1995) and is incompatible with any mechanism (e.g., viscous evolution/magnetospheric clearing) that produces a power law decline in disk quantities with time (Armitage, Clarke, & Tout 1999). Instead, it is necessary that the secular evolution of the disk diagnostics is at some stage supplemented by an additional mechanism that *rapidly* (i.e., on a $\sim 10^5$ year timescale) renders the inner disk (on scales less than a few AU) optically thin in the near infrared. Since the opacity is dominated by micron scale dust grains, this transition need not necessarily imply the dispersal of the inner *gas* disk, but must, at the least, involve the rapid depletion of such grains.

One possibility is that the disruption of the inner disk is associated with planet formation. This could be achieved either through the necessary precursor to planet formation (i.e., grain coagulation leading to the removal of opacity at a wavelength of a few microns) or else through the formation of a planet of sufficient mass that it dynamically cleared the inner disk of both gas and remaining grains (Rice et al. 2003b). In either case, the 'two timescale' behavior mentioned above means that the process must be regulated, perhaps by some threshold effect, such that rapid clearing sets in after the disk has spent a substantial prior period in the optically thick state. Although the detection of possible gaps and holes in discs is often assumed to be almost equivalent to a demonstration of planet formation, there are other mechanisms for clearing the inner disk in an appropriately rapid fashion. Clarke, Gendrin, & Sotomayor (2001) showed that the combination of viscous draining of the disk with photoevaporation of the outer disk by the ionizing continuum of the central star (Hollenbach et al. 1994) could replicate the required two timescale behavior: when the accretion flow through the outer disk drops to the low level where it matches the wind loss from a few AU, the inner disk is starved of resupply and empties on its (short) viscous timescale. Thus either planet formation (within a few AU) or photoevaporation of the disk can, in principle, reproduce the rapid clearing of the inner disk, but in both cases a disk is expected to remain at larger radii.

We thus have a general expectation that disks should clear from the inside out, essentially because all timescales are shorter at small radii. What evidence is there then of such 'doughnut sources' (i.e., those with inner holes)? Currently, the situation is rather confusing since, contrary to the expectations set out above, the Weak Line T Tauri stars (i.e., those lacking the spectral diagnostics of ongoing gas accretion) very rarely exhibit the millimeter emission indicative of cool outer disks, despite recent surveys probing to very low (few mJy) levels (Duvert et al. 2000; Mannings & Clarke 2004). The upper limits on the mass of millimeter emitting dust in these systems (around an earth mass) are not much greater than the detections of dust in the more massive debris disks (Wyatt, Dent, & Greaves 2003) and are one to two orders of magnitude lower than typical detections of millimeter emission in accreting T Tauri stars. This would suggest that the rapid clearing of disk material extends to the outer disk also, a requirement that is hard to achieve theoretically, given the very large dynamic range in timescales between the regions of the disk producing the near infrared and submillimeter emission. How could the disk's disappearance be so coordinated over such a large range of spatial scales?

In the last couple of years or so, however, several inner hole systems *have* been identified, though not, intriguingly, amongst Weak Line T Tauri stars (see Calvet et al. 2002; Rice et al. 2003b). Instead, it would appear that inner hole systems (i.e., disks that are optically thin in the near infrared and yet with evidence for excesses at longer wavelengths) are associated with ongoing accretion of gas on to the star at a range of levels (in the inner hole system with the highest accretion rate, GM Aurigae, the fact that the inner disk is optically thin in the near infrared either implies considerable depletion of micron size dust or else an accretion flow that is much closer to radial inflow than is the case for a conventional low viscosity protostellar disk). Evidently, then, the inner edge of the hole forms a leaky barrier, a feature that needs to be reproduced by disk clearing models. It may then be the case that in such systems, the outer disk and accretion onto the star decay together, thus explaining the lack of outer disks in stars without ongoing accretion. Such scenarios must however remain speculative, pending the assembly of better statistics on inner hole systems, an enterprise that will rely heavily on the sensitive mid-infrared capabilities of the *Spitzer Space Telescope*.

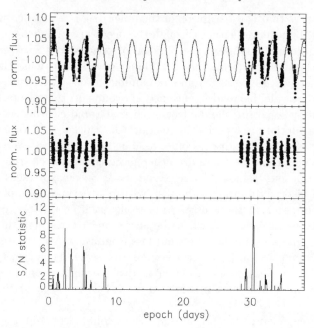

FIGURE 6. The feasibility of detecting planetary transits in pre-main sequence stars. The upper panel shows a simulated light curve for a 0.3 $M_\odot$ star with a 0.63 M_{jup} planet in a 2.8-day orbit, which is also subject to 10% photometric modulation from a starspot (with 2.3-day period) plus flickering on a 2-hour timescale at the 1% level. Notwithstanding these other variability sources and the gap in the data, the transit is detected. (Middle and lower panels show the data after subtraction of the detected starspot signal and the detection statistic for a single transit search). Courtesy of Suzanne Aigrain.

4.3. *Prospects of planet detections around young stars*

Given the theoretical uncertainties in the time required for planet formation by core accretion or disk instability models (see Pollack et al. 1996; Kornet, Bodenheimer, & Rozycka 2002; Boss 1997, 2004) and given also the uncertainty as to whether it is planet formation that disrupts the inner disk in the 'inner hole' systems described in Section 4.2.3 above (Clarke, Gendrin, & Sotomayor 2001; Rice et al. 2003b), it would obviously be a matter of extreme interest to be able to detect a planet around a pre-main sequence star. Even a single detection would start to place constraints on a process that is currently almost unconstrained by observations and would place an immediate upper bound on the time required for planet formation.

These considerations motivate a search that is considerably more challenging than the search for planets around older stars, since a number of techniques are problematical in the case of young stars, due to their variability and the possibility of complications from accretion and the presence of circumstellar material. For example, the precision of radial velocity measurements is considerably compromised in the case of young stars and likewise transit observations are complicated by the need to distinguish transits from the wealth of short term variations exhibited by young stars. Nevertheless, experience with filtering algorithms on synthetic data (Fig. 6; see Aigrain & Irwin 2004) suggests that a transit signal should be retrievable from stars with a realistic degree of periodic and aperiodic variability, and has motivated the upcoming campaign to seek planetary transits in a large sample of pre-main sequence stars in the Orion Nebula Cluster. The Space Interferometry Mission (SIM) will survey nearby star forming regions for astrometric signatures of planets and would be able to detect Jupiter-mass planets that have been postulated as agents of disk clearing in inner hole sources (Rice et al. 2003b).

Other detection methods, instead of being compromised by the messy circumstellar environments of young stars, exploit this feature to infer the presence of planets indirectly. Planets should distort circumstellar disks through the production of features such as gaps, inner holes and spiral waves which can in principle be revealed by high-resolution imaging (see Wolf et al. 2002 for simulations of the observability of planets in disks using ALMA). More speculatively, Clarke & Armitage (2003) have suggested that there may already be spectroscopic evidence for an embedded planet in two pre-main sequence stars undergoing FU Orionis outbursts, since periodic variations in the absorption line profiles are consistent with the signature of embedded hot Jupiters in these objects (Herbig, Petrov, & Duemmler 2003). If confirmed by further monitoring and modeling, this would push back the era of planet formation to very early times (i.e., less than a Myr). Evidently, the intense interest surrounding the possibility of observing planets in very young stars means that a variety of spectral/temporal/spatial bumps and wiggles will be tentatively interpreted as planets over the next few years. There is nevertheless a reasonable prospect that the next decade should yield some robust detections and that planet formation will become an observationally driven science.

REFERENCES

ADAMS, F. C. & FATUZZO, M. 1996 *ApJ* **464**, 256.
ADAMS, F. C., HOLLENBACH, D., LAUGHLIN, G., & GORTI, U. 2004 *ApJ*, **611**, 360.
ADAMS, F. C., LADA, C. J., & SHU, F. H. 1988 *ApJ* **326**, 865.
AIGRAIN, S. & IRWIN, M. 2004 *MNRAS* **350**, 331.
ALEXANDER, R. D., CLARKE, C. J., & PRINGLE, J. E. 2004 *MNRAS*, **354**, 71.
APPENZELLER, I., JANKOVICS, I., & OESTREICHER, R. 1985 *A&A* **141**, 108.
ARMITAGE, P., CLARKE, C., & PALLA, F. 2003 *MNRAS* **342**, 1139.
ARMITAGE, P., CLARKE, C., & TOUT, C. A. 1999 *MNRAS* **304**, 425.
ARMITAGE, P. J. & CLARKE, C. J. 1997 *MNRAS* **285**, 540.
ARONS, J. & MAX, C. E. 1975 *ApJ* **196**, L77.
BALLESTEROS-PAREDES, J., KLESSEN, R. S., & VAZQUEZ-SEMADENI, E. 2003 *ApJ* **592**, 188.
BALLY, J., O'DELL, C. R., & MCCAUGHREAN, M. J. 2000 *AJ* **119**, 2919.
BALLY, J., TESTI, L., SARGENT, A., & CARLSTROM, J. 1998 *AJ* **116**, 854.
BARAFFE, I., CHABRIER, G., ALLARD, F., & HAUSCHILDT, P. H. 1998 *A&A* **337**, 403.
BARAFFE, I., CHABRIER, G., ALLARD, F., & HAUSCHILDT, P. H. 2002 *A&A* **382**, 563.
BATE, M. R. 2000 *MNRAS* **317**, 1004.
BATE, M. R., BONNELL, I. A., & BROMM, V. 2002a *MNRAS* **332**, L65.
BATE, M. R., BONNELL, I. A., & BROMM, V. 2002b *MNRAS* **336**, 705.
BATE, M. R., BONNELL, I. A., & BROMM, V. 2003 *MNRAS* **339**, 577.
BATE, M. R., CLARKE, C. J., & MCCAUGHREAN, M. J. 1998 *MNRAS* **297**, 1163.
BELTRAN, M. T., CESARONI, R., NERI, R., CODELLA, C., FURUYA, R. S., TESTI, L., & OLMI, L. 2004 *ApJL* **601**, 187.
BONNELL, I. A. & BATE, M. R. 2002 *MNRAS* **336**, 659.
BONNELL, I. A., BATE, M. R., CLARKE, C. J., & PRINGLE, J. E. 1997 *MNRAS* **285**, 201.
BONNELL, I. A., BATE, M. R., & VINE, S. G. 2003 *MNRAS* **343**, 413.
BONNELL, I. A., BATE, M. R., & ZINNECKER, H. 1998 *MNRAS* **298**, 93.
BONNELL, I. A. & DAVIES, M. B. 1998 *MNRAS* **295**, 691.
BONNELL, I. A., VINE, S. G., & BATE, M. R. 2004 *MNRAS* **349**, 735.
BOSS, A. P. 1997 *ApJ* **591**, 581.
BOSS, A. P. 2004 *ApJ* **599**, 577.
BRANDNER, W., GREBEL, E., BARBA, R., WALBORN, N., & MONETI, A. 2001 *AJ* **122**, 858.

BRICENO, C., LUHMAN, K. L., HARTMANN, L., STAUFFER, J. R., & KIRKPATRICK, J. D. 2002 *ApJ* **580**, 317.

CALVET, N., D'ALESSIO, P., HARTMANN, L., WILNER, D., WALSH, A., & SITKO, M. 2002 *ApJ* **568**, 1008.

CASELLI, P. & MYERS, P. C. 1995 *ApJ* **446**, 665.

CHABRIER, G., BARAFFE, I., ALLARD, F., & HAUSCHILDT, P. 2000 *ApJ* **542**, 464.

CHINI, R., HOFFMEISTER, V., KIMESWINGER, S., NIELBOCK, M., NUERNBERGER, D., SCHMID-TOBRIECK, L., & STERZIK, M. 2004 *Nature* **429**, 155.

CLARKE, C. J. & ARMITAGE, P. J. 2003 *MNRAS* **345**, 691.

CLARKE, C. J. & BROMM, V. 2003 *MNRAS* **343**, 1224.

CLARKE, C. J., GENDRIN, A., & SOTOMAYOR, M. 2001 *MNRAS* **328**, 485.

CLARKE, C. J. & OEY, S. M. 2002 *MNRAS* **337**, 1299.

CRUTCHER, R. 1999 *ApJ* **520**, 706.

DALE, J. E., BONNELL, I. A., CLARKE, C. J., & BATE, M. R. 2004, *MNRAS*, submitted.

D'ANTONA, F. & CALOI, V. 2004 *ApJ*, **611**, 871.

DELGADO-DONATE, E. J., CLARKE, C. J., & BATE, M. R. 2003a *MNRAS* **342**, 926.

DELGADO-DONATE, E. J., CLARKE, C. J., & BATE, M. R. 2003b *MNRAS* **347**, 759.

DELGADO-DONATE, E. J., CLARKE, C. J., BATE, M. R., & HODGKIN, S. T. 2004 *MNRAS* **351**, 617.

DE WIT, W. J., TESTI, L., PALLA, F., VANZI, L., & ZINNECKER, H. 2004 *A&A*, **425**, 937.

DOPPMANN, G. W., JAFFE, D. T., & WHITE, R. J. 2003 *AJ* **126**, 3043.

DUQUENNOY, A. & MAYOR, M. 1991 *A&A* **248**, 485.

DUVERT, G., GUILLOTEAU, S., MENARD, F., SIMON, M., & DUTREY, A. 2000 *A&A* **355**,165.

EDGAR, R. G. & CLARKE, C. J. 2003 *MNRAS* **338**, 962.

EDGAR, R. G. & CLARKE, C. J. 2004 *MNRAS* **349**, 678.

EDWARDS, S., CABRIT, S., STROM, S. E., HEYER, I., STROM, K. M., & ANDERSON, E. 1987 *ApJ* **321**, 473.

ELMEGREEN, B. F. 2000a *ApJ* **530**, 277.

ELMEGREEN, B. F. 2000b *ApJ* **539**, 342.

FALGARONE, E., PUGET, J.-L., & PERAULT, M. 1992 *A&A* **257**, 715.

FEIGELSON, E., LAWSON, W., & GARMIRE, G. 2003 *ApJ* **599**, 1207.

GAMMIE, C. F. 2001 *ApJ* **553**, 174.

GAMMIE, C. F, LIN, Y.-T., STONE, J. M., & OSTRIKER, E. C. 2003 *ApJ* **592**, 203.

GOODWIN, S. P., WHITWORTH, A. P., & WARD-THOMPSON, D. 2004a *A&A* **419**, 543.

GOODWIN, S. P., WHITWORTH, A. P., & WARD-THOMPSON, D. 2004b *A&A* **423**, 169.

GIZIS, J. E., KIRKPATRICK, J. D., BURGASSER, A., REID, I. N., MONET, D. G., LIEBERT, J., & WILSON, J. C. 2001 *ApJ* **551**, 163.

GRADY, C., ET AL. 2003 *PASP* **115**, 1036.

GRADY, C., ET AL. 2004 *ApJ* **608**, 809.

GULLBRING, E., HARTMANN, L., BRICENO, C., & CALVET, N. 1998 *ApJ* **492**, 323.

HAISCH, K. E., LADA, E. A., & LADA, C. J. 2001 *ApJ* **553**, L153.

HERBIG, G. H., PETROV, P. P., & DUEMMLER, R. 2003 *ApJ* **595**, 384.

HARTIGAN, P., HARTMANN, L., KENYON, S., STROM, S. E., & SKRUTSKIE, M. F. 1990 *ApJ* **354**, 25.

HARTMANN, L., CALVET, N., GULLBRING, E., & D'ALESSIO, P. 1998 *ApJ* **495**, 385.

HEITSCH, F., MAC LOW, M.-M., & KLESSEN, R. 2001 *ApJ* **547**, 280.

HENNEY, W. J. & O'DELL, C. R. 1999 *ApJ* **118**, 2350.

HILLENBRAND, L. A. & WHITE, R. J. 2004 *ApJ* **604**, 741.

HOLLENBACH, D., JOHNSTONE, D., LIZANO, S., & SHU, F. 1994 *ApJ* **428**, 654.

JAYAWARDHANA, R., ARDILA, D., STELZER, B., & HAISCH, K. 2003 *AJ* **126**, 1515.

JAYAWARDHANA, R., MOHANTY, S., & BASRI, G. 2003 *ApJ* **592**, 282.

JOHNSTONE, D., HOLLENBACH, D., & BALLY, J. 1998 *ApJ* **499**, 758.

KENYON, S. J. & HARTMANN, L. 1995 *ApJS* **101**, 117.

KIM, W.-T., OSTRIKER, E. C., & STONE, J. M. 2003 *ApJ* **599**, 1157.

KLEIN, R., FISHER, R., KRUMHOLZ, M., & MCKEE, C. 2003 *Rev Mex AA* **15**, 92.

KLESSEN, R. 2001 *ApJ* **556**, 837.

KLESSEN, R., HEITSCH, F., & MAC LOW, M.-M. 2000 *ApJ* **535**, 887.

KOERNER, D. W., SARGENT, A. I., & BECKWITH, S. V. W. 1993 *Icarus* **106**, 2.

KORNET, K., BODENHEIMER, P., & ROZYCKA, M. 2002 *A&A* **396**, 977.

KROUPA, P. 2002 *Science* **395**, 82.

KROUPA, P., AARSETH, S., & HURLEY, J. 2001 *MNRAS* **321**, 699.

KROUPA, P. & BOUVIER, J. 2003 *MNRAS* **346**, 343.

KROUPA, P., TOUT, C. A., & GILMORE, G. 1990 *MNRAS* **244**, 76.

LADA, C. J., MUENCH, A. A., LADA, E. A., & ALVES, J. F. 2004 *AJ* **128**, 1254.

LARSON, R. B. 1969 *MNRAS* **145**, 271.

LARSON, R. B. 1981 *MNRAS* **194**, 809.

LARSON, R. B. 1998 *MNRAS* **301**, 569.

LI, Z-H. & NAKAMURA, F. 2004 *ApJ* **609**, L83.

LIU, M., NAJITA, J., & TOKUNAGA, A. 2003 *ApJ* **585**, 372.

LUHMANN, K., BRICENO, C., STAUFFER, J. R., ET AL. 2003a, *ApJ* **590**, 348.

LUHMANN, K., STAUFFER, J. R., MUENCH, A. A., ET AL. 2003b, *ApJ* **593**, 1093.

LYNDEN-BELL, D. & PRINGLE, J. E. 1974 *MNRAS* **168**, 603.

MAC LOW, M.-M., KLESSEN, R. S., BURKERT, A., & SMITH, M. D. 1998 *PhRvL* **80**, 275.

MAÍZ-APELLÁNIZ, J. & WALBORN, N. R. 2003. In *A Massive Star Odyssey: From Main Sequence to Supernova* (eds. K. van der Hucht, A. Herrero, & C. Esteban). IAU Symp. 212, p. 560. ASP.

MARCY, G. W. & BUTLER, R. P. 1998 *ARAA* **36**, 57.

MASON, B. D, GIES, D. R., HARTKPOF, W., ET AL. 1998 *AJ* **115**, 821.

MASSEY, P. 2002 *ApJS* **141**, 81.

MASUNAGA, H. & INUTSUKA, S. 1999 *ApJ* **510**, 822.

MCCAUGHREAN, M. J.& O'DELL, C. R. 1996 *AJ* **111**, 1977.

MORAUX, E. & CLARKE, C. 2004 *A&A*, submitted.

MUENCH, A. A., ALVES, J., LADA, E. A., & LADA, C. J. 2001 *ApJ* **558**, L51.

MUENCH, A. A., LADA, E. A., LADA, C. J., & ALVES, J. 2002 *ApJ* **573**, 366.

O'DELL, C. R., WEN, Z., & HU, X. 1993 *ApJ* **410**, 696.

OSSENKOPF, V., KLESSEN, R., & HEITSCH, F. 2001 *A&A* **379**, 1005.

OSTRIKER, E. C., GAMMIE, C. F., & STONE, J. M. 1999 *ApJ* **513**, 259.

PADOAN, P., BALLY, J., BILLAWALA, Y., JUVELA, M., & NORDLUND, A. 1999 *ApJ* **525**, 318.

PADOAN, P., JUVELA, M., BALLY, J., & NORDLUND, A. 1998 *ApJ* **504**, 300.

PADOAN, P. & NORDLUND, A. 1999 *ApJ* **526**, 279.

PETTINI, M., RIX, S. A., STEIDEL, C. C., SHAPLEY, A. E., & ADELBERGER, K. L. 2003. In *A Massive Star Odyssey: From Main Sequence to Supernova* (eds. K. A. van der Hucht, A. Herrero, C. Esteban). Proc. IAU Symp. 212, p. 671. ASP.

POLLACK, J. B., HUBICKYI, O., BODENHEIMER, P., LISSAUER, J. L., PODOLAK, M., & GREENZWEIG, Y. 1996 *Icarus* **124**, 62.

PRATO, L., SIMON, M., MAZEH, T., ZUCKER, S., & MCLEAN, I. 2002 *ApJ* **579**, 99.

PRICE, D. J. & MONAGHAN, J. J. 2004 *MNRAS* **348**, 123.

REIPURTH, B. 2000 *AJ* **120**, 3177.

REIPURTH, B. & CLARKE, C. 2001 *AJ* **122**, 432.

RICE, W. K. M., ARMITAGE, P. J., BATE, M. R., & BONNELL, I. A. 2003a *MNRAS* **339**, 1025.

RICE, W. K. M., WOOD, K., ARMITAGE, P. J., WHITNEY, B. A., & BJORKMAN, J. E. 2003b *MNRAS* **342**, 79.

ROBBERTO, M., ET AL. 2004 *ApJ* **606**, 952.

ROMANIELLO, M., ROBBERTO, M. & PANAGIA, N. 2004 *ApJ* **608**, 220.

SALPETER, E. E. 1955 *ApJ* **121**, 161.

SARGENT, A. & BECKWITH, S. 1987 *ApJ* **323**, 294.

SCALLY, A. & CLARKE, C. J. 2001 *MNRAS* **325**, 449.

SCALLY, A. & CLARKE, C. J. 2002 *MNRAS* **334**, 156.

SCALO, J. 1986 *Fundam. Cosmic Phys.* **11**, 1.

SICILIA-AGUILAR, A., HARTMANN, L. W., BRICENO, C., MUZEROLLE, J., & CALVET, N. 2004 *ApJ* **128**, 805.

SKRUTSKIE, M. F., DUTKEVITCH, D., STROM, S. E., EDWARDS, S., STROM, K. M., SHURE, M. A. 1990 *AJ* **99**, 1187.

SMITH, L. & GALLAGHER, J. S. 2001 *MNRAS* **326**, 1027.

SMITH, N., BALLY, J., & MORSE, J. A. 2003 *ApJ* **587**, L105.

STERZIK, M. F. & DURISEN, R. H. 1995 *A&A* **304**, 9.

STERZIK, M. F. & DURISEN, R. H. 1998 *A&A* **339**, 95.

STOERZER, H. & HOLLENBACH, D. 1999 *ApJ* **515**, 669.

STONE, J. M., OSTRIKER, E. C., & GAMMIE, C. F. 1998 *ApJ* **508**, L99.

TESTI, L., PALLA, F., & NATTA, A. 1999 *A&A* **342**, 515.

VAN ALBADA, T. 1968 *Bull. Astr. Inst. Neth.* **20**, 57.

VAZQUEZ-SEMADENI, E., BALLESTEROS-PAREDES, J., & KLESSEN, R. S. 2003 *ApJ* **585**, 131.

WADA, K., MEURER, G., & NORMAN, C. 2002 *ApJ* **577**, 197.

WEIDNER, C. & KROUPA, P. 2004 *MNRAS* **348**, 187.

WHITEHOUSE, S. C. & BATE, M. R. 2004 *MNRAS* **353**, 1078.

WHITWORTH, A. 1979 *MNRAS* **186**, 59.

WILLIAMS, J. P. & BLITZ, L. 1998 *ApJ* **494**, 657.

WOLF, S., GUETH, F., HENNING, T., & KLEY, W. 2002 *ApJ* **566**, L97.

WOLFIRE, M. G. & CASSINELLI, J. P. 1987 *ApJ* **319**, 850.

WYATT, M. C., DENT, W. R. F., & GREAVES, J. S. 2003 *MNRAS* **342**, 876.

YORKE, H. W. & SONNHALTER, C. 2002 *ApJ* **569**, 846.

ZUCKER, S. & MAZEH, T. 2001 *ApJ* **562**, 1038.

Star formation in clusters

By SØREN S. LARSEN

ESO/ST-ECF, Karl-Schwarzschild Strasse 2, D-85748 Garching bei München, Germany

The *Hubble Space Telescope* is very well tailored for observations of extragalactic star clusters. Obvious reasons are *HST*'s ability to recognize clusters as extended objects and measure sizes out to distances of several Mpc. Equally important is the wavelength range offered by the instruments on board *HST*—in particular the blue and near-UV coverage—which is essential for age-dating young clusters. *HST* observations have helped establish the ubiquity of young massive clusters (YMCs) in a wide variety of star-forming environments, ranging from dwarf galaxies and spiral disks to nuclear starbursts and mergers. These YMCs have masses and structural properties similar to those of old globular clusters in the Milky Way and elsewhere, and the two may be closely related. Several lines of evidence suggest that a large fraction of all stars are born in clusters, but most clusters disrupt rapidly and their stars disperse to become part of the field population. In most cases studied to date, the luminosity functions of young cluster systems are well fit by power laws $dN(L)/dL \propto L^{-\alpha}$ with $\alpha \approx 2$, and the luminosity of the brightest cluster can (with few exceptions) be predicted from simple sampling statistics. *Mass* functions have only been constrained in a few cases, but appear to be well approximated by similar power laws. The absence of any characteristic mass scale for cluster formation suggests that star clusters of all masses form by the same basic process, without any need to invoke special mechanisms for the formation of "massive" clusters. It is possible, however, that special conditions *can* lead to the formation of a few YMCs in some dwarfs where the mass function is discontinuous. Further studies of mass functions for star clusters of different ages may help test the theoretical prediction that the power-law mass distribution observed in young cluster systems can evolve towards the approximately log-normal distribution seen in old globular cluster systems.

1. Introduction

The wide range of topics covered in this volume—*from planets to cosmology*—bear testimony to the fact that, in spite of much recent progress in competing technologies such as ground-based adaptive optics, for many purposes the capabilities offered by *Hubble* remain unique. For studies of extragalactic stellar populations, the combination of relatively wide-field imaging at diffraction-limited resolution, even in the optical and near-UV, offered by *HST* is invaluable, and will remain unsurpassed for the foreseeable future. *HST* has contributed much to our understanding of star formation in clusters, both in a Galactic and extragalactic context. Rather than attempting to cover everything, I will concentrate mainly on star clusters beyond the Local Group, partly because that is what I am most familiar with, and partly because Local Group galaxies are covered elsewhere in this volume (Grebel). Much of the work on young clusters done in the first decade of *HST*'s lifetime has been reviewed by Whitmore (2003), and although some overlap is unavoidable, the primary aim of this review is to discuss more recent results and try to look ahead.

Almost since the day it was launched, *HST* has had a tremendous impact on the field of extragalactic star clusters. Some major ground-breaking discoveries were, in fact, made even prior to the 1993 repair mission. Prior to *HST*, young star clusters had been identified in a few galaxies other than the Milky Way, including most Local Group members and a few galaxies slightly beyond the Local Group. But with *HST*, it became possible to undertake systematic surveys in larger samples of galaxies, and to better characterize the properties of young clusters in different environments. Of course, ground-based capabilities have also evolved during the lifetime of *HST*. Larger-format CCD detectors with excellent blue and UV quantum efficiency have become available, a better understanding of "dome" seeing has led to improved image quality, and 8–10 m ground-based telescopes have made it possible to obtain high-quality spectra of the faint objects detected in *HST* images.

2. Why star clusters?

While star clusters have been the subject of substantial interest for many years, it may be worth recalling some of the main motivations for studying them.

First, there are a number of problems which make clusters interesting in their own right. These involve both their formation, subsequent dynamical evolution, and ultimate fate. At first glance, clusters appear deceptively simple: they are aggregations of a few hundred to about a million individual stars and generally constitute a gravitationally bound system (although the latter may not be true for some of the youngest systems). Yet, constructing realistic models of their structure and dynamical evolution has proven to be a major challenge, and it is only now becoming possible to carry out reasonably realistic N-body simulations including the effects of stellar evolution, external gravitational fields, and the rapidly varying gravitational potential in the early phases of cluster evolution during which gas is expelled from the system (Joshi et al. 2000; Giersz 2001; Kroupa & Boily 2002). The models must be tested observationally, and *HST* data currently represent the only way to reliably measure structural parameters for extragalactic star clusters.

Second, there is growing evidence that a significant fraction of all stars form within clusters, although only a small fraction of these stars eventually end up in *bound* clusters (Lada & Lada 2003; Fall 2004). Therefore, the problem of understanding *star* formation is intimately linked to that of understanding *cluster* formation, and a theory of one cannot be complete without the other. It is of interest to investigate how the properties of star clusters might depend on environment, as this might provide important clues to any differences in the star formation process itself. In particular, *HST* has made important contributions towards establishing the presence of "young massive clusters" (YMCs†) in a variety of environments, which appear to be very similar to young versions of the old *globular* clusters, which are ubiquitous around all major galaxies. Globular cluster formation was once thought to be uniquely related to the physics of the early Universe (e.g., Peebles & Dicke 1968; Fall & Rees 1985) but it now seems to be an ongoing process which can be observed even at the present epoch.

Third, star clusters are potentially very useful as tracers of the stellar populations in their host galaxies. Clusters can be identified and studied at much greater distances than individual stars. In most cases, they are composed of stars which, to a very good approximation, formed at the same time and have the same metallicity. This is in contrast to the integrated light from the galaxies, which may originate from an unknown mix of stellar populations with different ages and metallicities. Although the effects of stellar evolution alone cause a cluster to fade by 5–6 magnitudes (in V-band) over 10 Gyrs (Bruzual & Charlot 2003), in principle it is possible to detect clusters which have formed during the entire lifetime of galaxies, out to distances of several Mpc. In particular, *globular clusters* have been used extensively in attempts to constrain the star formation histories of early-type galaxies.

† Also known as "super star clusters," "populous clusters," or "young globular clusters."

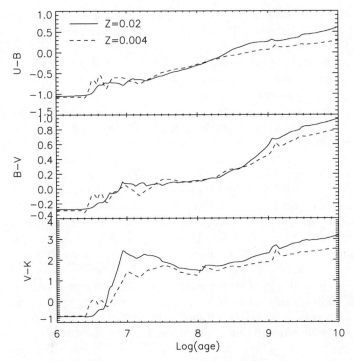

FIGURE 1. Evolution of broad-band colors as a function of age and metallicity according to Bruzual & Charlot (2003) simple stellar population models.

3. *HST* and extragalactic star clusters

HST is almost ideally tailored for studies of extragalactic star clusters. Three main reasons for this are:

• Angular resolution: clusters typically have half-light radii of 2–4 pc (see Section 5), and can thus be recognized as extended objects out to distances of 10–20 Mpc with the $\sim 0''.05$ resolution offered by WFPC2 or ACS. With careful modeling of the point spread function (PSF), this limit may be pushed even further.

• Field size: At 10 Mpc, the ACS $200'' \times 200''$ field of view corresponds to about 10 kpc $\times$ 10 kpc, making it possible to cover a significant fraction of a typical galaxy in a single pointing.

• Spectral range: For studies of young stellar populations, optical and near-UV spectral coverage is essential, as discussed below.

There is currently no alternative to *HST* on the horizon which offers a similar combination of capabilities. The *James Webb Space Telescope* (*JWST*), while offering vastly improved efficiency in the IR, will offer no significant gain in resolution over *HST*, and will be limited to longer wavelengths. Ground-based adaptive optics (AO) can provide similar, or even better resolution than *HST*, but only within a small ($\sim 20''$) isoplanatic field of view. Furthermore, AO lacks the stable PSF of *HST*, which is critical for many purposes (e.g., when measuring structural parameters for star clusters at the limit of the resolution), and is in any case limited to the IR (at least for now). The *GALEX* mission offers wide-field UV imaging, but with a spatial resolution that is inferior by far to that of *HST* (about $5''$).

The need for optical and near-UV imaging in particular deserves some additional comments. Figure 1 shows simple stellar population (SSP) model calculations (Bruzual &

Charlot 2003) for the evolution of the $U-B$, $B-V$, and $V-K$ broad-band colors of a single-burst stellar population. The models are shown for metallicities $Z = 0.02$ (Solar) and $Z = 0.004$ between ages of 10^6 years and 10^{10} years. As seen from the figure, the $U-B$ color is an excellent age indicator in the range from 10^7 to a few times 10^8 years, increasing by more than 0.5 mag and with little metallicity dependence over this age range. The $B-V$ color, in contrast, remains nearly constant over the same age range, and offers little leverage for age determinations. In practice, there are complicating problems such as dust extinction, which in general will make it difficult to obtain accurate age estimates from a single color. Using a combination of two colors, such as $U-B$ and $B-V$, will make it possible to constrain both age and reddening, while at the same time being relatively insensitive to metallicity effects. The relation between age and location of a cluster in the $(U-B, B-V)$ two-color plane has been calibrated with clusters in the Large Magellanic Cloud through the so-called 'S'-sequence (Elson & Fall 1985; Girardi et al. 1995). For ages younger than about 10^7 years, line emission becomes important (Anders & Fritze-v. Alvensleben 2003), while the age-metallicity degeneracy (Worthey 1994) becomes a difficulty at older ages. A more recent discussion of photometric age indicators, with emphasis on the importance of blue and UV data, is in Anders et al. (2004b). Use of, e.g., the $V-K$ color can help put further constraints on the metallicity and may also help constrain the ages of stellar populations in the range $\sim$200 Myr to $\sim$500 Myr (Maraston et al. 2002), although the models are more uncertain and depend strongly on the stellar evolutionary tracks used in the construction of the SSP models (Girardi 2000).

High-resolution, wide-field imaging in the blue and/or UV will be especially important for attempts to constrain not only the luminosity function, but also the mass function of clusters. For a long time, WFPC2 was the "workhorse" on *HST*, and it remains the only wide-field imager on board *HST* with U-band imaging capability. However, the sensitivity of WFPC2 in the U-band is rather low and the detectors are steadily degrading. The Wide Field Camera 3, with its panchromatic coverage, would be an ideally suited instrument for such studies.

4. Setting the stage: early developments

Even within the Local Group, it has long been known that the traditional distinction between open and globular clusters that can be applied fairly easily in the Milky Way breaks down in some other galaxies. The "blue globular" clusters in the Large Magellanic Cloud are classical examples, which are not easily classified as either open or globular clusters. The most massive of these objects have masses up to $\sim$$10^5$ $M_\odot$ (Elson & Fall 1985; Fischer et al. 1992; Richtler 1993; Hunter et al. 2003), similar to the median mass of old globular clusters and about an order of magnitude more massive than any young open cluster known in the Milky Way. Yet, these objects have young ages, and are still being produced today by the LMC. Similar clusters have been found in M33 (Christian & Schommer 1982, 1988).

A good indication of the status of research in extragalactic young star clusters shortly prior to *HST* is provided by Kennicutt & Chu (1988; hereafter KC88). These authors compiled observations of what they refer to as "young populous clusters" (PCs) in 14 galaxies for which data was available at that time. Half of the galaxies studied by KC88 were Local Group members (Milky Way, LMC, SMC, M33, M31, NGC 6822 and IC 1613). As noted by KC88, the widely variable completeness of the surveys and the different definitions of such clusters makes comparing observations of PCs in different galaxies very difficult. KC88 adopted a (somewhat arbitrary) definition of a young PC as an object with an estimated mass $> 10^4 M_\odot$ and a color $B-V < 0.5$. They noted a conspicuous deficiency of populous clusters in the Milky Way and M31, the two only large Sb/Sbc-type spirals in the sample, and suggested that this might be linked to the deficiency of giant H II regions in the same two galaxies. By comparing the relative numbers of PCs and giant H II regions in their sample of galaxies, KC88 concluded that PCs may indeed form inside such regions, but not all giant H II regions produce bound clusters. This is very much in line with recent indications that only a small fraction of star clusters of *any* mass remain bound (Fall 2004). The galaxies which did contain PCs were all late-type, though not *all* late-type galaxies were found to contain PCs. A significant exception is the Local Group dwarf irregular IC 1613, which contains few if any star clusters at all (van den Bergh 1979; Hodge 1980), in spite of some on-going star formation. The near-absence of star clusters in IC 1613 may be as important a clue to the nature of the cluster formation process as the abundant cluster systems in starbursts and merger galaxies (Section 5).

To a large extent, research in old globular clusters (GCs) remained detached from that of YMCs until fairly recently. It was well known that early-type galaxies typically have many more GCs per unit host galaxy luminosity (Harris & van den Bergh 1981) than spirals and irregulars, a fact that was recognized as a problem for the idea that early-type galaxies form by mergers of gas-rich spirals (van den Bergh 1982). Schweizer (1987) proposed that this problem might be solved if new GCs form *during* the merger. This idea was further explored by Ashman & Zepf (1992), who predicted that the resulting merger product should contain two distinct GC populations: one metal-poor population inherited from the progenitor galaxies, and a new metal-rich population formed in the merger. The two GC populations should be identifiable in the color distributions of the resulting GC systems. Two highly influential discoveries soon followed: Bimodal color distributions were discovered in several GC systems around early-type galaxies (Zepf & Ashman 1993; Secker et al. 1995; Whitmore et al. 1995), and highly luminous, compact young star clusters were found in ongoing or recent mergers like the Antennae and NGC 7252 (Whitmore et al. 1993; Whitmore & Schweizer 1995). In retrospect, it had already been known for a long time that even the metallicity distribution of the Milky Way GC system is strongly bimodal (Zinn 1985). The mean metallicities of the two modes in the Milky Way are, in fact, quite similar to those seen in early-type galaxies. The Milky Way is unlikely to be the result of a major merger, and there are also other indications that not all properties of GC systems in early-type galaxies can be explained by a naive application of the merger model. Alternative scenarios were proposed to explain the presence of multiple GC populations (e.g., Forbes et al. 1997; Côté et al. 1998), but it is beyond the scope of this paper to discuss any of these in detail. Comprehensive discussions can be found e.g., in Harris (2001) and Kissler-Patig (2000). Nevertheless, the discovery of young globular cluster-like objects in ongoing mergers was a tantalizing hint that it might be possible to study the process of globular cluster formation close-up at the present epoch, and not just from the fossil record.

5. Extragalactic star clusters in different environments

Tables 1–5 are an attempt to collect a reasonably complete list of galaxies where YMCs have been identified (up to May 2004), along with some pertinent references. For each galaxy, the main facilities used for the observations are listed, although in many cases it is impossible to give a comprehensive listing. Standard abbreviations (ACS, FOC, GHRS, STIS, NICMOS, WFPC, WFPC2) are used for *HST* instruments. Other abbreviations are WIYN (Wisconsin Indiana Yale NOAO 3.5 m), UKIRT (United Kingdom

Galaxy	Instrument	References
NGC 3125	STIS	Chandar et al. (2004)
NGC 3310*	FOC, WFPC2, NICMOS	Meurer et al. (1995), de Grijs et al. (2003a)
NGC 3991*	FOC	Meurer et al. (1995)
NGC 4670*	FOC	Meurer et al. (1995)
NGC 5253*	STIS, WFPC2, FOC	Maíz-Apellániz (2001), Tremonti et al. (2001), Harris et al. (2004), Turner & Beck (2004), Vanzi & Sauvage (2004), Meurer et al. (1995)
NGC 6745	WFPC2	de Grijs et al. (2003a)
NGC 7469	NICMOS	Scoville et al. (2000)
NGC 7673*	WIYN, WFPC2	Homeier & Gallagher (1999), Homeier et al. (2002)
IC 883	NICMOS	Scoville et al. (2000)
M 82	Pal 200-inch,WFPC, WFPC2	van den Bergh (1971), O'Connell et al. (1995), de Grijs et al. (2003b)
TOL1924-416*	FOC	Meurer et al. (1995)
Zw 049.057	NICMOS	Scoville et al. (2000)
VII Zw 031	NICMOS	Scoville et al. (2000)
IR 15250+3609	NICMOS	Scoville et al. (2000)
IR 17208-0014	NICMOS	Scoville et al. (2000)
NGC 1614	NICMOS, UKIRT	Alonso-Herrero et al. (2001a), Kotilainen et al. (2001)
NGC 7714	UKIRT JHK	Kotilainen et al. (2001)

TABLE 1. Observations of young star clusters in starburst galaxies

Infra-Red Telescope), NTT (ESO 3.5 m New Technology Telescope), CFHT (3.6 m Canada-France-Hawaii Telescope), DK154 (Danish 1.54 m at ESO, La Silla) and NOT (2.56 m Nordic Optical Telescope). The level of detail provided in the different studies varies enormously—in some cases, identifications of YMCs are only a byproduct of more general investigations of galaxy properties (e.g., Meurer et al. 1995; Scoville et al. 2000), while other studies are dedicated analyses of cluster systems in individual galaxies. Galaxies marked with a star ($\star$) are used later (Section 6.2) when discussing luminosity functions. In the following I briefly discuss a few illustrative cases from each table and then move on to discuss more general properties of young cluster systems.

5.1. *Starburst galaxies*

The richest populations of YMCs are often found in major mergers (Section 5.2). However, there are also examples of YMCs in starbursts which are not directly associated with major mergers, although in some cases they may be stimulated by more benign interactions or accretion of companion satellites (Table 1). In the case of NGC 7673, for example, Homeier & Gallagher (1999) argue that the morphological features of the galaxy point toward a minor merger, while the starburst in M 82 may have been triggered by tidal interactions with M 81. M 82 is also noteworthy for being the first galaxy in which the term 'super star cluster' was used. It was introduced by van den Bergh (1971), who was careful to point out that the nomenclature was not intended to imply that these objects are necessarily bound. The presence of SSCs in M 82 was confirmed by O'Connell et al. (1995), who identified about 100 clusters in WFPC images. NGC 5253, which is located about 600 kpc from its nearest neighbor, M 83, is an example of a starburst which is unlikely to be triggered by an interaction (Harris et al. 2004).

One of the first surveys to provide a systematic census of star clusters in a sample of starburst galaxies was the work by Meurer et al. (1995), who observed nine galaxies with *HST*'s Faint Object Camera (FOC). Meurer et al. noted that a high fraction, on average about 20%, of the UV luminosity in these starbursts originated from clusters or

Galaxy	Instrument	References
NGC 253*	WFPC2, ROSAT	Watson et al. (1996), Forbes et al. (2000)
NGC 1808	NTT K-band	Tacconi-Garman et al. (1996)
NGC 4303	WFPC2, NICMOS, STIS	Colina & Wada (2000), Colina et al. (2002)
NGC 5236	WFPC2	Harris et al. (2001)
NGC 6240	WFPC2, NICMOS	Pasquali et al. (2004), Scoville et al. (2000)
NGC 1079	FOC	Maoz et al. (1996)
NGC 1326*	WFPC2	Buta et al. (2000)
NGC 1433	FOC	Maoz et al. (1996)
NGC 1512*	FOC, WFPC2, NICMOS	Maoz et al. (1996), Maoz et al. (2001a)
NGC 1097*	WFPC2	Barth et al. (1995)
NGC 2903	CFHT/AO NICMOS	Alonso-Herrero et al. (2001b)
NGC 2997	FOC, WFPC2	Maoz et al. (1996)
NGC 3310*	KPNO 4m, WFPC2	Elmegreen et al. (2002)
NGC 4314*	WFPC	Benedict et al. (1993)
NGC 5248*	FOC, WFPC2, NICMOS	Maoz et al. (1996), Maoz et al. (2001a)
NGC 6951*	WFPC2	Barth et al. (1995)
NGC 7552*	FOC	Meurer et al. (1995)

TABLE 2. Observations of young star clusters in nuclear and circumnuclear starbursts

compact objects, and saw a hint of a trend for this fraction to increase with the underlying UV surface brightness. They also measured cluster sizes similar to those of Galactic globular clusters, and found the luminosity functions to be well represented by a power law $dN(L)/dL \propto L^{-\alpha}$ with $\alpha \approx 2$.

YMCs have been identified in several nuclear and circumnuclear starburst regions, and are often associated with barred spiral galaxies (Table 2). Maoz et al. (1996) studied five circumnuclear starbursts and found that as much as 30%–50% of the UV light came from compact, young star clusters with half-light radii < 5 pc and estimated masses up to about 10^5 $M_\odot$. Again they found the luminosity functions to be well approximated by power laws with slope $\alpha \approx 2$. Buta et al. (2000) found a much steeper slope ($\alpha = 3.7 \pm 0.1$) in their study of the circumnuclear starburst in NGC 1326, but noted that their sample might be contaminated by individual supergiant stars. In some cases, the ring-like structure of the nuclear starburst is not quite so evident. Watson et al. (1996) discovered four luminous clusters in the central starburst region of NGC 253, the brightest of which has $M_V = -15$, an inferred mass in excess of 1.5×10^6 $M_\odot$ and a half-light radius of 2.5 pc. However, these clusters may be part of a compact ring-like structure with a radius of about 50 pc (Forbes et al. 2000). Most of the clusters in the nuclear starburst of M83 are also located within a semicircular annulus (Harris et al. 2001), but again the ring is more poorly defined.

5.2. Mergers

Many of the most spectacular YMC populations have been found in merger galaxies (Table 3). NGC 1275 was one of the first galaxies in which *HST* data confirmed the existence of YMCs, although at least one object in this galaxy was already suspected to be a massive cluster based on ground-based data (Shields & Filippenko 1990). With the Planetary Camera on *HST*, Holtzman et al. (1992) identified about 60 cluster candidates with absolute magnitudes up to $M_V = -16$. Using WFPC2 data, Carlson et al. (1998) identified about 3,000 clusters, of which about 1,200 have blue integrated colors and estimated ages between 0.1 and 1 Gyr. The young clusters had estimated masses and sizes similar to those of old globular clusters, although Brodie et al. (1998) found that the Balmer line equivalent widths measured on spectra of five clusters were too strong

Galaxy	Instrument	References
NGC 520	UKIRT JHK	Kotilainen et al. (2001)
NGC 1275*	WFPC, WFPC2, Keck/LRIS	Holtzman et al. (1992), Carlson et al. (1998), Brodie et al. (1998)
NGC 1741*	FOC, WFPC2, GHRS	Johnson et al. (1999)
NGC 2207*/IC 2163	WFPC2	Elmegreen et al. (2001)
NGC 2366*	WFPC2, STIS, CFHT Hα, JHK	Drissen et al. (2000)
NGC 2623	NICMOS	Scoville et al. (2000)
NGC 3256*	WFPC2	Zepf et al. (1999)
NGC 3395/96	STIS	Hancock et al. (2003)
NGC 3597*	ESO 2p2, NTT, WFPC2	Lutz (1991), Holtzman et al. (1996), Carlson et al. (1999), Forbes & Hau (2000)
NGC 3690*	FOC	Meurer et al. (1995)
NGC 3921*	WFPC2	Schweizer et al. (1996)
NGC 4038/39*	WFPC, WFPC2, GHRS	Whitmore & Schweizer (1995), Whitmore et al. (1999)
NGC 6052*	WFPC2	Holtzman et al. (1996)
NGC 6090	NICMOS	Dinshaw et al. (1999), Scoville et al. (2000)
NGC 6240	WFPC2	Pasquali et al. (2003)
NGC 7252*	WFPC, WFPC2	Whitmore et al. (1993), Miller et al. (1997)
NGC 7727	?	Crabtree & Smecker-Hane (1994)
Arp 220	NICMOS	Scoville et al. (1998), Scoville et al. (2000)
II ZW 96	Univ. Haw. 2p2 BRHK	Goldader et al. (1997)
The Mice	ACS	de Grijs et al. (2003c)
Tadpole	ACS	de Grijs et al. (2003c)
HCG 31	WIYN, WFPC2	Johnson & Conti (2000)
VV 114E/W	NICMOS	Scoville et al. (2000)
UGC 5101	NICMOS	Scoville et al. (2000)
UGC 10214	ACS	Tran et al. (2003)
IR 10565+2448W	NICMOS	Scoville et al. (2000)
IR 15206+3342	WHT, WFPC2	Arribas & Colina (2002)
IR 22491-1808W	NICMOS	Scoville et al. (2000)
Mrk 273S	NICMOS	Scoville et al. (2000)
Stephan's Quintet	WFPC2	Gallagher et al. (2001)
Tidal Tails	WFPC2	Knierman et al. (2003)

TABLE 3. Observations of young star clusters in mergers

to be consistent with standard SSP models, unless a stellar mass function truncated at 2–3 $M_\odot$ was adopted. With accurate modeling of the *HST* point-spread function and high-dispersion spectroscopy with 8–10 m class telescopes, it might be possible to constrain the virial masses of some of the brightest clusters, and thereby provide independent constraints on their stellar IMF.

While NGC 1275 may have experienced a recent merger/accretion event (Holtzman et al. 1992), it is hardly one of the classical "Toomre" mergers (Toomre & Toomre 1972). One of the nearest ongoing, major mergers is the "Antennae" NGC 4038/39, where *HST* observations have revealed a rich population of luminous, compact young star clusters with typical half-light radii ∼4 pc (Whitmore & Schweizer 1995; Whitmore et al. 1999). The brightest of them reach $M_V \approx -14$ and have estimated masses close to 10^6 $M_\odot$ (Zhang & Fall 1999). Similar rich populations of YMCs have been found in many other mergers, like NGC 3256 where Zepf et al. (1999) identified about 1,000 compact bright, blue objects on WFPC2 images within the central 7 kpc × 7 kpc region. Again, the young clusters contribute a very significant fraction (15%–20%) of the blue light within the starburst region. Zepf et al. (1999) estimated half-light radii of 5–10 pc for the clusters in NGC 3256, somewhat larger than for the Antennae, but note that 1 PC pixel corresponds to a linear scale of 8 pc at the distance of NGC 3256, so that the clusters

Galaxy	Instrument	References
NGC 1140*	WFPC	Hunter et al. (1994)
NGC 1156*	NOT, WFPC2	Larsen & Richtler (1999), Larsen (2002)
NGC 1313*	DK154, WFPC2	Larsen & Richtler (1999), Larsen (2002)
NGC 1569*	Pal 200-inch, WFPC, WFPC2, STIS	Arp & Sandage (1985), O'Connell et al. (1994), de Marchi et al. (1997), Hunter et al. (2000), Maoz et al. (2001b), Maíz-Apellániz (2001), Origlia et al. (2001), Anders et al. (2004a), Gilbert & Graham (2003)
NGC 1705*	Las Campanas 2p5, WFPC, WFPC2, STIS	Melnick et al. (1985), O'Connell et al. (1994), Maíz-Apellániz (2001), Billett et al. (2002), Vázquez et al. (2004)
NGC 3077*	WFPC2, CFHT JHK	Harris et al. (2004), Davidge (2004)
NGC 4194	STIS	Weistrop et al. (2004)
NGC 4214*	WFPC2	Maíz-Apellániz (2001), Billett et al. (2002)
NGC 4449*	WFPC2	Gelatt et al. (2001), Maíz-Apellániz (2001)
ESO-338-IG04	WFPC2	Östlin et al. (1998)
HE 2-10	WFPC2, GHRS	Conti & Vacca (1994), Johnson et al. (2000), Beck et al. (2001)
I Zw 18*	FOC	Meurer et al. (1995)
UGC 7636	KPNO 4m C,T1	Lee et al. (1997)
POX 186	NTT RI	Doublier et al. (2000)
SBS 0335-052	NTT JHK	Vanzi et al. (2000)

TABLE 4. Observations of young star clusters in dwarf and irregular galaxies

are only marginally resolved. Interestingly, only a shallow trend of cluster size versus luminosity was found, with radius r scaling with luminosity L roughly as $r \propto L^{0.07}$.

NGC 7252 is a somewhat more advanced system than NGC 3256 or the Antennae. Miller et al. (1997) date the cluster system at between 650 Myr and 750 Myr. Remarkably, both photometry and dynamical measurements yield a mass of about 8×10^7 $M_\odot$ for the most massive object (W3; Maraston et al. 2004), making it about an order of magnitude more massive than any old globular cluster in the Milky Way. With a half-light radius of 17.5 ± 1.8 pc, this object is much larger than a normal star cluster, and may be more closely associated with the "Ultra Compact Dwarf Galaxies" in Fornax (Hilker et al. 1999; Drinkwater et al. 2003).

5.3. *Dwarf/Irregular galaxies*

The bright "central condensations" in NGC 1569 were noted already by Mayall (1935) on plates taken with the 36-inch Crossley reflector at Lick Observatory, though Arp & Sandage (1985) were probably the first to recognize them as likely star clusters. At a distance of only $\sim$2 Mpc (Makarova & Karachentsev 2003), these clusters appear well resolved on *HST* images with half-light radii of about 2 pc (O'Connell et al. 1994; de Marchi et al. 1997). One of the clusters, NGC 1569-A, is actually a double cluster, and STIS spectroscopy has shown that one component exhibits Wolf-Rayet features, while the other component is devoid of such features, which suggests an age difference of a few Myrs between the two components (Maoz et al. 2001b). Using high-dispersion spectroscopy from the NIRSPEC spectrograph on the Keck II telescope, Gilbert & Graham (2003) derived dynamical mass estimates of about 0.3×10^6 $M_\odot$ for each of the two components of NGC 1569-A, and 0.18×10^6 $M_\odot$ for NGC 1569-B, again very similar to the typical masses of old globular clusters, and consistent with the clusters having "normal" stellar mass functions (see also Section 6.4).

A peculiar feature of the NGC 1569 cluster population is that the next brightest clusters after NGC 1569-A and NGC 1569-B are more than two magnitudes fainter (O'Connell

Galaxy	Instrument	References
M 51*	Lick 3m UBV, WFPC2, NICMOS	Larsen (2000), Bik et al. (2003), Bastian et al. (2005)
M 81	WFPC2	Chandar et al. (2001)
M 101	WFPC2	Bresolin et al. (1996)
NGC 2403*	Loiano 1p5, WFPC2, NOT	Battistini et al. (1984), Drissen et al. (1999), Larsen & Richtler (1997)
NGC 2997*	DK154, WFPC2	Larsen & Richtler (1999), Larsen (2002)
NGC 3081	WFPC2	Buta et al. (2004)
NGC 3621*	DK154, WFPC2	Larsen & Richtler (1999), Larsen (2002)
NGC 3627*	WFPC2	Dolphin & Kennicutt (2002)
NGC 5236*	UITP,† DK154, WFPC2	Bohlin et al. (1990), Larsen & Richtler (1999), Larsen (2002)
NGC 7793*	DK154, WFPC2	Larsen & Richtler (1999), Larsen (2002)
NGC 6946*	NOT, WFPC2	Larsen & Richtler (1999), Larsen (2002)

† UITP = prototype of the Ultraviolet Imaging Telescope flown on Spacelab.

TABLE 5. Observations of young star clusters in spiral galaxy disks

et al. 1994). An even more dramatic discontinuity in the luminosity function is seen in NGC 1705, which has only a single bright cluster, and in NGC 4214 there is a gap of about 1.5 mag from the brightest two clusters down to number three (Billett et al. 2002). Interestingly, while the clusters in NGC 1569 and NGC 1705 are young ($\sim 10^7$ years), the two clusters in NGC 4214 are both about 250 Myrs old (Billett et al. 2002), demonstrating that massive clusters are capable of surviving for substantial amounts of time at least in some dwarf galaxies.

These observations are detailed in Table 4.

5.4. *Spiral galaxy disks*

Most of the YMCs discussed in the preceding sections are located in environments that are peculiar in some way, or at least different from what we see in the solar neighborhood. Thus, it is tempting to speculate that the absence of YMCs in the Milky Way indicates that their formation somehow requires special conditions. There is, however, increasing evidence that YMCs can form even in the disks of spiral galaxies. Table 5 lists a number of nearby spirals in which YMCs have been identified. A few (e.g., M 51) are clearly involved in interactions, but none of them are disturbed to a degree where they are not clearly recognizable as spirals. The nuclear starburst in M 83 was already mentioned in Section 5.1, but there is also a rich population of young star clusters throughout the disk (Bohlin et al. 1990; Larsen & Richtler 1999), the most massive of which have masses of several times 10^5 $M_\odot$. An even more extreme cluster is in NGC 6946, with a dynamical mass estimate of about 1.7×10^6 $M_\odot$ (Larsen et al. 2001). The disks of spiral galaxies can evidently form star clusters with masses as high as those observed in any other environment, including merger galaxies like the Antennae and starbursts like M 82.

Most of the spirals in Table 5 are type Sb or later, but one exception is NGC 3081. In this barred S0/Sa-type spiral, Buta et al. (2004) detected a number of luminous young clusters in the inner Lindblad resonance ring at 5 kpc. Buta et al. (2004) found rather large sizes for these clusters, with estimated half-light radii of about 11 pc. This is much larger than the typical sizes of Milky Way open and globular clusters and indeed of YMCs found in most other places, and raises the question whether these objects might be related to the "faint fuzzy" star clusters which are located in an annulus of similar radius in the lenticular galaxy NGC 1023 (Larsen & Brodie 2000; Brodie & Larsen 2002), but have globular cluster-like ages.

6. General properties of cluster systems

Just how similar are the properties of star clusters in different environments, and what might they tell us about the star formation process? Objects like NGC 1569-A appear extreme compared to Milky Way open clusters or even to young LMC clusters: O'Connell et al. (1994) estimate that NGC 1569-A has a half-light surface brightness over 65 times higher than the R136 cluster in the LMC, and 1200 times higher than the mean rich LMC cluster after allowing for evolutionary fading. Do such extreme objects constitute an altogether separate mode of star/cluster formation, or do they simply represent a tail of a distribution, extending down to the open clusters that we encounter locally? And are YMCs really young analogs of the old GCs observed in the Milky Way and virtually all other major galaxies?

6.1. *Luminosity and mass functions*

One of the best tools to address these questions is the cluster mass function (MF). In the Milky Way and the Magellanic Clouds, the MF of young star clusters is well approximated by a power law $dN(M)/dM \propto M^{-\alpha}$ where $\alpha \approx 2$ (Elmegreen & Efremov 1997; Hunter et al. 2003). This is deceptively similar to the *luminosity* functions derived in many young cluster systems, but it is important to recognize that luminosity functions are not necessarily identical, or even similar to the underlying MFs (unless the age distribution is a delta function). Unfortunately, MFs are difficult to measure directly. The only practical way to obtain mass estimates for large samples of clusters is from photometry, but because the mass-to-light ratios are strongly age-dependent, masses cannot be estimated without reliable age information for each individual cluster. As discussed in Section 3, this is best done by including U-band imaging, which is costly to obtain in terms of observing time. So far, MFs have only been constrained for a few well-studied systems. In the Antennae, Zhang & Fall (1999) found a power-law shape with exponent $\alpha \approx 2$ over the mass range 10^4 $M_\odot$ to 10^6 $M_\odot$, similar to the MF of young LMC clusters. Bik et al. (2003) find $\alpha = 2.1 \pm 0.3$ over the range 10^3 $M_\odot$ to 10^5 $M_\odot$ for M 51, and de Grijs et al. (2003a) find $\alpha = 2.04 \pm 0.23$ and $\alpha = 1.96 \pm 0.15$ in NGC 3310 and NGC 6745.

The many studies which have found similar power-law *luminosity* functions are of course consistent with these results, but should not be taken as proof that the MF is as universal as the LF. Conversely, any differences in the LFs observed in different systems would not necessarily imply that the MFs are different. There are some hints that slight LF variations may be present: Elmegreen et al. (2001) find LF slopes of $\alpha = 1.58 \pm 0.12$ and $\alpha = 1.85 \pm 0.05$ in NGC 2207 and IC 2163, while Larsen (2002) and Dolphin & Kennicutt (2002) find somewhat steeper slopes ($\alpha = 2.0 - 2.5$) in several nearby spiral galaxies. While Whitmore et al. (1999) find $\alpha = 2.12 \pm 0.04$ for the full sample of Antennae clusters, there is some evidence for a steepening at brighter magnitudes with $\alpha = 2.6 \pm 0.2$ brighter than $M_V = -10.4$. However, measurements of LF slopes are subject to many uncertainties, as completeness and contamination effects can be difficult to fully control, and it is not presently clear how significant these differences are. More data is needed.

Another important question is how the MF evolves over time. While the evidence available so far indicates that the MF in most young cluster systems is well approximated by a uniform power law with slope $\alpha \approx 2$ down to the detection limit, old GC systems show a quite different behavior. Here, the luminosity function is well fit by a roughly log-normal distribution with a peak at $M_V \sim -7.3$ (about 10^5 $M_\odot$ for an age of 10–15 Gyr) and dispersion ~ 1.2 mag (e.g., Harris & van den Bergh 1981). Thus, *old* globular clusters appear to have a characteristic mass of about $\sim 10^5$ $M_\odot$, while there is no characteristic mass for *young* clusters. This difference might seem to imply fundamentally different formation mechanisms, however, model calculations for the Milky Way GC system by

Fall & Zhang (2001) indicate that this difference can be accounted for by dynamical evolution of the cluster system. This makes the low-mass clusters disrupt more quickly and thereby causes an initial power-law mass distribution to eventually approach the bell-shaped MF seen in old GC systems. Simulations by Vesperini et al. (2003) and Vesperini & Zepf (2003) suggest that an initial power-law MF can evolve towards a bell-shaped distribution also in ellipticals.

It is puzzling, however, that the "faint fuzzy" star clusters in NGC 1023 do *not* show a turn-over in the MF at $\sim 10^5 \ M_\odot$ even though they appear as old as the normal GCs in NGC 1023 (which do show the usual turn-over). Instead, there is a steady increase in the number of faint fuzzies at least down to the detection limit at $M_V \sim -6$ (Larsen & Brodie 2000). It appears counter-intuitive that these diffuse objects should be more stable against disruption than compact GCs, although it may be significant that the faint fuzzies seem to be on roughly circular orbits in the disk of NGC 1023 (Burkert et al., in preparation).

Deriving a MF for an intermediate-age sample of clusters would provide an important observational constraint on models for dynamical evolution. In the ~ 1 Gyr fossil starburst M 82B, the analysis by de Grijs et al. (2003b) indicates a turn-over in the MF at a mass of about $10^5 \ M_\odot$, making the MF rather similar to that of old globular clusters. This would suggest that the erosion of the MF is already well advanced at an age of $\sim 10^9$ years in this system. On the other hand, Goudfrooij et al. (2003) find no turn-over in the mass distribution of 3-Gyr-old clusters in the merger remnant NGC 1316 down to their completeness limit at $M_B \approx -6$, or about 1 mag below the mass corresponding to the turn-over observed in old GC systems (accounting for evolutionary fading from 3 Gyr to 10 Gyrs). Thus, while it appears plausible that the MF observed in old GC systems may indeed have evolved from an initial power-law distribution as seen in young cluster systems, more observational constraints would be highly desirable. Observations with ACS or WFC3 would play a crucial role here, since high spatial resolution is required to detect the faintest clusters and separate them from stars and background galaxies.

6.2. *Size-of-sample effects*

Because power laws have no characteristic scale, it is hard to make a meaningful division between low-mass open clusters and higher-mass "super" clusters in cluster systems with a power-law MF. The lack of a characteristic mass suggests that there is no fundamental difference between the physical processes behind formation of clusters of various masses. Nevertheless, there are evidently differences in the numbers of YMCs (according to any-one's preferred definition) from one galaxy to another. But how significant are these differences? And in particular, is there an upper limit to the mass of a star cluster that can form in a given galaxy?

Size-of-sample effects may play an important role in explaining the apparent differences between cluster systems in different galaxies. As demonstrated by Whitmore (2003), Billett et al. (2002) and Larsen (2002), there is a strong correlation between the luminosity of the brightest cluster in a galaxy and the total number of young clusters down to some magnitude limit. Moreover, this relation has the same form as one would expect if the luminosity function is a power law where the maximum luminosity is simply dictated by sampling statistics (Fig. 10 in Whitmore 2003). Monte-Carlo simulations indicate that the *scatter* around the expected relation (about 1 mag) is also consistent with sampling statistics (Larsen 2002). In other words, current data are consistent with a universal, power-law luminosity function for young clusters in most galaxies, with the brightest clusters simply forming the tail of a continuous distribution. A similar analysis has yet

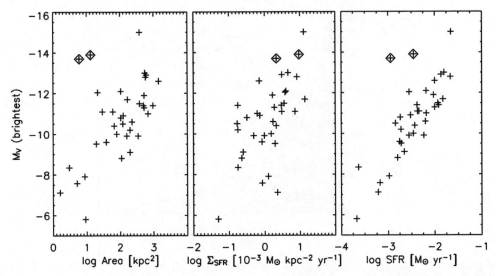

FIGURE 2. Magnitude of brightest cluster versus total surveyed area (left), area-normalized star formation rate (center), and total star formation rate (right). NGC 1569 and NGC 1705 are marked with diamonds.

to be carried out for the *mass* distributions of star clusters in a significant sample of galaxies.

While the comparison of maximum cluster luminosity versus total cluster population is suggestive, it has its difficulties. Estimating the total cluster population to some magnitude limit is subject to uncertainties, due to the different completeness limits and detection criteria applied in various studies. Larsen (2002) attempted to circumvent this problem by assuming that the total cluster population scales with the galaxy SFR, but here the difficulty is to determine the proper normalization and account for the possibility that the scaling may not be linear (Meurer et al. 1995; Larsen & Richtler 2000).

Figure 2 shows the magnitude of the brightest cluster M_V^{br} in the sample of galaxies studied by Larsen (2002) versus total surveyed area A (left panel), area-normalized star formation Σ_{SFR} (center) and total star formation rate, SFR $= A \times \Sigma_{\mathrm{SFR}}$ (right). All three panels show some degree of correlation, emphasizing the difficulty of disentangling size-of-sample effects from physical effects. The most obvious interpretation of the correlation between M_V^{br} and A is a purely *statistical* one, i.e., larger galaxies have more clusters on average, and therefore M_V^{br} becomes brighter, by the size-of-sample effect. The Σ_{SFR} vs. M_V^{br} relation, on the other hand, is suggestive of a *physical* explanation: Σ_{SFR} correlates with the gas density, for example (Kennicutt 1998), and the higher gas densities and pressures in galaxies with high Σ_{SFR} might provide conditions which are conducive for YMC formation (Elmegreen & Efremov 1997). An important clue may lie in the fact that the third plot, SFR vs. M_V^{br}, shows the tightest correlation of all. This suggests that global galaxy properties (SFR) are more important than local ones (Σ_{SFR}) for a galaxy's ability to form massive clusters.

An alternative metric is to simply compare the luminosities of the two brightest clusters in galaxies. If the luminosity function is an untruncated power law, this magnitude difference (ΔMag) should behave in a predictable way which can then be compared with observations. From an observational point of view, this approach has a number of attractive features: only the two brightest clusters in a galaxy need be detected, distance uncertainties are irrelevant, and heterogeneous data (e.g., use of different bandpasses in

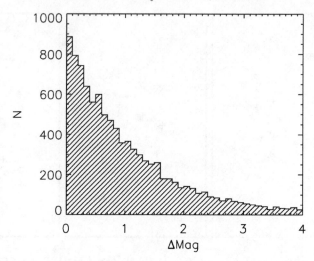

FIGURE 3. Simulated distribution of the difference ΔMag, defined as the magnitude difference between two random clusters drawn from a power-law luminosity distribution, for 10,000 experiments. This is equivalent to simulating the distribution of magnitude differences between the brightest and second-brightest cluster drawn from a power-law luminosity distribution in a sample of galaxies. The median ΔMag is 0.76 mag, while 25%, 75% and 90% of the experiments have ΔMag less than 0.31 mag, 1.50 mag and 2.51 mag.

different galaxies) do not constitute a problem. An obvious disadvantage is that this metric does not "catch" cases like NGC 1569, where the magnitude difference between the two brightest clusters will not reveal the gap down to number 3. Also, a large sample of galaxies is needed to get meaningful results.

Figure 3 shows a simulation of the distribution of ΔMag, obtained in a series of 10,000 experiments where cluster populations were drawn at random from a power-law distribution with exponent $\alpha = 2$. Fig. 4 shows the distribution of observed ΔMag values for a sample of 57 galaxies, consisting of the spirals and dwarf galaxies surveyed by Larsen & Richtler (1999) and Billett et al. (2002) and data from the sources listed in Tables 1–5 (galaxies marked with asterisks) where photometry is given for individual clusters. The observed ΔMag distribution in Fig. 4 is qualitatively similar to that in Fig. 3, peaked towards ΔMag = 0 and with a tail extending up to larger ΔMag values. Quantitatively, a Kolmogov-Smirnov test returns only a 3% probability that the distribution in Fig. 4 is drawn from that in Fig. 3, but the agreement improves greatly for a somewhat steeper power-law slope: For $\alpha = 2.2$ and $\alpha = 2.4$, the K-S test returns a probability of 30% and 77% that the observed ΔMag distribution is consistent with the equivalent of Fig. 3. Since this comparison is only sensitive to the very upper end of the LF, it may indicate that the slope at the bright end of the LF is typically somewhat steeper than $\alpha = 2$.

One remaining issue is the apparently discontinuous luminosity function in some dwarf galaxies, where the brightest clusters are much *too luminous* for the total number of clusters. It has been suggested that these cases may represent a mode of star formation which is distinct from that operating in larger galaxies, possibly caused by transient high-pressure disturbances (Elmegreen 2004). Again, it may be worth asking how significant these exceptions are. The median ΔMag from Fig. 3 is 0.76 mag, but in 25% of the cases there is a difference of ΔMag = 1.50 mag or more, and a ΔMag > 2.5 mag is found in 10% of the cases. Thus, it is not entirely unlikely to find a gap of 2 mag from the brightest to the second-brightest cluster.

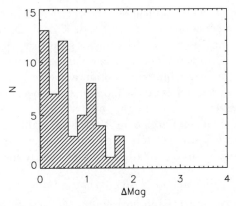

FIGURE 4. Observed distribution of ΔMag values for 51 galaxies.

One might argue that it does seem unlikely to form *two* or more very massive clusters by chance. However, the luminosity (or mass) function is a statistical tool which may not apply within small regions, but only when averaged over an entire galaxy. A dwarf galaxy like NGC 1569 may be considered as essentially a single starburst region, whereas a larger galaxy contains a multitude of starforming regions of various sizes. The experiment on which Fig. 3 is based does not apply within a region that has been selected *a priori* to contain a massive cluster. For example, if one selects a small subregion around the brightest cluster in a large galaxy, it would be very unlikely to find the second-brightest cluster in the galaxy within that subregion by chance. A related question is whether the formation of massive clusters is correlated—is there an increased probability of finding one massive cluster forming next to another one? The answer appears to be affirmative. In addition to the case of NGC 1569-A, there are several examples of binary clusters with roughly equal masses in the Large Magellanic Cloud (Dieball et al. 2002). Closer to home, the famous "double cluster" h and χ Per is another example. The double cluster, incidentally, is also among the most luminous open clusters known in the Milky Way.

6.3. *Cluster sizes*

Another point hinting at a universal cluster formation mechanism is the observation that most star clusters seem to have about the same size. The initial WFPC data for young clusters in the Antennae indicated a mean half-light radius of about 18 pc (Whitmore & Schweizer 1995), rather large compared to the ∼3 pc typical of old globular clusters. This was used as an argument against the idea that the young clusters in the Antennae are young globular clusters (van den Bergh 1995), but with WFPC2 data the mean size was revised to 4 ± 1 pc (Whitmore et al. 1999). Similar results have since been found for star clusters in many other galaxies (see examples in preceding sections). The discrepancy between size measurements of Antennae clusters carried out on WFPC and WFPC2 images illustrates the importance of having sufficient spatial resolution, however.

Naively, one might expect clusters to form with a constant density rather than a constant size, but intriguingly, there is little to suggest any significant size-mass correlation for star clusters. This has now been demonstrated both in young cluster systems in starburst regions (Zepf et al. 1999; Carlson & Holtzman 2001), in spiral galaxies (Larsen 2004), for old globular clusters in the Milky Way (van den Bergh et al. 1991) and elsewhere (Kundu & Whitmore 2001), and even in nuclear star clusters (Böker et al. 2004). In the absence of any such relation, it is clear that the most massive clusters will also tend to have very high densities, and explaining these high densities is therefore just a special case of the more general problem of why star clusters form with a nearly constant size, rather than with a constant density. Ashman & Zepf (2001) proposed that the lack of a size-mass relation may be related to a higher star-formation efficiency in high-mass clusters, causing them to expand less than lower-mass clusters after the residual gas is blown away, but more work is required to obtain a deeper understanding of this issue.

The lack of a size-mass relation is all the more puzzling because there *are* real variations in the sizes of star clusters. The half-light radii of old globular clusters correlate with galactocentric distance (van den Bergh et al. 1991; McLaughlin 2000), and some lenticular galaxies have "faint fuzzy" star clusters with much larger sizes than normal open and globular clusters (Larsen & Brodie 2000). This issue clearly merits further investigation.

6.4. *Efficiency of cluster formation and disruption*

Many of the studies cited in Section 5 have found that a high fraction of the luminosity in starburst regions is emitted by clusters or compact sources, although a direct inter-comparison is difficult because of the different bandpasses (UV through IR) used and different detection limits. Meurer et al. (1995) warn that some of their compact sources may not be individual star clusters, and similar caution applies in other systems, especially as observations are pushed to greater distances—where the nature of cluster candidates cannot be verified because of insufficient resolution. Nevertheless, there are strong indications that most stars tend to form in groups or clusters, although only a small fraction of all stars eventually end up in *bound* clusters. In the Milky Way, Lada & Lada (2003) estimate that the vast majority (70%–90%) of star formation in nearby molecular clouds takes place in embedded clusters, while only 4%–7% of these embedded clusters survive to become bound clusters with ages similar to that of the Pleiades ($\sim 10^8$ years). Similar results have been found in other galaxies: In NGC 5253, Tremonti et al. (2001) found their data to be consistent with a scenario where all stars are initially born in clusters, of which most disperse on a short time scale (~ 10 Myrs). In the Antennae, Fall (2004) estimates that at least 20%, and possibly all, of the star formation takes place in clusters, although he also concludes that most are unbound and short-lived. This also seems consistent with the finding by KC88 (Section 4), that only a small fraction of giant H II regions in late-type galaxies form massive, bound clusters, while the rest are forming unbound associations.

It is unclear how these findings relate to the apparent variations in specific cluster luminosity T_L with host galaxy star formation rate (Larsen & Richtler 2000). These authors found that the fraction of U-band light from star clusters relative to their host galaxies increases with the area-normalized SFR, from 0.1%–1% for most normal spiral galaxies up to the very large fractions (20%–50%) found in starbursts. If nearly all stars initially form in clusters, T_L may reflect a survival—rather than a formation efficiency. However, other factors, such as dust extinction and the details of the star formation history of the host galaxy, could also affect T_L. Within an on-going, strong starburst, a large fraction of the light comes from clusters, leading to a high T_L. After the burst, fading alone would not affect T_L (since field stars and clusters will fade by the same amount), but cluster disruption would cause T_L to gradually decrease by an amount that depends on the disruption timescale (Boutloukos & Lamers 2003).

Cluster disruption occurs on several timescales (Whitmore 2004). Initially, the star formation efficiency (SFE) within the parent molecular cloud is a critical factor in determining whether or not a given embedded cluster remains bound. Unless the SFE exceeds 30%–50% (Hills 1980; Boily & Kroupa 2003), the cluster is likely to become unbound once the gas is expelled. Even when the SFE is high, a large fraction of the stars may be unbound and eventually disperse away (Kroupa et al. 2001). The high "infant mortality" for clusters may represent a combination of rapid disruption of clusters which are unbound altogether, and clusters which retain a bound core containing only a small fraction of the initial mass. Interestingly, this effect appears to be largely independent of cluster mass (Fall 2004). On longer timescales, clusters continue to disrupt due to two-body relaxation, tidal shocks and other effects (Boutloukos & Lamers 2003), which may be instrumental in shaping the globular-cluster mass function (Section 6.1).

6.5. *Dynamical masses and the distribution of stellar masses*

A somewhat controversial question, which is related to the disruption timescales, concerns the universality of the stellar mass function (SMF) in YMCs. Note that the term "IMF" (initial mass function) is deliberately avoided here, since the present-day mass function in a star cluster may differ from the initial one. If the SMF is biased towards high-mass stars, compared, e.g., to a Kroupa (2002)-type function, the clusters might be rapidly disrupted (Goodwin 1997). Direct observations of individual stars in YMCs are usually beyond reach even with *HST*, especially at the low-mass end of the SMF. However, by measuring structural parameters on *HST* images and using ground-based high-dispersion spectra to estimate the line-of-sight velocity dispersions of the cluster stars, dynamical mass estimates can be obtained by simple application of the virial theorem. If the cluster ages are known, the masses thus derived can be compared with SSP model calculations for various SMFs. An increasing amount of such data is now becoming available, but the results remain ambiguous. Based on observations by Ho & Filippenko (1996a, 1996b), Sternberg (1998) found some evidence for differences in the SMF slopes of NGC 1569-A (SMF at least as steep as a Salpeter law down to 0.1 $M_\odot$) and NGC 1705-1 (SMF may be shallower than Salpeter or truncated between 1 $M_\odot$ and 3 $M_\odot$), and Smith & Gallagher (2001) also concluded that M 82-F has a top-heavy SMF. However, Gilbert & Graham (2003) found M/L ratios consistent with "normal" SMFs for three clusters in NGC 1569, Maraston et al. (2004) reported "excellent" agreement between the dynamical mass of the W3 object in NGC 7252 and SSP model predictions, and Larsen & Richtler (2004) and Larsen et al. (2004) found standard Kroupa (2002)-like SMFs for seven YMCs in a sample of dwarfs and spiral galaxies. Other authors have found a mixture of normal and top-heavy SMFs (Mengel et al. 2002; McCrady et al. 2003). At present, it is not clear to what extent these differences are real, or could be a result of different measurement techniques, crowding and resolution effects, as well as the inherent uncertainties in the analysis (e.g., assumption of virial equilibrium, effects of mass segregation, different macroturbulent velocities in the atmospheres of the cluster supergiants and the template stars). However, this aspect of YMC research would have been virtually impossible without the symbiosis between *HST* and high-dispersion spectrographs large ground-based telescopes.

7. Summary and outlook

Over the past decade, research in extragalactic young star clusters has evolved into a mature field. This is in no small part due to the complement of instruments available on *HST*, although the impact of parallel developments in ground-based instrumentation should not be underestimated.

One major advance has been the establishment of the ubiquity of young massive, globular-like clusters in a wide variety of star-forming environments. These objects can have sizes and masses which make them virtually identical to the old globular clusters observed in the Milky Way and indeed around all major galaxies.

With few exceptions, the luminosity functions of young cluster systems are well approximated by a power law with exponent ~ 2. It appears that random sampling from such a luminosity function can account, to a large extent, for differences in the numbers of YMCs and in the luminosity of the brightest star clusters observed in different galaxies. So far, no case has been found in which the luminosity of the brightest cluster is limited by anything other than sampling statistics. In other words, *YMCs are present whenever clusters form in large numbers*. Studies of the *mass* functions of young star clusters are more difficult, but the few studies that have been made seem to indicate that the mass functions are also well approximated by power laws.

It remains unclear to what extent dwarf galaxies like NGC 1569 and NGC 1705, with only a few very bright clusters, pose a problem for the idea of a universal cluster luminosity (or mass) function and a universal cluster formation mechanism. Billett et al. (2002) noted that massive star clusters are rare even in actively star-forming dwarfs, but when they do form, they are accompanied by a high level of star-formation activity. Billett et al. suggested that large-scale flows and gravitational instabilities in the absence of shear may favor the formation of massive clusters in dwarf galaxies.

By focusing on the most extreme starburst environments, one naturally finds the most extreme cluster populations. However, a complete picture can only be formed by examining the whole range of environments—from very quiescent, over normal star-forming galaxies, to starbursts—and good progress is being made towards this goal. An example of the first extreme is IC 1613, whose extremely low (but non-vanishing) star formation rate has produced only a very feeble cluster system. Even among "normal" spiral galaxies there are substantial variations in the SFR, and this translates directly to corresponding differences in the richness of the cluster systems. With their exceedingly rich young cluster systems, starbursts like the Antennae and M 82 are at the other extreme.

While special conditions may lead to the formation of a few massive clusters in some dwarfs, YMC formation apparently does not *require* special triggering mechanisms. This is supported by the Zhang et al. (2001) study of young clusters in the Antennae. No correlation between strong gradients in the velocity field and the formation sites of star clusters was found, although it was noted that *some* clusters might have been triggered by cloud-cloud collisions. By a slight extrapolation of this argument, globular clusters could also have formed by normal star formation processes in the early Universe.

A large fraction of all stars, possibly the majority, form in star clusters. However, due to an initially rapid disruption of clusters on timescales of 10^7 years which appears to be largely independent of mass, only a small fraction of these stars eventually end up as members of *bound* star clusters. Clusters which survive the initial phase of rapid destruction continue to dissolve on longer timescales, and this process may cause an initial power-law mass function to evolve towards the approximately log-normal MF observed in old globular cluster systems.

Further progress is likely to come from multi-wavelength campaigns, which will allow detailed analyses of the age and mass distributions of cluster systems. Is there a universal cluster mass function, as hinted at by the many studies which have found similar power-law luminosity functions? Is the "infant mortality" rate the same everywhere? Can we see the signatures of dynamical evolution? A good place to look for such signatures might be in galaxies with rich cluster systems formed over an extended period of time. Also, what are the properties of young clusters in early-type (Sa/Sb) spirals and other environments (e.g., low surface-brightness galaxies) which remain poorly studied?

Over its lifetime, *HST* has been upgraded several times, each time essentially leaving us with a new, much more capable observatory. The Wide Field Planetary Camera 2 has now been in operation for more than 10 years, and has produced many spectacular results. We have only just started to see the potential of the Advanced Camera for Surveys. The Wide Field Camera 3, with its panchromatic wavelength coverage from the near-ultraviolet to the infrared, has the potential to once more boost the power of *Hubble* by a significant factor.

REFERENCES

ALONSO-HERRERO, A., ENGELBRACHT, C. W., RIEKE, M. J., ET AL. 2001 *ApJ* **546**, 952.

ALONSO-HERRERO, A., RYDER, S. D., & KNAPEN, J. H. 2001 *MNRAS* **322**, 757.

ANDERS, P., BISSANTZ, N., FRITZE-V. ALVENSLEBEN, U., DE GRIJS, R. 2004 *MNRAS* **347**, 196.

ANDERS, P., DE GRIJS, R., FRITZE-V. ALVENSLEBEN, U., & BISSANTZ, N. 2004 *MNRAS* **347**, 17.

ANDERS, P. & FRITZE-V. ALVENSLEBEN, U. 2003 *A&A* **401**, 1063.

ARP, H. & SANDAGE, A. 1985 *AJ* **90**, 1163.

ARRIBAS, S. & COLINA, L., 2002 *ApJ* **573**, 576.

ASHMAN, K. M. & ZEPF, S. E. 1992 *ApJ* **384**, 50.

ASHMAN, K. M. & ZEPF, S. E. 2001 *AJ* **122**, 1888.

BARTH, A. J., HO, L. C., FILIPPENKO, A. V., & SARGENT, W. L. 1995 *AJ* **110**, 1009.

BASTIAN, N., GIELES, M., LAMERS, H. J. G. L. M., ET AL. 2005 *A&A*, **431**, 905.

BATTISTINI, P., BONOLI, F., FEDERICI, L., ET AL. 1984 *A&A* **130**, 162.

BECK, S. C., TURNER, J. L., & GORJIAN, V. 2001 *AJ* **122**, 1365.

BENEDICT, G. F., HIGDON, J. L., JEFFREYS, W. H., ET AL. 1993 *AJ* **105**, 1369.

BIK, A., LAMERS, H. J. G. L. M., BASTIAN, N., ET AL. 2003 *A&A* **397**, 473.

BILLETT, O. H., HUNTER, D. A., & ELMEGREEN, B. G. 2002 *AJ* **123**, 1454.

BOHLIN, R. C., CORNETT, R. H., HILL, J. K., & STECHER, T. P. 1990 *ApJ* **363**, 154.

BOILY, C. M. & KROUPA, P. 2003 *MNRAS* **338**, 673.

BÖKER, T., SARZI, M., MCLAUGHLIN, D. E., ET AL. 2004 *AJ* **127**, 105.

BOUTLOUKOS, S. G. & LAMERS, H. J. G. L. M. 2003 *MNRAS* **338**, 717.

BRESOLIN, F., KENNICUTT, R. C., & STETSON, P. B. 1996 *AJ* **112**, 1009.

BRODIE, J. P. & LARSEN, S. S. 2002 *AJ* **124**, 1410.

BRODIE, J. P., SCHRODER, L. L., HUCHRA, J. P., ET AL. 1998 *AJ* **116**, 691.

BRUZUAL, G. A. & CHARLOT, S. 2003 *MNRAS* **344**, 1000.

BUTA, R., BYRD, G. G., & FREEMAN, T. 2004 *AJ* **127**, 1982.

BUTA, R., TREUTHARDT, P. M., BYRD, G. G., & CROCKER, D. A. 2000 *AJ* **120**, 1289.

CARLSON, M. N. & HOLTZMAN, J. A. 2001 *PASP* **113**, 1522.

CARLSON, M. N., HOLTZMAN, J. A., GRILLMAIR, C. J., ET AL. 1999 *AJ* **117**, 1700.

CARLSON, M. N., HOLTZMAN, J. A., WATSON, A. M., ET AL. 1998 *AJ* **115**, 1778.

CHANDAR, R., LEITHERER, C., & TREMONTI, C. A. 2004 *ApJ* **604**, 153.

CHANDAR, R., TSVETANOV, Z., & FORD, H. C. 2001 *AJ* **122**, 1342.

CHRISTIAN, C. A. & SCHOMMER, R. A. 1982 *ApJS* **49**, 405.

CHRISTIAN, C. A. & SCHOMMER, R. A. 1988 *AJ* **95**, 704.

COLINA, L., GONZALEZ-DELGADO, R., MAS-HESSE, J. M., & LEITHERER, C. 2002 *ApJ* **579**, 545.

COLINA, L. & WADA, K. 2000 *ApJ* **529**, 845.

CONTI, P. S. & VACCA, W. D. 1994 *ApJ* **423**, L97.

CÔTÉ, P., MARZKE, R. O., & WEST, M. J. 1998 *ApJ* **501**, 554.

CRABTREE, D. R. & SMECKER-HANE, T. 1994 *BAAS* **26**, 1499.

DAVIDGE, T. J. 2004 *AJ* **127**, 1460.

DE GRIJS, R., ANDERS, P., BASTIAN, N., ET AL. 2003 *MNRAS* **343**, 1285.

DE GRIJS, R., BASTIAN, N., LAMERS, H. J. G. L. M. 2003 *MNRAS* **340**, 197.

DE GRIJS, R., LEE, J. T., CLEMENCIA MORA HERRERA, M., ET AL. 2003 *New Astronomy* **8**, 155.

DE MARCHI, G., CLAMPIN, M., GREGGIO, L., ET AL. 1997 *ApJ* **479**, L27.

DIEBALL, A., MÜLLER, H., & GREBEL, E. K. 2002 *A&A* **391**, 547–564

DINSHAW, N., EVANS, A. S., EPPS, H., ET AL. 1999 *ApJ* **525**, 702.

DOLPHIN, A. E. & KENNICUTT, R. C. 2002 *AJ* **124**, 158.

DOUBLIER, V., KUNTH, D., COURBIN, F., & MAGAIN, P. 2000 *A&A* **353**, 887.

DRINKWATER, M., GREGG, M. D., HILKER, M., ET AL. 2003 *Nature* **423**, 519.

DRISSEN, L., ROY, J.-R., MOFFAT, A. F. J., & SHARA, M. M. 1999 *AJ* **117**, 1249.

DRISSEN, L., ROY, J.-R., ROBERT, C., ET AL. 2000 *AJ* **119**, 688.

ELMEGREEN, B. G. 2004. In *The Formation and Evolution of Massive Young Star Clusters* (eds. H. J. G. L. M. Lamers, A. Nota, & L. Smith). ASP Conf. Ser. Vol. 322, p. 277. ASP.

ELMEGREEN, B. G. & EFREMOV, YU. N. 1997 *ApJ* **480**, 235.

ELMEGREEN, D. M., CHROMEY, F. R., McGRATH, E. J., & OSTENSON, J. M. 2002 *AJ* **123**, 1381.

ELMEGREEN, D. M., KAUFMAN, M., ELMEGREEN, B. G., ET AL. 2001 *AJ* **121**, 182.

ELSON, R. A. W. & FALL, S. M. 1985 *ApJ* **299**, 211.

FALL, S. M. 2004. In *The Formation and Evolution of Massive Young Star Clusters* (eds. H. J. G. L. M. Lamers, A. Nota, & L. Smith). ASP Conf. Ser. Vol. 322, p. 399. ASP.

FALL, S. M. & REES, M. J. 1985 *ApJ* **298**, 18.

FALL, S. M. & ZHANG, Q. 2001 *ApJ* **561**, 751.

FISCHER, P., WELCH, D. L., CÔTÉ, P., ET AL. 1992 *AJ* **103**, 857.

FORBES, D. A., BRODIE, J. P., & GRILLMAIR, C. J. 1997 *AJ* **113**, 1652.

FORBES, D. A., & HAU, G. K. T. 2000 *MNRAS* **312**, 703.

FORBES, D. A., POLEHAMPTON, E., STEVENS, I. R., ET AL. 2000 *MNRAS* **312**, 689.

GALLAGHER, S. C., CHARLTON, J. C., HUNSBERGER, S. D., ET AL. 2001 *AJ* **122**, 163.

GELATT, A. E., HUNTER, D. A., & GALLAGHER, J. S. 2001 *PASP* **113**, 142.

GIERSZ, M. 2001 *MNRAS* **324**, 218.

GILBERT, A. M. & GRAHAM, J. R. 2003. In *Extragalactic Globular Clusters and their Host Galaxies* ed. T. J. Bridges). IAU Joint Discussion 6.

GIRARDI, L. 2000. In *Massive Stellar Clusters* (eds. A. Lançon & C. Boily). ASP Conf. Ser. Vol. 211, p. 133. ASP.

GIRARDI, L., CHIOSI, C., BERTELLI, G., & BRESSAN, A. 1995 *A&A* **298**, 87.

GOLDADER, J. D., GOLDADER, D. L., JOSEPH, R. D., ET AL. 1997 *AJ* **113**, 1569.

GOODWIN, S. P. 1997 *MNRAS*, **286**, 669.

GOUDFROOIJ, P., ALONSO, M. V., MARASTON, C., & MINNITI, D. 2003 *MNRAS* **328**, 237.

HANCOCK, M., WEISTROP, D., EGGERS, D., & NELSON, C. H. 2003 *AJ* **125**, 1696.

HARRIS, J., CALZETTI, D., GALLAGHER, J. S., III, ET AL. 2001 *AJ* **122**, 3046.

HARRIS, J., CALZETTI, D., GALLAGHER, J. S., III, ET AL. 2004 *ApJ* **603**, 503.

HARRIS, W. E. 2001. In *Star Clusters, Saas-Fee Advanced Course 28* (eds. L. Labhardt & B. Binggeli). Lecture Notes 1998, Swiss Society for Astrophysics and Astronomy, p. 223. Springer-Verlag.

HARRIS, W. E. & VAN DEN BERGH, S. 1981 *AJ* **86**, 1627.

HILKER, M., INFANTE, L., VIEIRA, G., ET AL. 1999 *A&AS* **134**, 75.

HILLS, J. G. 1980 *ApJ* **235**, 986.

HO, L. C. & FILIPPENKO, A. V. 1996a *ApJ* **466**, L83.

HO, L. C. & FILIPPENKO, A. V. 1996b *ApJ* **472**, 600.

HODGE, P. W. 1980 *ApJ* **241**, 125.

HOLTZMAN, J., FABER, S. M., SHAYA, E. J., ET AL. 1992 *AJ* **103**, 691.

HOLTZMAN, J. A., WATSON, A. M., MOULD, J. R., ET AL. 1996 *AJ* **112**, 416.

HOMEIER, N. L. & GALLAGHER, J. S. 1999 *ApJ* **522**, 199.

HOMEIER, N. L., GALLAGHER, J. S., III, & PASQUALI, A. 2002 *A&A* **391**, 857.

HUNTER, D. A., ELMEGREEN, B. G., DUPUY, T. J., & MORTONSON, M. 2003 *AJ* **126**, 1836.

HUNTER, D. A., O'CONNELL, R. W., & GALLAGHER, J. S., III 1994 *AJ* **108**, 84.

HUNTER, D. A., O'CONNELL, R. W., GALLAGHER, J. S., & SMECKER-HANE, T. A. 2000 *AJ* **120**, 2383.

JOHNSON, K. E. & CONTI, P. S. 2000 *AJ* **119**, 2146.

JOHNSON, K. E., LEITHERER, C., VACCA, W. D., & CONTI, P. S. 2000 *AJ* **120**, 1273

JOHNSON, K. E., VACCA, W. D., LEITHERER, C., ET AL. 1999 *AJ* **117**, 1708.

JOSHI, K. J., RASIO, F. A., & PORTEGIES ZWART, S. 2000 *ApJ* **540**, 969.

KENNICUTT, R. C. 1998 *ARA&A* **36**, 189.

KENNICUTT, R. C. & CHU, Y.-H. 1988 *AJ* **95**, 720.

KISSLER-PATIG, M., 2000 *Reviews in Modern Astronomy* **13**, 13.

KNIERMAN, K. A., GALLAGHER, S. D., CHARLTON, J. C., ET AL. 2003 *AJ* **126**, 1227.

KOTILAINEN, J. K., REUNANEN, J., LAINE, S., & RYDER, S. D. 2001 *A&A* **366**, 439.

KROUPA, P. 2002 *Science* **295**, 82.

KROUPA, P., AARSETH, S., & HURLEY, J. 2001 *MNRAS* **321**, 699.

KROUPA, P. & BOILY, C. M. 2002 *MNRAS* **336**, 1188.

KUNDU, A. & WHITMORE, B. C. 2001 *AJ* **121**, 2950.

LADA, C. J. & LADA, E. A. 2003 *ARA&A* **41**, 57.

LARSEN, S. S. 2000 *MNRAS* **319**, 893.

LARSEN, S. S. 2002 *AJ* **124**, 1393.

LARSEN, S. S. 2004 *A&A* **416**, 537.

LARSEN, S. S. & BRODIE, J. P. 2002 *AJ* **120**, 2938.

LARSEN, S. S., BRODIE, J. P., ELMEGREEN, B. G., ET AL. 2001 *ApJ* **556**, 801.

LARSEN, S. S., BRODIE, J. P., & HUNTER, D. A. 2004 *AJ*, **128**, 2295.

LARSEN, S. S. & RICHTLER, R. 1997 In *Galactic Halos: A UC Santa Cruz Workshop* (ed. D. Zaritsky). ASP Conf. Ser. Vol. 136, p. 67. ASP.

LARSEN, S. S. & RICHTLER, R. 1999 *A&A* **345**, 59.

LARSEN, S. S. & RICHTLER, R. 2000 *A&A* **354**, 836.

LARSEN, S. S. & RICHTLER, R. 2004 *A&A*, **427**, 495.

LEE, M. G., KIM, E., & GEISLER, D. 1997 *AJ* **114**, 1824.

LUTZ, D., 1991 *A&A* **245**, 31.

MAÍZ-APELLÁNIZ, J. 2001 *ApJ* **563**, 151.

MAKAROVA, L. N. & KARACHENTSEV, I. D. 2003 *Astrophysics* **46**, 144.

MAOZ, D., BARTH, A. J., HO, L. C., ET AL. 2001 *AJ* **121**, 3048.

MAOZ, D., FILIPPENKO, A. V., HO, L. C., ET AL. 1996 *ApJS* **107**, 215.

MAOZ, D., HO, L. C., & STERNBERG, A. 2001 *ApJ* **554**, L139.

MARASTON, C., BASTIAN, N., SAGLIA, R. P., ET AL. 2004 *A&A* **416**, 467.

MARASTON, C., KISSLER-PATIG, M., BRODIE, J. P., ET AL. 2002 *AP&SS* **281**, 137.

MAYALL, N. U. 1935 *PASP* **47**, 319.

MCCRADY, N., GILBERT, A. M., & GRAHAM, J. R. 2003 *ApJ* **596**, 240.

MCLAUGHLIN, D. E. 2000 *ApJ* **539**, 618.

MELNICK, J., MOLES, M., & TERLEVICH, R. 1985 *A&A* **149**, L24.

MENGEL, S., LEHNERT, M. D., THATTE, N., & GENZEL, R. 2002 *A&A* **383**, 137.

MEURER, G. R., HECKMAN, T. M., LEITHERER, C., ET AL. 1995 *AJ* **110**, 2665.

MILLER, B. W., WHITMORE, B. C., SCHWEIZER, F., & FALL, S. M. 1997 *AJ* **114**, 2381.

O'CONNELL, R. W., GALLAGHER, J. S., III, HUNTER, D. A. 1994 *ApJ* **433**, 65.

O'CONNELL, R. W., GALLAGHER, J. S., III, HUNTER, D. A. & WESLEY, C. N. 1995 *ApJ* **446**, L1.

ORIGLIA, L., LEITHERER, C., ALOISI, A., ET AL. 2001 *AJ* **122**, 815

ÖSTLIN, G., BERGVALL, N., & RÖNNBACK, J. 1998 *A&A* **335**, 85.

PASQUALI, A., DE GRIJS, R., & GALLAGHER, J. S. 2003 *MNRAS* **345**, 161.

PASQUALI, A., GALLAGHER, J. S., DE GRIJS, R. 2004 *A&A* **415**, 103.

PEEBLES, P. J. E. & DICKE, R. H. 1968 *ApJ* **154**, 891.

RICHTLER, T. 1993. In *The Globular Cluster-Galaxy Connection* (eds. G. H. Smith & J. P. Brodie). ASP Conf. Ser. Vol. 48, p. 375. ASP.

SCHWEIZER, F., 1987. In *Nearly Normal Galaxies: From the Planck Time to the Present*. p. 18. Springer-Verlag.

SCHWEIZER, F., MILLER, B. W., WHITMORE, B. C., & FALL, S. M. 1996 *AJ* **112**, 1839.

SCOVILLE, N. Z., EVANS, A. S., DINSHAW, N., ET AL. 1998 *ApJ* **492**, L107.

SCOVILLE, N. Z., EVANS, A. S., THOMPSON, R., ET AL. 2000 *AJ* **119**, 991.

SECKER, J., GEISLER, D., MCLAUGHLIN, D. E., & HARRIS, W. E. 1995 *AJ* **109**, 1019.

SHIELDS, J. C. & FILIPPENKO, A. V. 1990 *ApJ* **353**, L7.

SMITH, L. J. & GALLAGHER, J. S. 2001 *MNRAS* **326**, 1027.

STERNBERG, A. 1998 *ApJ* **506**, 721.

TACCONI-GARMAN, L. E., STERNBERG, A., & ECKART, A. 1996 *AJ* **112**, 918.

TOOMRE, A. & TOOMRE, J. 1972 *ApJ* **178**, 623.

TRAN, H. D., SIRIANNI, M., FORD, H. C., ET AL. 2003 *ApJ* **585**, 750.

TREMONTI, C. A., CALZETTI, D., LEITHERER, C., & HECKMAN, T. M. 2001 *ApJ* **555**, 322.

TURNER, J. L. & BECK, S. C. 2004 *ApJ* **602**, L85.

VAN DEN BERGH, S. 1971 *A&A* **12**, 474.

VAN DEN BERGH, S. 1979 *ApJ* **230**, 95.

VAN DEN BERGH, S. 1982 *PASP* **94**, 459.

VAN DEN BERGH, S. 1995 *ApJ* **450**, 27.

VAN DEN BERGH, S., MORBEY, C., & PAZDER, J. 1991 *ApJ* **375**, 594.

VANZI, L., HUNT, L. K., THUAN, T. X., & IZOTOV, Y. I. 2000 *A&A* **363**, 493.

VANZI, L. & SAUVAGE, M. 2004 *A&A* **415**, 509.

VÁZQUEZ, G. A., LEITHERER, C., HECKMAN, T. M., ET AL. 2004 *ApJ* **600**, 162.

VESPERINI, E. & ZEPF, S. E. 2003 *ApJ* **587**, L97.

VESPERINI, E., ZEPF, S. E., KUNDU, A., & ASHMAN, K. M. 2003 *ApJ* **593**, 760.

WATSON, A. M., GALLAGHER, J. S., III, HOLTZMAN, J. A., ET AL. 1996 *AJ* **112**, 534.

WEISTROP, D., EGGERS, D., HANCOCK, M., ET AL. 2004 *AJ* **127**, 1360.

WHITMORE, B. C. 2003. In *A Decade of HST Science* (eds. M. Livio, K. Noll & M. Stiavelli. p. 153. Cambridge University Press.

WHITMORE, B. C. 2004. In *The Formation and Evolution of Massive Young Star Clusters* (eds. H. J. G. L. M. Lamers, L. J. Smith, & A. Nota). ASP Conf. Ser. Vol. 322, p. 419. ASP.

WHITMORE, B. C. & SCHWEIZER, F. 1995 *AJ* **109**, 960.

WHITMORE, B. C., SCHWEIZER, F., LEITHERER, C., BORNE, K., & ROBERT, C. 1993 *AJ* **106**, 1354.

WHITMORE, B. C., SPARKS, W. B., LUCAS, R. 1995 *ApJ* **454**, L73.

WHITMORE, B. C., ZHANG, Q., LEITHERER, C., ET AL. 1999 AJ **118**, 1551.

WORTHEY, G., 1994 *ApJS* **95**, 107.

ZEPF, S. E. & ASHMAN, K. M. 1993 *MNRAS* **264**, 611.

ZEPF, S. E., ASHMAN, K. M., ENGLISH, J., ET AL. 1999 *AJ* **118**, 752.

ZHANG, Q. & FALL, S. M. 1999 *ApJ* **527**, L81.

ZHANG, Q., FALL, S. M., & WHITMORE, B. C. 2001 *ApJ* **561**, 727.

ZINN, R. 1985 *ApJ* **293**, 424.

HST abundance studies of low metallicity stars

By J. W. TRURAN,[1] C. SNEDEN,[2] F. PRIMAS,[3]
J. J. COWAN,[4] AND T. BEERS[5]

[1] Department of Astronomy & Astrophysics, University of Chicago

[2] Department of Astronomy and McDonald Observatory, University of Texas

[3] European Southern Observatory, Garching, Germany

[4] Department of Physics & Astronomy, University of Oklahoma

[5] Department of Physics and Astronomy, Michigan State University

1. Introduction

Abundance studies of the oldest stars provide critical clues to—and constraints upon—the characteristics of the earliest stellar populations in our Galaxy. Such constraints include those upon: light element production and BBN; the early star-formation and nucleosynthesis history of the Galaxy; the characteristics of heavy-element nucleosynthesis mechanisms; and the ages of early stellar populations from nuclear chronometers. Discussions of many of these issues are to be found in a number of review papers (Wheeler et al. 1989; McWilliam 1997; Truran et al. 2002; Gratton, Sneden, & Caretta 2004).

While much of the available data has been obtained with ground-based telescopes, there is much to learn with *HST*. Studies in the wavelength region accessible with *HST* can, in fact, address issues ranging from the origin of the light elements Li, Be, and B to the production mechanisms responsible for the synthesis of the heaviest elements through thorium and uranium. In the following two sections, we will review specifically first boron abundance studies at low Z and then abundances of the heavy elements Ge, Zr, Os, Pt, Au, and Pb, at low Z.

2. Boron abundances in halo stars

Knowledge of lithium, beryllium, and boron abundances in stars play a major role in our understanding of Big Bang nucleosynthesis, cosmic-ray physics, and stellar interiors.

In the standard model for the origin and evolution of the light elements, only ^{7}Li is produced in significant amounts from Big Bang (primordial) nucleosynthesis. Spallation reactions on CNO nuclei during the propagation of cosmic rays (CR) in the Galaxy (with an extra α–α fusion channel for ^{6}Li and ^{7}Li) account for the other light-element isotopes, ^{6}Li, ^{9}Be, and ^{10}B (cf. Reeves et al. 1973). The origin of ^{11}B is not yet completely clear, as the exact contribution from ν-spallation in supernovae remains uncertain.

While lithium is within easy reach of ground-based telescopes (its main transition falls at 6707 Å), Be and B have their main atomic transitions in the near-UV (3130 Å) and UV (2500 Å), thus requiring, respectively, ground-based instruments equipped with high-quality UV-sensitive detectors and space-based observing facilities. The recent availability of instruments like UVES at the ESO VLT and HIRES at the Keck telescope, and before that the advent of the *Hubble Space Telescope* equipped with medium-to-high resolution spectrographs (first with the Goddard High Resolution Spectrograph, and now with STIS) have had a strong impact on this specific field of research.

Our recent analysis of high resolution, high S/N, near-UV VLT spectra for a large sample (50 stars) of Galactic stars spanning a wide range of metallicities (Primas et al. 2004) has identified a very well-defined and narrow correlation between Be and Fe for

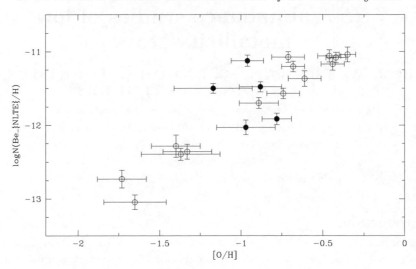

FIGURE 1. The abundance of beryllium as a function of [O/H].

[Fe/H] ≤ -1.5, but a Be-dispersed region at higher metallicities. Theories which include stochastic supernova-related processes (Parizot & Drury 2000; Ramaty et al. 2000) seem to be able to fit the existing Be data and make clear predictions about the expected observational scatter.

However, the mean ingredient of spallation reactions is O, not Fe. In Figure 1, we show the Be abundances plotted versus [O/H], the latter derived by Nissen et al. (2002) using the forbidden oxygen line at 630 nm (considered to be the best oxygen indicator because of its insensitivity to NLTE effects). Although the strong dispersion emerging in the Be versus Fe plane is significantly reduced, some scatter still remains. Furthermore, if one takes into account the error bars associated to the O abundances, all objects with [O/H] between -1.3 and -0.6 may be considered having the same oxygen content. Hence, the large difference detected in their Be abundances cannot be ascribed to their O content (nor to stellar processes since the Li abundances, also determined from the same set of VLT spectra, have not been affected yet by stellar depletion).

In order to disentangle the abundance patterns of these stars and to further constrain the mechanism(s) responsible for the observed Be abundances, one clearly needs boron, which is the main scientific goal of our Cycle 12 *HST* observations, currently being taken. Our test will consist of investigating how B compares to Be, for the five objects represented with filled circles in Figure 1. Galactic cosmic-ray spallation predicts a B/Be ratio close to 17 (e.g., Walker et al. 1985): if B is a pure CR spallation product, we expect to observe a B/Be ratio matching the one predicted by the theory. Otherwise, this new set of B abundances may hold important new clues on the contribution to the production of boron from ν-spallation or from some other mechanism.

3. Heavy-element abundances in halo stars

Abundance studies of extremely metal-deficient stars have revealed significant trends that constrain nucleosynthesis mechanisms, the characteristics of early stellar generations, and early-galactic chemical evolution. These include trends in iron-group elements (chromium through zinc) with decreasing metallicity (Cayrel et al. 2004) and the dominance of *r*-process elements (effective absence of *s*-process elements) at metallicities below [Fe/H] ~ -2.5 (Truran et al. 2002).

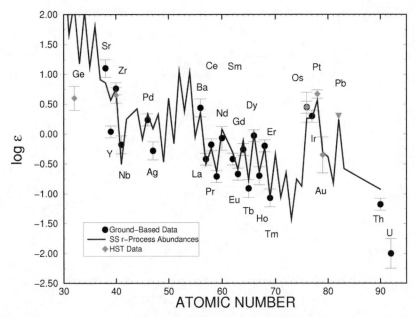

FIGURE 2. Neutron-capture elements in the halo star BD +17°3248 obtained from ground-based and *HST* observations are compared with the solar system *r*-process abundance curve. The upper limit on the lead abundance is denoted by an inverted triangle (Cowan et al. 2002).

We recall that the elements whose abundance histories have been revealed in *HST* studies are Ge, Zr, Os, Pt, Au, and Pb. The isotopes of these elements are generally understood to be products of one of the two processes of neutron capture. Ge and Zr are primarily *s*-process products; Os, Pt, and Au are mostly produced by the *r*-process; and Pb is a complicated product of both processes. Observations of their abundance histories can therefore provide important constraints on the characteristics of these processes and their operation in stellar environments.

A representative example of the manner in which *HST* observations complement those at other wavelengths is shown in Figure 2. Here, the abundances of *r*-process elements in the region past iron are shown as a function of atomic number for the *r*-process rich ([*r*-process/Fe] $\approx$ 1), low metallicity ([Fe/H] = -2.7) field halo star BD +17°3248, are compared with the solar-system *r*-process pattern (Cowan et al. 2002; Truran et al. 2002). The *HST* data for Os, Pt, Au, and Pb are shown as solid diamonds, while other *r*-process nuclei are shown as solid circles. We call attention to the remarkable agreement with solar system *r*-process abundances for the heavier nuclei, while the abundance pattern in the mass regime below A $\approx$ 130–140 does not exhibit this consistency.

The importance of the *HST* abundance data for the elements Os, Pt, Au, and Pb shown here arises from the fact that these all lie in (proximity to) the *r*-process abundance peak at mass A $\approx$ 190. These data, together with that for other well-studied cases of *r*-process-rich stars (Figure 3), unambiguously establish the robustness of the *r*-process mechanism responsible for the production of the "heavy" (A $\geq$ 130–140) *r*-process isotopes (Truran et al. 1981, 2002).

The agreement between the solar system *r*-process curve and the heavy (Z $\geq$ 56) elemental abundances in CS 22892−052 has now been seen in several other halo stars Lawler et al. (2004), and suggests a robust *r*-process operating over many billions of years. It also implies a well-defined range of astrophysical conditions (e.g., neutron number densities) and/or that not all supernovae are responsible for the *r*-process—instead, perhaps

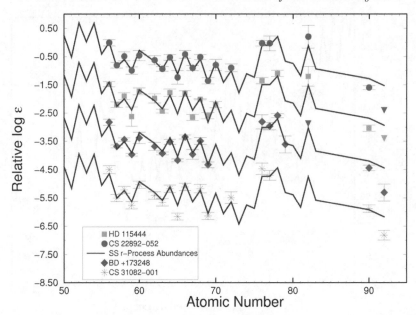

FIGURE 3. Heavy-element abundance patterns for the four stars CS 22892−052, HD 155444, BD +17°3248, and CS 31082−001 are compared with the scaled solar system *r*-process abundance distribution (solid line). See Sneden et al. 2000; Westin et al. 2000; Cowan et al. 2003.

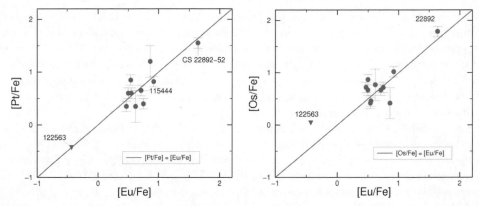

FIGURE 4. Abundance correlation between [Pt/Fe] (left panel) and [Os/Fe] (right panel) and [Eu/Fe].

only a narrow mass range. We note, however, that the well-studied stars showing this pattern are all *r*-process-rich and much less is known about *r*-process-poor stars, such as HD 122563.

4. Abundance Trends in Halo Stars

Our new *HST* abundance observations of a sample of 11 halo stars have provided new information not previously attainable concerning both the very light *n*-capture elements (such as Ge) and the $3^{\rm rd}$ *r*-process peak elements, including Pt and Os. While Ge was not detected in CS 22892−052, it was found in many of the other target stars. Zr was also detected in these stars using *HST*. The results of those observations will be forthcoming Cowan et al. (2005). We show here the elemental abundance trends of [Pt/Fe] and [Os/Fe]

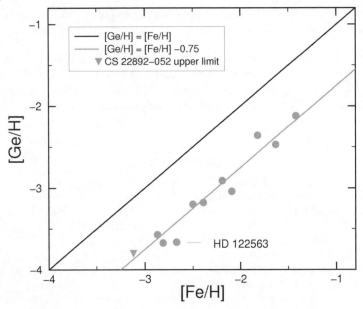

FIGURE 5. The behavior of Ge/Fe is displayed as a function of [Fe/H]. The fact that [Ge/Fe] traces [Fe/H] suggests a possible contribution to Ge from explosive nucleosynthesis in early stellar populations.

with respect to [Eu/Fe] in Figure 4. The consistency between the solar system curve and the halo star abundances (for the elements from Ba and above) has, in the past, mostly been predicated on the rare-earth elements detected from the ground. Now with the *HST* detections of elements such as Os and Pt—with dominant spectral transitions in the near UV—that agreement seems to extend through the $3^{\rm rd}$ r-process peak. We see a direct comparison of these two latter elements with the r-process element Eu—the abundances of this element were obtained with ground-based observations. It is clear in Figure 4 that the abundances of both Pt and Os seem to be correlated—there is a 45° angle straight-line relationship—with Eu in these metal-poor halo stars. This strongly suggests a similar synthesis origin for all three of those n-capture elements in the halo star progenitors.

The abundance data for germanium also provides a potentially interesting challenge to nucleosynthesis theorists. Our data in Figure 5 reveals that the abundance of Ge traces that of iron, over the range $-3 <$ [Fe/H] < -1. This is in conflict with the behavior expected if Ge were a pure s-process element. We note, however, the fact that ground-based observations of [Zn/Fe] indicate that the abundance of zinc relative to that of iron increases with decreasing metallicity at metallicities [Fe/H] < -2 (Cayrel et al. 2004). The possibility therefore exists that this behavior may reflect explosive nucleosynthesis contributions that would be identifiable in the [Ge/Fe] behavior as well (Truran et al. 2004).

5. Conclusions

HST observations of elemental abundances in stars of low metallicity have taught us much concerning nucleosynthesis, stellar evolution, and Galactic chemical evolution. Boron abundances at low metallicity challenge and constrain theoretical models for its origin. Trends in Os, Pt, Au, and Pb confirm the robustness of r-process synthesis over the mass range $140 \leq A \leq 190$ and strengthen our confidence in the use of the actinide chronometers ^{238}U and ^{232}Th as measures of the age of the Universe.

This research has been supported in part by NSF grants AST-0307279 (JJC), and AST-0307495 (CS), and by STScI grants GO-8111 and GO-08342. JWT and TCB acknowledge support of the NSF Physics Frontier Center/JINA: Joint Institute for Nuclear Astrophysics under grant PHY 02-16783 and JWT acknowledges DOE support under grant DE-FG 02-91ER 40606.

REFERENCES

CAYREL, R., ET AL. 2004 *A&A* **416**, 1117.

COWAN, J. J., ET AL. 2002 *ApJ* **572**, 861.

COWAN, J. J., ET AL. 2005 *ApJ* **627**, 238.

GRATTON, R., SNEDEN, C., & CARETTA, E. 2004 *ARA&A*, 42, 385.

LAWLER, J. E., SNEDEN, C., & COWAN, J. J. 2004 *ApJ* **608**, 850.

MCWILLIAM, A. 1997 *ARA&A* **35**, 503.

NISSEN, P. E., PRIMAS, F., ASPLUND, M., & LAMBERT, D. L. 2002 *A&A* **390**, 235.

PARIZOT, E. & DRURY, L. 2000 *A&A* **356L**, 66.

PRIMAS, F., GARCÌA PÈREZ, A., NISSEN, P. E., ASPLUND, M., & LAMBERT, D. L. 2005, in preparation.

RAMATY, R., TATISCHEFF, V., THIBAUD, J. P., KOZLOVSKY, B., & MANDZHAVIDZE, N. 2000 *ApJ* **534**, 747.

REEVES, H., AUDOUZE, J., FOWLER, W. A., & SCHRAMM, D. N. 1973 *ApJ* **179**, 909.

TRURAN, J. W. 1981 *A&A* **97**, 391.

TRURAN, J. W. & BOITTIN, N. 2005, in preparation.

TRURAN, J. W., COWAN, J. J., PILACHOWSKI, C. A., & SNEDEN, C. 2002 *PASP* **114**, 1293.

WALKER, T. P., VIOLA, V. E., & MATHEWS, G. J. 1985 *ApJ* **229**, 745.

WHEELER, J. C., SNEDEN, C., & TRURAN, J. W. 1989 *ARA&A* **27**, 279.

Physical conditions and feedback: *HST* studies of intense star-forming environments

By J. S. GALLAGHER,[1] L. J. SMITH,[2] AND
R. W. O'CONNELL[3]

[1]Department of Astronomy, University of Wisconsin, Madison, WI, USA

[2]Department of Physics & Astronomy, University College London, London, UK

[3]Department of Astronomy, University of Virginia, Charlottesville, VA, USA

Starbursts represent a different style of star-forming activity: not only is star formation more intense, but it also tends to produce more stars in compact, massive star clusters. This concentration of stars into small regions and their influence on the surroundings sets a requirement for high angular resolution observations over a range of wavelengths that only *HST* can meet. These points are illustrated through a discussion of some of the current issues regarding the nature and impact of super star clusters in nearby starburst galaxies.

1. Introduction

Starbursts are not simply scaled-up versions of the disks of normal spiral and irregular galaxies. The composite *HST* WFPC2 image of the classic starburst galaxy M82 in Figure 1 illustrates some of the differences. Star formation is localized in a well-defined central zone, where it is concentrated in clumps, beyond which there is virtually no star-forming activity (O'Connell & Mangano 1978). The well-known superwind extends above and below the plane out to kiloparsecs beyond the main starburst zone (Shopbell & Bland-Hawthorn 1998 and references therein). In M82 we can observe the combined effects of stellar feedback and a weak interaction with M81 in sufficient detail to test our models of galactic star formation. This is critical for understanding how the cycling of baryonic matter through stars relates to the overall structure of a galaxy, including its dark matter halo; e.g., through its influence in varying the luminosity part of the Tully–Fisher relationship (van Driel, van den Broek & Baan 1995). Nearby starburst systems also provide useful insights into the operation of the extreme starbursts found in young galaxies observed at high redshifts, such as Lyman break galaxies (LBGs; e.g., Giavalisco 2002; Shapley et al. 2003).

Within the M82 clumps a significant fraction of young stars are born in compact star clusters with half-light radii of <5 pc, an early *HST* result (O'Connell et al. 1995). These clusters are unresolved from the ground under normal seeing conditions even in the nearby M82 galaxy, where 1 arcsec subtends 18 pc. Although pioneering ground-based studies (e.g., Arp & Sandage 1985; Melnick, Moles, & Terlevich 1985; Kennicutt & Chu 1988) pointed out examples of luminous compact star clusters in the pre-*HST* era, WFPC imaging with the aberrated *HST* opened the field of high angular resolution optical imaging by revealing the extremely compact nature of even the nearest of these clusters, R136a in the LMC (Hunter et al. 1995). Further early *HST* studies demonstrated the association of youthful luminous compact star clusters with starbursts, showed that in some cases they form in large numbers, and established their globular cluster-like sizes and central stellar densities (see Whitmore 2003 for a review).

Following the correction to the *HST* optics, this research area again expanded. As a result, super star clusters (SSCs)—young star clusters with the mass and size of globular clusters—have been found in a variety of starbursts and, less frequently, in normal spiral galaxies. Nearby examples studied with WFPC2 include SSCs found in a range

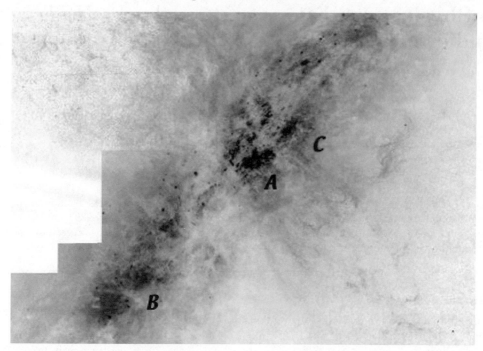

FIGURE 1. Composite image of WFPC2 frames of the starburst galaxy M82 (north to upper left). This picture contains F555W and F814W broad band images as well as narrow band images in the light of the Hα and [N II] emission lines. Here we see only the inner part of the stellar disk that runs from the lower left to the upper right, and the base of the superwind emerging from the starburst region. Major starburst clumps identified by O'Connell & Mangano (1978) are labeled. As discussed by de Grijs, Bastian, & Lamers (2003), clump B is a postburst region whose dominant stellar population has an age of ∼1 Gyr. The star-like objects sprinkled across the galaxy are compact star clusters, including SSCs.

of host galaxy types (Whitmore 2003). These include the starburst nuclei of the spirals NGC 253 (Watson et al. 1996), and M83 (Harris et al. 2001); the starburst dwarfs such as NGC 1569 (Hunter et al. 2000) and NGC 5253 (Gorjian 1996; Harris et al. 2004); the disks of spirals (e.g., NGC 6946; Larsen et al. 2001); starbursts such as in M82 (de Grijs, O'Connell, & Gallagher 2001) and circumnuclear rings (Maoz et al. 2001); and the spectacular Antennae merger (Whitmore et al. 1999). However, even with *HST*, resolution of SSCs begins to become difficult at the 20 Mpc distance of the Antennae where the *HST* diffraction pattern corresponds to 5 pc. SSCs are likely to be present in more distant objects, including the extraordinary brightest member of the Perseus cluster of galaxies, NGC 1275 (Holtzman et al. 1992; Carlson et al. 1998), and major mergers, such as NGC 6240 (Pasquali, de Grijs, & Gallagher 2003), but we lack the angular resolution to be certain that the clusters have extremely compact sizes of true SSCs.

In §2 we consider the physical properties of SSCs. The question of the survival of the SSCs and related objects is briefly discussed in §3, while in §4 we turn to the issue of how extreme clustering of young stars affects galaxies. We illustrate this example of a small-large spatial scale connection within starbursts using STIS observations of an SSC in M82 along with combined *HST* and ground-based images to gain a new perspective on this galaxy's superwind. Finally, we discuss the substantial capabilities of *HST* to make further contributions to research on extreme modes of star formation.

2. Physical properties of super star clusters

Even though SSCs are small, their presence has far-reaching impacts on the host galaxy, especially if the SSCs themselves are spatially concentrated in clumps. An understanding of the astrophysics of SSCs requires that we measure their basic physical parameters. At minimum this includes the size, total mass, luminosity, and age, but we would also like to know about other properties, such as the stellar-mass function, binary-star fraction, and metallicity. It is only with the advent of *HST* that reliable sizes became available for SSCs, indicating that their half-light radii typically fall in the range of 1–5 pc. These overlap with the radii of globular star clusters, which sparked the debate as to whether we are seeing young versions of globular clusters. The answer thus far is *yes*, qualified mainly by uncertainties about the dynamical evolution of SSCs (e.g., Fall & Zhang 2001).

Ages of SSCs can be determined by matching simple stellar population models to observed photometry (e.g., de Grijs, Bastian, & Lamers 2003). In this approach reddening presents problems, and a wide wavelength range extending from below the Balmer jump into the near infrared (NIR) helps to disentangle the effects of reddening from age. A second practical problem arises due to crowding of the clusters in clumps and their tendency to have close neighbors. Only *HST* offers the combination of wavelength agility, angular resolution, and photometric stability to reliably photometer individual SSCs packed in starburst clumps, and can do so for distances up to ≈ 100 Mpc. In principle, better age estimates can be derived by fitting spectrophotometric data, especially those with sufficient spectral purity to allow line strength and profile measurements in the optical for clusters older then 20–30 Myr (see Gallagher & Smith 1999; Smith & Gallagher 2001), and in the UV (e.g., de Mello, Leitherer, & Heckman 2000). However, ages are more problematic when only NIR observations are available. This mainly is due to the dominance of cool stars in the NIR whose spectra are not very sensitive to the age of the stellar population and where stochastic effects can be produced by a few luminous stars (see Lançon & Mouchine 2002).

Direct measurements of dynamical masses require that an appropriate mass-weighted radius and stellar velocity dispersion is measured. This approach was pioneered by Ho & Filippenko (1996a) who used the Keck Observatory HIRES echelle spectrograph to find velocity dispersions and WFPC2 data to derive sizes. Dynamical masses now are available for several SSCs, and the results show that the most massive SSCs populate the globular cluster mass range of 10^5–10^6 $M_\odot$ (Ho & Filippenko 1996a,b; Smith & Gallagher· 2001; Mengel et al. 2002; McCrady, Gilbert, & Graham 2003).

The combination of photometry, age, and mass allows us to compare SSCs with theoretical stellar population models that predict M/L for a given IMF. This shows that the SSC M82-F appears to lack low-mass stars, and therefore is a poor candidate for survival, while NGC 1569-A is rich in low-mass stars and thus should be internally stable over a cosmic time period (Smith & Gallagher 2001). A key issue in interpreting these results is the effect of dynamical mass segregation on cluster properties. If the massive stars in a cluster are concentrated within the half-mass radius, then the luminosity profile yields a substantial underestimate of the *half mass* radius, and thus of the total mass. Significant mass segregation also reduces the likelihood that a cluster will remain gravitationally bound. Thus, the internal structure of SSCs remains relevant to understanding whether SSCs can routinely evolve into globular clusters with ages of >10 Gyr. Thus, the survival of an SSC with too low of a mass for its age and luminosity (such as M82-F), is unlikely, even if the region where it formed had a normal (on the average) IMF.

A major investment of *HST* resources has been made in determining if large populations of SSCs which form during starbursts could yield globular cluster systems (see

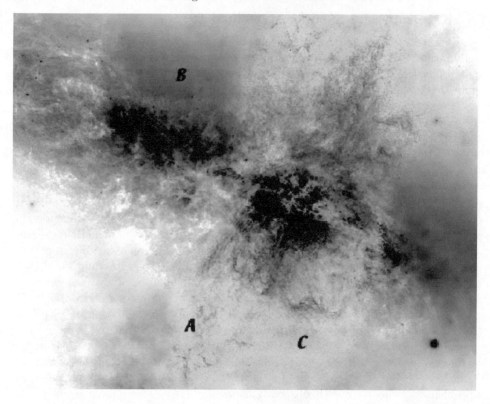

FIGURE 2. Inner region of M82 is shown in this combination of broad- and narrow-band images from *HST* and the WIYN 3.5-m telescope made by M. Westmoquette. They contain a mix of broad-band and Hα-region narrow-band images to show the locations of stellar light and regions where optical line emission is strong, which generally have complex, filamentary structures. Extraplanar nebulae are labeled by their associated starburst clumps as identified by O'Connell & Mangano (1978). Multiple "chimneys" defined by ionized gas emanate from clumps A and C, while the smooth diffuse light above the older region B appears to be a large reflection nebula. A diffuse reflection component also can be seen as the sharply defined light cone to the left of the label 'A.' The complex small scale chimneys from the clumps appear to interact above the clumps, and merge to form major components of the optically visible wind, as shown in the next figure.

Holtzman et al. 1992). One approach to this issue is based around the determination of cluster luminosity functions. *HST* observations show that young star clusters generally follow the same type of M_{cl}^{-2} power-law mass distributions as open star clusters (Whitmore et al. 1999). The production of the log-normal luminosity-stellar mass distributions seen for globular clusters then requires evolution of the luminosity function. Theoretical studies suggest this is feasible if less massive clusters are most prone to destruction. *HST*-based studies of disk galaxy star cluster systems indicates the expected trend, with time scale for dissolution scaling approximately as $M_{cl}^{0.6}$ (Boutloukos & Lamers 2003). However, a complication arises from the evidence for very rapid dissolution of SSCs in dense regions—∼10–20 Myr in some cases (Harris et al. 2001; Tremonti et al. 2001). Thus, this issue remains under discussion.

3. Young SSCS in M82 and their surroundings

An early hint of the existence of SSCs came from the extraordinarily strong Wolf-Rayet (W-R) spectral features seen in the spectra of some regions of galaxies. Such features

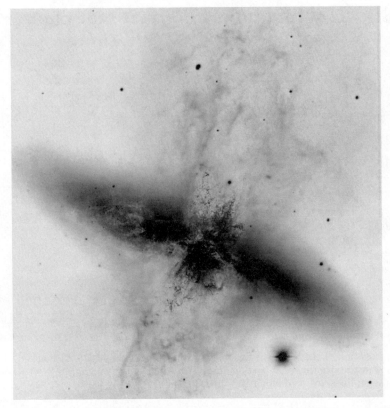

FIGURE 3. Wider angle view of the optical combined WFPC2 and WIYN image of M82. North is up and east to the left. Note the smooth cone surrounding the northern wind, probably a result of reflection by extraplanar dust. This image also shows the off-center nature of the burst, and how small features at the base of the wind blend into a large-scale superwind. These data are discussed by Gallagher et al. (2004).

implied the existence of hundreds or even thousands of W-R stars, which could only occur if large numbers of stars formed in a short time period (Conti 1993). These observations also indicated a key feature of young SSCs: the sustained input of mechanical energy in small volumes of space, combined with huge ultraviolet luminosities. A 3 Myr-old SSC with a size of ∼10 pc is perfectly capable of producing a kpc-scale giant H II region. It also can be a substantial contributor to driving galactic outflows (Gilbert & Graham 2004; Metz et al. 2004; Silich, Tenorio-Tagle, & Ródríguez-González 2004). Like galaxy nuclei, SSCs connect events occurring in very small regions to galactic scale processes. We therefore need to understand the tiny SSCs, which represent an extremum in the formation of gravitationally-bound units within galaxies, if we are to understand galaxies themselves.

M82 provides excellent illustrations of these types of connections. Figures 2 and 3 show an image of M82 produced by combining *HST* WFPC2 data with ground-based CCD images taken in similar band passes using the WIYN 3.5-m telescope. Figure 2 displays the region near clump A where a clear connection exists between the starburst clump and a localized outflow marked by ionized gas. As suggested by the models of Tenorio-Tagle, Silich, & Muñoz-Tuñón (2003), the presence of multiple star clusters leads to local collimation and directed outflows from the clump. As a result of combining the superb angular resolution of *HST* with the narrow-band-filter surface brightness sensitivity of WIYN, we immediately see in Figure 3 that the wind in M82 is

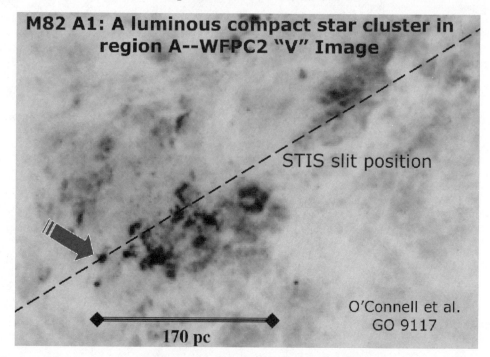

FIGURE 4. Location of the STIS slit for observations of the SSC M82−A1 marked by the arrow. The slit runs just above the main part of clump A to the right (west) of M82−A1 where extended optical emission lines are seen along the slit (next figure).

complex. As suggested by Shopbell & Bland-Hawthorn (1998), as well as Ohyama et al. (2002), the wind in M82 is not the simple form of a Chevalier-Clegg bipolar outflow, such as are seen in many of the nearby starburst nuclei (e.g., Cecil, Bland-Hawthorn, & Veilleux 2002). Gallagher et al. (2004) discuss the combined WIYN+*HST* images in more detail.

Figure 4 shows the location of the STIS slit in our program (O'Connell et al.; GO9117) to observe the luminous SSC M82−A1. Our objectives were to obtain optical spectroscopy covering the $\lambda\lambda 3500$–5200 Å and Hα spectral regions. Figure 5 displays part of the resulting spectrum for the Hα region. This spectrum has several notable features: (1) A compact H II region is associated with the SSC M82−A1, but its spatial extent is difficult to determine due to the effects of interstellar obscuration in M82. (2) Spatially extended emission is seen along most of the STIS slit, especially where it passes close to clump A. (3) The ratio of the [S II] doublet implies an electron density of $\sim 10^3$, implying an impressive pressure in the extended ionized gas of $P/k \approx 10^7$. (4) The diffuse emission line widths at half maximum intensity of ~ 100 km s^{-1} are large compared to diffuse gas found in normal galaxies, but about half the line widths of narrow components in the main superwind (Bland & Tully 1988). Evidently the warm ionized gas we are observing is not participating in the outflow near the base of the wind.

These observations raise a variety of questions, which will be addressed by Smith et al. (2004): What is the origin of a compact H II region that survives despite the combined stellar winds from what we believe is not an extremely young SSC? Why is the warm ionized ISM in relatively quiescent state close to the base of the superwind?

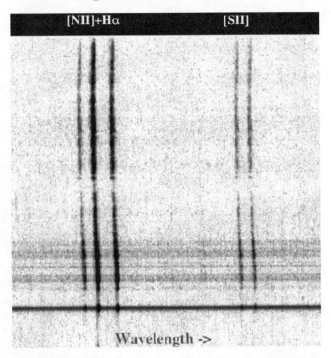

FIGURE 5. High angular resolution combined with good spectral purity make STIS spectroscopy a powerful tool for exploring the M82 starburst. In the lower part of the figure is the SSC M82−A1 that is surrounded by a small nebula. Extending beyond the cluster towards clump A we see moderately broad emission lines. The smooth velocity field perhaps is surprising given the presence of the superwind in M82. The intensities of the emission lines indicate near solar gas metallicity and high pressures in the ionized regions. These data are interpreted in Smith et al. (2004).

Is the high pressure a signature of the presence of a pervasive hot phase of the ISM, consistent with the *Chandra* observations of diffuse emission from high temperature gas in the central region of M82 (Griffiths et al. 2000; Strickland et al. 2004)? Answers to these questions could be supplied by further STIS studies, especially of the bases of superwinds in M82 and other nearby starbursts, where we can use the resolution of *HST* to full advantage. These types of data can be compared with predictions of increasingly sophisticated models of winds from groups of SSCs, such as those of Tenorio-Tagle et al. (2003).

The high pressure, hot ISM in this region carries some potentially important implications for the operation of the starburst. This type of ISM is characterized by short time scales. The sound speed is high, and cooling clouds must rapidly reach high densities if they are to remain close to pressure equilibrium with their surroundings. In addition, the trend towards making stars in compact clusters indicates that local star formation efficiencies exceed 30% and are likely twice this, as compared with normal efficiencies of <10%. It therefore is possible that galaxies in the starburst mode, where a significant fraction of stars are born in compact regions (Meurer et al. 1995; Homeier, Gallagher, & Pasquali 2002), are several times more efficient in using their molecular material than are the disks of normal spirals. All of these factors could enhance the effectiveness of star formation under starburst conditions; starbursts may feed on themselves to sustain high SFRs.

4. The *HST* spectroscopic frontiers

4.1. *The optical*

Optical/IR astronomers are accustomed to using spectrographs in seeing-limited modes, where one operates far from the theoretical resolution of the grating and the slit size is independent of aperture. However, for a fixed spectral resolution and a spectrograph working at the diffraction limit of a telescope, the signal per solid angle resolution element from a uniform intensity source is independent of aperture and scales linearly with wavelength for a constant f_λ source. Thus, for example, the limiting surface brightness for a spectrograph+telescope combination with 10% throughput observing at the diffraction limit corresponds to an R-band (Kron-Cousins system) surface brightness detection limit of $\mu_R \sim 17.5$ mag arcsec^{-2} in a $\sim 10^3$ s exposure with 8 Å spectral resolution; only bright continuum sources are in range.

However, as our data show, many SSCs reach the threshold where STIS provides unique optical observations—both of the clusters themselves and potentially of the intervening ISM—through the presence of interstellar absorption lines such as Na I or Ca II. Furthermore, emission lines in a wide range of intensely star-forming galaxies are within reach, allowing spectroscopic studies of wind excitation and velocity fields to be made with adequate resolution in the substantial selection of wind and superwind starbursts located within 100 Mpc. These types of data allow us to effectively address the apparently puzzling nature of M82. Among nearby starbursts, it stands out in the physical extent and brightness of the optically visible wind (Veilleux et al. 2003), and in its large, inferred wind velocities far from the plane and over relatively large spatial regions. In these regards, M82 could be more closely related to LBGs, which also show evidence for widespread high velocity outflows of multi-phase interstellar gas (Pettini et al. 2001), than the more common nearby nuclear starbursts.

Unique high-angular resolution optical spectroscopy with *HST* offers important payoffs for understanding the operation and impact of starbursts. This, in turn, affects a spectrum of key astrophysical processes, including the mixing of metals into the IGM, and understanding feedback in both its positive and negative modes in starbursts. However, little of this type of work has been done. That such studies are useful is illustrated by emission-line image ratios derived from WFPC2 narrow-band images of nearby starbursts by D. Calzetti and collaborators (see Calzetti et al. 2004). These data show the presence of likely shock structures that are easily missed in blurred ground-based observations.

4.2. *The ultraviolet*

As discussed elsewhere in this volume, *HST* is widely recognized for its many fundamental contributions to ultraviolet spectroscopy. This includes important studies of the FUV spectra of SSCs and starbursts that have been carried out mainly with FOS and STIS. For SSCs with ages of <10 Myr, the UV spectrum, when accessible, is a uniquely powerful age indicator. It offers added benefits of sensitivity to the characteristics of stellar winds and diagnosis of the intervening ISM through the properties of resonant interstellar absorption lines spanning a range in ionization potentials. Among the many results from these types of data is the Tremonti et al. (2001) comparison of FUV spectra from the field and star clusters in the starburst core of NGC 5253. This indicates the field is composed of stars that are systematically older than those seen in the compact clusters, opening the possibility that the field is fed through the rapid dissolution of compact star clusters, presumably including most SSCs. The more extensive investigation of FUV spectra in starbursts by Chandar, Leitherer, & Tremonti (2004) shows wonderful comparisons between the spectra of nearby starbursts relative to those of LBGs. This study is built around the authors' observations of WR clusters in NGC 3125 taken as part of the Chandar et al. GO9036 *HST* program. The spectra show remarkably intense He II λ1640 Å emission and illustrate the importance of WR stars in the FUV spectra of rapidly star-forming galaxies. A proper understanding of the FUV spectra of galaxies, and especially those at high redshifts, rests on having a detailed knowledge of how the spectra of massive stars evolve. *HST* FUV spectroscopy remains the key source of empirical information for this enterprise.

5. Summary

The discovery of compact star clusters, extending up to the massive SSCs, as common products of intense star formation is an important result from *HST*. These objects have masses of $\sim 10^5 -> 10^6$ $M_\odot$ and are produced in ~ 1 Myr. The resulting intense radiative and mechanical luminosities cause SSCs to have the impacts of extremely massive stars; they can photoionize kpc-scale nebulae and energize galactic winds. Yet their compact sizes of ~ 10 pc put SSCs beyond the resolution of seeing-limited ground-based telescopes for galaxies outside of the Local Group. *HST* has been and continues to be essential for developing a physical understanding of these most extreme products of star formation. In this regard, the SSCs are important in their own right, e.g., as the best present-day analogs to young globular star clusters, and as factors in the operation of feedback in rapidly star-forming galaxies.

Much remains to be done with *HST* and with follow-on observatories offering similar capabilities. This includes photometric imaging taking advantage of the superior resolution and near-UV sensitivity of the ACS and the narrow-band filters of WFC3. In addition, high angular resolution optical spectroscopy with STIS already is a key tool for defining the astrophysical states and impacts of SSCs. Mid- and far-UV spectroscopy also remain to be fully exploited and hopefully can be extended in the future with the installation of COS.

We would like to express our appreciation to NASA, STScI and the many people and taxpayers who made the *HST* possible. We also thank our collaborators, and especially Richard de Grijs and Mark Westmoquette, for their contributions to the M82 STIS project. This research was supported in part by NASA through grant GO-9117 administered by STScI, and by the University of Wisconsin–Madison Graduate School through funds supplied by the Wisconsin Alumni Research Foundation.

REFERENCES

ARP, H. A. & SANDAGE, A. R. 1985 *AJ* **90**, 1163.
BLAND, J. & TULLY, R. B. 1988 *Nature* **334**, 43.
BOUTLOUKOS, S. G. & LAMERS, H. J. G. L. M. 2003 *MNRAS* **338**, 717.
CALZETTI, D., ET AL. 2004 *AJ* **127**, 1405.
CARLSON, M. N., ET AL. (WFPC2 TEAM) 1998 *AJ* **115**, 1778.
CECIL, G., BLAND-HAWTHORN, J., & VEILLEUX, S. 2002 *ApJ* **576**, 745.
CHANDAR, R., LEITHERER, C, & TREMONTI, C. 2004 *ApJ* **604**, 153.
CONTI, P. 1993. In *Massive Stars: Their Lives in the Interstellar Medium* (eds. J. P. Cassinelli, E. B. Churchwell. ASP Conf. Ser. 35, p. 449. ASP.
DE GRIJS, R., O'CONNELL, R. W., & GALLAGHER, J. S. 2001 *AJ* **121**, 768.
DE GRIJS, R., BASTIAN, N., & LAMERS, H. J. G. L. M. 2003 *MNRAS* **340**, 197.
DE MELLO, D. F., LEITHERER, C., & HECKMAN, T. M. 2000 *ApJ* **530**, 251.
FALL, S. M. & ZHANG, Q. 2001 *ApJ* **561**, 751.

GALLAGHER, J. S. & SMITH, L. J. 1999 *MNRAS* **304**, 540.

GALLAGHER, J. S., ET AL. 2004, in preparation.

GERSSEN, J., ET AL. 2004 *AJ* **127**, 75.

GIAVALISCO, M. 2002 *ARAA* **40**, 579.

GILBERT, A. M. & GRAHAM, J. R. 2004. In *Recycling Intergalactic and Interstellar Matter* (eds. P.-A. Duc, J. Braine, & E. Brinks, IAU Symp. 217, p. 316. ASP.

GORJIAN, V. 1996 *AJ* **112**, 1886.

GRIFFITHS, R. E., ET AL. 2000 *Science* **290**, 1325.

HARRIS, J., CALZETTI, D., GALLAGHER, J. S., CONSELICE, C. J., & SMITH, D. A. 2001 *AJ* **122**, 3046.

HARRIS, J., CALZETTI, D., GALLAGHER, J. S., SMITH, D., & CONSELICE, C. J. 2004 *ApJ* **603**, 503.

HO, L. C. & FILIPPENKO, A. V. 1996a *ApJ* **466**, L83.

HO, L. C. & FILIPPENKO, A. V. 1996b *ApJ* **472**, 600.

HOLTZMAN, J. A., ET AL. (WFPC TEAM) 1992 *AJ* **103**, 691.

HOMEIER, N., GALLAGHER, J. S., & PASQUALI, A. 2002 *A&A* **391**, 857.

HUNTER, D. A., ET AL. 1995 *ApJ* **448**, 179.

HUNTER, D. A., O'CONNELL, R. W., GALLAGHER, J. S., & SMECKER-HANE, T. 2000 *AJ* **120**, 2383.

KENNICUTT, R. C. & CHU, Y.-H. 1988 *AJ* **95**, 720.

LANÇON, A. & MOUCHINE, M. 2002 *A&A* **393**, 167.

LARSEN, S., ET AL. 2001. In *Extragalactic Star Clusters* (eds. D. Geisler, E. Grebel, D. Minniti). IAU Symp. 207, p. 700. ASP.

MAOZ, D., BARTH, A. J., HO, L. C., STERNBERG, A., & FILIPPENKO, A. V. 2001 *AJ* **121**, 3048.

MCCRADY, N., GILBERT, A. M., & GRAHAM, J. R. 2003 *ApJ* **596**, 240.

MELNICK, J., MOLES, M., & TERLEVICH, R. 1985 *A&A* **149**, L24.

MENGEL, S., LEHNERT, M. P., THATTE, N., & GENZEL, R. 2002 *A&A* **383**, 137.

METZ, J. M., ET AL. 2004 *ApJ* **605**, 725.

MEURER, G. R., ET AL. 1995 *AJ* **110**, 2665.

O'CONNELL, R. W. & MANGANO, J. J. 1978 *ApJ* **221**, 62.

O'CONNELL, R. W., GALLAGHER, J. S., HUNTER, D. A., & COLLEY, W. N. 1995 *ApJ* **446**, L1

OHYAMA, Y., ET AL. 2002 *PASJ* **54**, 891.

PASQUALI, A., DE GRIJS, R., & GALLAGHER, J. S. 2003 *MNRAS* **345**, 161.

PETTINI, M., ET AL. 2001 *ApJ* **554**, 981.

SHAPLEY, A., STEIDEL, C. C., PETTINI, M., & ADELBERGER, K. L. 2003 *ApJ* **588**, 65.

SHOPBELL, P. L. & BLAND-HAWTHORN, J. 1998 *ApJ* **493**, 129.

SILICH, S., TENORIO-TAGLE, G., & RÓDRÍGUEZ-GONZÁLEZ, A. 2004 *ApJ* **610**, 226.

SMITH, L. J. & GALLAGHER, J. S. 2001 *MNRAS* **326**, 1027.

SMITH, L. J., ET AL. 2004, in preparation.

STRICKLAND, D. K., HECKMAN, T. M., COLBERT, E. J. M., HOOPES, C. G., & WEAVER, K. A. 2004 *ApJ* **606**, 829.

TENORIO-TAGLE, G., SILICH, S., & MUÑOZ-TUÑÓN, C. 2003 *ApJ* **597**, 279.

TREMONTI, C. A., CALZETTI, D., LEITHERER, C., & HECKMAN, T. M. 2001 *ApJ* **555**, 322.

VAN DRIEL, W., VAN DEN BROEK, A. C., & BAAN, W. 1995 *ApJ* **444**, 80.

VEILLEUX, S., SHOPBELL, P. L., RUPKE, D. S., BLAND-HAWTHORN, J. & CECIL, G. 2003 *AJ* **126**, 2185.

WATSON, A., ET AL. (WFPC2 TEAM) 1996 *AJ* **112**, 534.

WHITMORE, B. C. 2003. In *A decade of Hubble Space Telescope science* (eds. M. Livio, K. Noll, M. Stiavelli). Space Telescope Science Institute Symp. Ser. 14, p. 153. Cambridge University Press.

WHITMORE, B. C., ET AL. 1999 *AJ* **118**, 1551.

Quasar hosts: Growing up with monstrous middles

By KIM K. MCLEOD

Department of Astronomy, Wellesley College, Wellesley, MA 02481, USA

The *Hubble Space Telescope* has shown us the homes of nearby quasars in revealing detail, and has dealt us surprising answers to some of our long-standing questions about quasar host galaxy morphology. However, like all cutting-edge instruments, *HST* has taught us that the very questions we were asking were not necessarily the most interesting ones. Exploring the latter will require a combination of ground- and space-based work over the remaining lifetime of *HST*, and beyond. Such studies promise to give us insight into the formation and evolution of galaxies like our own over the whole history of the Universe.

1. Introduction

HST and quasar host galaxy studies have grown up together over the past 30 years. Indeed, "the imaging of low-redshift quasars at high angular resolution ($\sim 0\overset{''}{.}1$) is one of the principal scientific goals for which the *Hubble Space Telescope* was designed" (Bahcall, Kirhakos, & Schneider 1994). The nice demonstration by Kristian (1973) that nearby quasars are, in fact, surrounded by "fuzz" in deep 200-inch photographs provided timely input for the design of *HST* and its instruments, the specifications for which were outlined by the Large Space Telescope Science Working Group in 1974 (*HST* website). While *HST* has changed the way we look at quasar hosts, the ultimate goal of our studies has not changed over the decades. Then, as now, we strive to understand the roles played by quasars in galaxy evolution. The coeval peaks at $z = 2$–3 in quasar activity and the star-formation history of the Universe, together with the recently discovered relation between black hole mass and galaxy mass, both indicate that the quasars' role is not to be ignored.

In this paper, I will give the non-specialist a quick primer on the history of quasar host galaxy studies (§2), some insight into the challenges of imaging quasar host galaxies (§3), a description of how *HST* has changed our view (§4), a status report on our current understanding (§5), and a speculative list of future projects (§6). The bulk of this paper is based on imaging studies of quasars with redshift $z \lesssim 0.4$, to which I henceforth refer as "low-z." For a more comprehensive recent review on the host galaxies of nearby quasars, I refer the reader to e.g., Dunlop (2004). I ignore here the host galaxies of e.g., BL Lacs, radio galaxies, and LINERS, but point out that their stories often parallel that for the quasars.

2. A decade-by-decade history, Astro 101 style

One way to track the progress in our understanding of quasar hosts is to examine it in the tradition of Astronomy 101 ("'Scopes for Dopes"?). In 1974, quasar host galaxies had not yet made their way into introductory textbooks. Ten years later, we were told in *Realm of the Universe* (Abell 1984) that "In a few quasars, we can actually observe the underlying galaxies in which they are embedded... The total amount of light in the faint glow around one of these quasars and its radial extension are about right for a normal galaxy." These results were based on imaging with photographic plates.

73

In the decade that followed, CCDs helped the cause by providing the linear response that is critical for untangling the light of the point-like quasar nucleus from the surrounding galaxian nebulosity. According to the textbook *Universe* (Kaufmann 1994) "...painstaking observations have revealed some basic properties of these host galaxies. *Radio-quiet quasars seem to be located in spiral galaxies, whereas radio-loud quasars seem to be located in ellipticals* [my italics] ...Astronomers look forward to using the repaired *Hubble Space Telescope* to examine quasars and the host galaxies in greater detail."

By this year, 2004, our students (at least those who actually read the textbook...) learned in *Voyages to the Stars and Galaxies* (Fraknoi, Morrison, & Wolf 2004) that "Observations with the *Hubble Space Telescope* provided even stronger evidence. Quasars turn out to be located at the centers of galaxies. Hints that this is true had been obtained with ground-based telescopes, but space observations were required to make a convincing case[!]...*Quasars have been found in the cores of both spiral and elliptical galaxies* [my italics]."

To a specialist in this field, the situation as represented here is both over- and understated at the same time. There was really no question in the eyes of ground-based observers that pre-*HST* observations had convincingly demonstrated the existence of host galaxies. We didn't need *HST* to prove that. On the other hand, *HST* put us in our collective places by showing that our previous statements about the specifics of galaxy types had been premature and overzealous. Before *HST*, we are now ashamed to admit, the identification of spirals with radio-quiet objects, and ellipticals with radio-loud objects, was largely based upon argument by analogy with the lower-luminosity cousins of these objects, namely Seyferts and radio galaxies.

As I explain below, *HST* observations were key in distinguishing even this most basic morphological property of the hosts. However, it is worth remembering that quasar host galaxies need not be classifiable into a standard type. Many hosts show signs of tidal distortion and interaction. The extent to which an interaction, or even a major merger, is required to fuel quasar activity is still debated.

3. Three evil letters: PSF

The main problems we face in probing the nature of quasar hosts are *contrast* and *resolution*. The former is illustrated in Fig. 1, which gives a graphic rationale for observing in the near-IR for nearby objects. The latter issue affects both detection of the galaxy itself and the detection of features within it. For a quasar at $z = 0.4$, the largest galaxies have diameters of $< 10''$ even in deep images (corresponding to the isophote D_{25}), whereas scale lengths and spiral arms appear on a sub-arcsecond scale. Thus, attempting ground-based imaging of quasar hosts can be a humbling experience.

Because of these difficulties, host galaxy imaging studies have always required some form of point-spread-function (PSF) analysis to separate the contribution of the nuclear point source from that of the extended host. Figure 2 gives a good picture of the technical challenges of PSF subtraction. These challenges are similar in many ways to those faced in trying to image extrasolar planets, a big difference being that here we are looking for extended, potentially smooth features that mimic scattered light.

Since the earliest studies with photographic plates, for example in the back-to-back papers of Hutchings et al. (1981) and Wyckoff, Gehren, & Wehinger (1981), a fruitful approach has been to use observations of stars to represent the PSF, and to generate radial intensity profiles of both quasar and "PSF star." The quasar profile is then modeled as a combination of PSF + galaxy, with the galaxy being either an exponential disk or a deVaucouleurs $r^{1/4}$ elliptical. This technique has proven useful in estimating overall

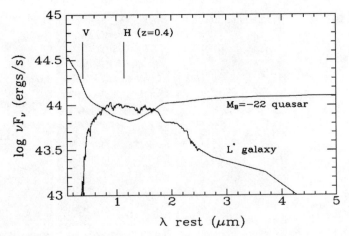

FIGURE 1. The contrast problem. Spectral energy distributions of a quasar and a typical galaxy are shown in the rest frame. The galaxy has luminosity L^*slightly brighter than, e.g., the Milky Way. The quasar represents a low-power nucleus, in this case one at the nominal quasar-Seyfert dividing line. The vertical bars show the rest-frame wavelengths sampled by observations in the visible and near-IR for a low-z quasar. In the visible, the nucleus can outshine the entire host by tens to hundreds of times. In the near-IR the contrast is improved. Adapted from McLeod & Rieke (1995); $H_0 = 80$ km s^{-1} Mpc^{-1} is used throughout.

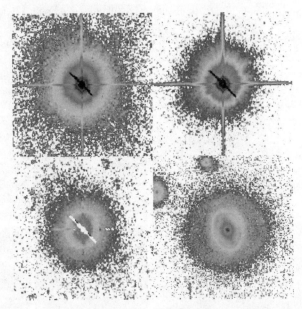

FIGURE 2. Why PSF subtraction is required. This false-color mosaic created from WFPC2 images shows a quasar (upper left); a star observed with the same instrument configuration, representing the PSF (upper right); the quasar with the star normalized and subtracted (lower left); and a galaxy on the same WFPC2 image as the quasar, illustrating what a $\sim1.4L^*$galaxy would look like at the quasar's redshift. Without PSF subtraction, the light from the nucleus completely overwhelms that from the galaxy; it is hard to tell the difference between the quasar and the star. After PSF subtraction, this quasar's host is easily seen and resembles the galaxy at lower right. The scattered light halo, diffraction spikes, and even the radial features of the PSF have been fairly effectively removed. However, information from the $0\rlap{.}''5$ saturated core and the central bleeding region is forever lost. Each box is $\approx 20''$ across. The author thanks John Bahcall for these data, and for pointing *HST* at quasars.

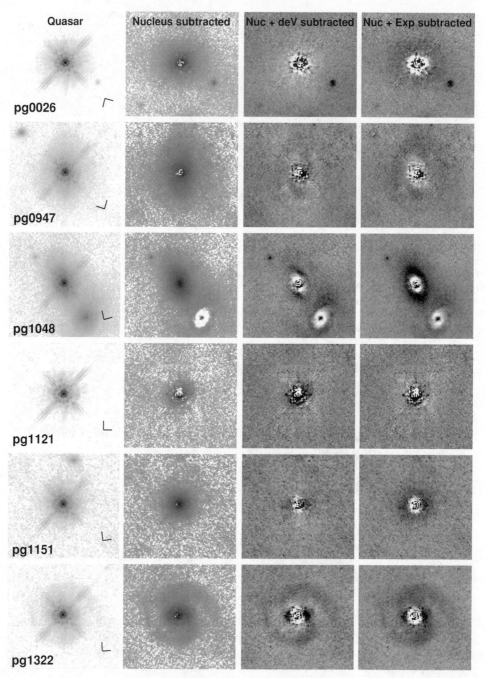

FIGURE 3. PSF removal in two dimensions. These are low-z quasars observed with NICMOS, shown in boxes $\approx 10''$ on a side. The leftmost panel on each row shows the quasar image itself, and illustrates again the difficulty in discerning host properties without PSF removal. In the second panel, the nucleus has been subtracted, and though the centers remain noisy, the Airy ring and scattered light have been effectively removed. The third and fourth panels show the quasar with nucleus+deVaucouleurs model and nucleus+exponential model removed, respectively. In some cases this view highlights spiral arm structure. Taken from McLeod & McLeod (2001).

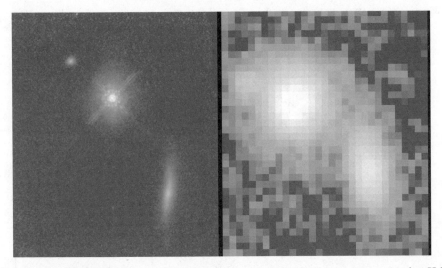

FIGURE 4. Two 2.3 m telescopes, with very different views of a low-z quasar in the H-band. Left: *HST*/NICMOS image. Right: Steward 90 inch. This figure shows clearly why *HST* has changed the way we do business in this business, and makes us wonder how we learned *anything* from the ground-based view!

host galaxy luminosities, but unreliable for determining galaxy type in all but the very nearest objects.

With more recent CCD and near-IR imaging, where sensitivity is improved and the linear response has been a great boon to the PSF removal business, techniques have extended to two dimensions. One approach is to estimate host galaxy flux in a model-independent way by subtracting a PSF image scaled so as to leave either zero flux in the central pixel, or to leave a profile that is monotonically decreasing. Another approach is to convolve the galaxy model+point source with the observed PSF, and perform a least-squares fit to the observed quasar image. The parameters allowed to vary in the fit include the nuclear intensity and position, the background level, and the galaxian intensity, position, scale length, ellipticity, and position angle. A sample of such fits as applied to NICMOS images is shown in Fig. 3.

4. Before and after the arrival of *HST*

4.1. *Before...*

By the time WFPC2 was ready to roll, several large (dozens of objects) ground-based studies of low-z quasars had been completed in the visible by, e.g., Smith et al. (1986), Hutchings, Janson, & Neff (1989), and Veron-Cetty & Woltjer (1990), with the somewhat confused state of affairs summarized nicely in the last of these. One of the uncertainties that these studies always must address results from the observations being carried out at different rest wavelengths for quasars at different redshifts. In interpreting the results, one must remember that there can be different contributions to the host light from star formation and from scattered quasar emission lines. The analyses indicated that quasar hosts tend to be drawn from the bright end of the galaxy luminosity function, and that there is a possible difference between radio-quiet and radio-loud quasars in the sense that the latter typically have more luminous hosts. We now know that the situation is more complicated, as we discuss below.

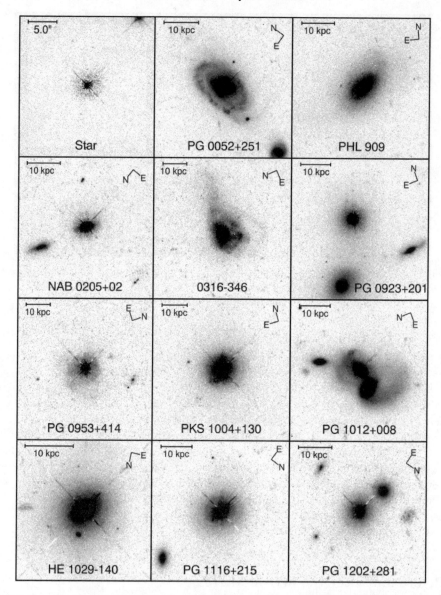

FIGURE 5. Sample of quasar hosts with nuclei removed, from Bahcall et al. (1997) Fig. 2. Note the range of types represented. Some galaxies are smooth, some are undergoing strong tidal interactions, and when spiral arms are present, we can *finally* see them! Boxes are $\approx 20''$ across.

The development of near-IR arrays offered us a complementary view of hosts. Ground-based near-IR images show the mass-tracing red stellar population of the host while being less biased by star formation. They also probe the hosts to fainter isophotes, owing in part to longer effective exposure times. The results of the IR studies ran somewhat contrary to those of the CCD studies. The Dunlop et al. (1993) program found no luminosity difference between hosts of carefully-matched samples of quasars with different radio properties, whereas the McLeod & Rieke (1994ab) program showed that the host masses allowed are dependent on the luminosity of the nucleus, foreshadowing the events that were soon to follow.

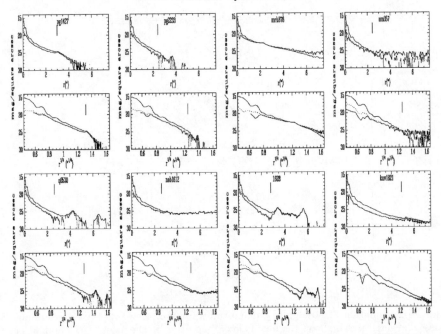

FIGURE 6. Sample radial profiles from NICMOS images. The three curves shown on each plot are, top to bottom: the "raw" quasar profile; the quasar with just enough of the PSF subtracted to leave a monotonic result; and the quasar with the PSF subtracted to leave zero flux in the center. The vertical line shows the radius inside which the profile is free from nearby sources and excess noise. For each quasar, the profiles are plotted twice: once versus r, on which an exponential disk would appear as a straight line; and once versus $r^{1/4}$, on which a deVaucouleurs law would appear as a straight line. The cases in which the choice of galaxy type looks ambiguous here are generally the ones whose two-dimensional fits are also nondefinitive, in the sense that the χ^2 for a given type is heavily dependent on the weighting scheme adopted. From McLeod & McLeod (2001).

4.2. *After...*

With the resolution of *HST*, our views of low-z quasars have improved dramatically, as illustrated in Fig. 4. A sampling of images from the large WFPC2 program of Bahcall, Kirhakos, & Schneider (1994), shown in Fig. 5, highlights the power of *HST* to reveal detailed structures of low-z hosts. The WFPC2 images have, for the first time, shown clearly delineated spiral arms in some objects, providing proof that those particular hosts are indeed disk systems. It has also shown spectacular cases of galaxy interactions (e.g., Bahcall, Kirhakos, & Schneider 1994; Hutchings & Morris 1995; Boyce et al. 1998; McLure et al. 1999).

Together, *HST* and near-IR imaging have proven to be a winning combination. Using host galaxy masses derived from ground-based near-IR imaging, together with the lack of high spatial frequency features in many of these objects in the WFPC2 images, we got a strong hint that luminous, radio-quiet, quasars can in fact inhabit *massive, early-type hosts* (McLeod & Rieke 1995).

With the subsequent insertion of NICMOS into *HST*, we achieved a best-of-both-worlds marriage of near-IR light and excellent resolution that has been put to good use on many low-z quasars (e.g., McLeod, Rieke, & Storrie-Lombardi 1999; McLure, Dunlop,

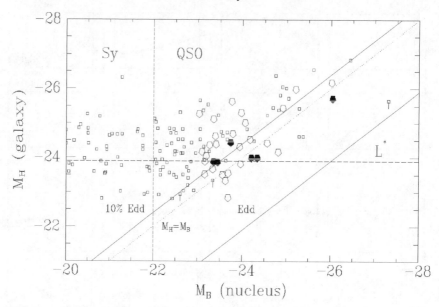

FIGURE 7. Compilation of host galaxy magnitudes for samples of $z < 0.4$ quasars and Seyfert galaxies. The vertical axis shows the host galaxy magnitude in the near-IR H band, after removal of the nucleus. The horizontal axis shows the B magnitude of the nucleus alone. QSOs shown as pentagons have host magnitudes derived from *HST* images. Also shown are the QSO/Seyfert boundary (dashed vertical line) and the position of an L* galaxy (dashed horizontal line). Note that the lower right corner of the diagram is empty! The diagonal lines show the fraction of the Eddington limit at which the nuclei are radiating, assuming that they follow the black hole-bulge relation discussed in the text. All magnitudes are given in the rest frame. From McLeod & McLeod (2001).

& Kukula 2000; McLeod & McLeod 2001; Dunlop et al. 2003 and references therein). The NICMOS camera has a very efficient non-destructive readout mode that is helpful to the host-seeking astronomer. With it, we can easily build up a deep image that reveals the faint isophotes of the host, yet avoids the saturation problem that affects long WFPC2 images.

Thanks to the relative stability of *HST*'s PSF, we have achieved good results from the fitting. A successful observing procedure has been to observe a quasar in a dither pattern for most of a single orbit, but then to take a few minutes at the end of the orbit to observe a nearby bright PSF star on the same dither pointings. As it turns out, the discrimination of exponential v. $r^{1/4}$ is still not reliable in all cases, even when tidal features are not present. In practice the light in the center of a peaky $r^{1/4}$ profile is degenerate in the fits with the central point source component. For a similar reason, even *HST* resolution is generally not high enough to warrant attempts to fit a bulge+disk. Sample profiles from *HST* images are shown in Fig. 6.

5. Where we are now

5.1. *Galaxy types—discard preconceptions before entering*

By providing the first decently high-resolution look at $z \lesssim 0.4$ quasars, the *Hubble Space Telescope* has helped us to rewrite the textbooks on quasar host galaxies. A compilation of host galaxy magnitudes is shown in Fig. 7, where we can take the near-IR (H-band) luminosity to serve as a good proxy for host galaxy mass. We see that Seyferts can inhabit

0054+144(RQQ)M4M

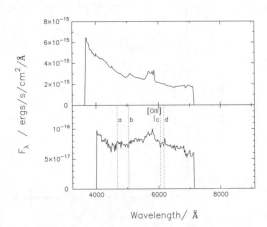

Wavelength/ Å

FIGURE 8. Observed-frame spectrum of a low-z quasar (top) and its host (bottom). The contour plot is $30''$ on a side, and shows how the image can be used to determine a judicious slit placement far from the nucleus. Line "a" marks the position of the 4000 Å break. The spectrum was taken at the Kitt Peak 4m. From Hughes et al. (2000).

galaxies a few times more or less massive than L^*, and we know these are typically spirals. However, as we move into the quasar range, we see that there are no longer small galaxies acting as hosts. Instead, there is a luminosity/host-mass limit in the sense that the more luminous nuclei require more massive hosts.

Thanks to the *HST* data, we (the collective we, with contributions from all of the authors cited already) now know the morphological types of many of these galaxies. For radio-quiet objects, the fraction of quasars inhabiting spirals drops as the nuclear luminosity increases. By the time we reach the objects on the high-luminosity end of Fig. 7, at least $\sim 1/2$ of the radio quiet objects are seen to be in *massive, early-type galaxies.* I have deliberately used this italicized term in preference to using "ellipticals" because in some cases I do not believe that the data are good enough to rule out a disk component that one would have in an armless S0 galaxy. Unfortunately, improving the statistics at the bright end is difficult because these are precisely the objects for which the nuclei make the fits difficult.

The radio-loud quasars tend to have very luminous nuclei, and by the luminosity/host-mass limit, very massive hosts. In fact, they do seem to inhabit elliptical galaxies. Thus, we conclude that *the determinant for host galaxy type is not explicitly the radio loudness, but rather the nuclear luminosity.* The physical implications of this are discussed in the next section.

The deep near-IR quasar host images have also proven useful for informing followup spectroscopy of the fuzz, as shown in Fig. 8 taken from Hughes et al. (2000). Thanks to the images, these authors have been able to place the slit at greater distance ($\sim 5''$) than had been previously possible in the pioneering efforts of Boroson & Oke (1984). This, along with spectrometer improvement, has greatly reduced amounts of nuclear scattered light entering the slit, and has made possible not only fuzz detection, but also stellar-population analysis. The synthesis models run on the very nearest quasars indicate that at $z \sim 0.2$, radio-loud and radio-quiet quasars appear to inhabit galaxies dominated by an evolved (8–12 Gyr) population (Nolan et al. 2001), typical of giant ellipticals. This is consistent with the morphological analysis showing early-type hosts.

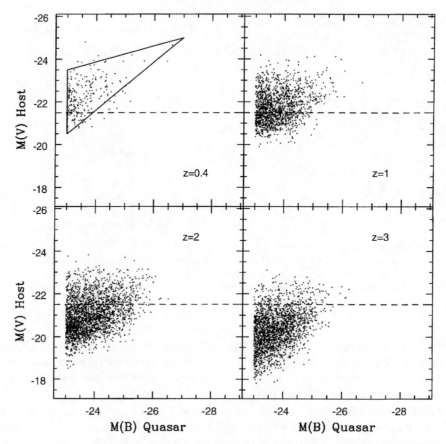

FIGURE 9. Model predictions for the distribution of galaxy v. quasar absolute magnitude as a function of redshift based on hierarchical galaxy formation models. The dashed line shows the magnitude of an L^* galaxy. The triangle in the low-z bin represents the observed locations of quasars shown in Fig. 7. From Kauffmann & Haehnelt (2000).

5.2. *The black hole knows about the bulge*

While the quasar host observers were using *HST* to uncover the luminosity/host-mass limit, other teams were probing the cores of nearby, nonactive galaxies with unprecedented resolution. Their now-famous result, specified by Magorrian et al. (1998) and discussed by Doug Richstone and Laura Ferrarese in this symposium, shows that nonactive galaxies exhibit a correlation between the mass of the galaxy's spheroid component (i.e., an elliptical galaxy, or a bulge of a spiral) and the mass of the apparently ubiquitous central black hole.

We have interpreted the luminosity/host-mass limit within this framework by assuming that the quasars follow the same relation. If we take the relation to be $f = M_{\text{black hole}}/M_{\text{spheroid}} \sim 0.006$, then we derive the following relation between the B-band magnitude of the nucleus and the H-band magnitude of the host galaxy:

$$M_B = M_H - 2.1 - 2.5 \left[\log_{10}(\epsilon) + \log_{10} \left(\frac{\Upsilon_V}{7.2 \ M_\odot/L_\odot} \right) + \log_{10} \left(\frac{f}{0.006} \right) - \log_{10} \left(\frac{BC}{12} \right) \right]$$

(McLeod, Rieke, & Storrie-Lombardi 1999). Here, ϵ represents the fraction of the Eddington limit at which the quasar is radiating; Υ_V is the V-band mass-to-light ratio of the galaxy; and BC is the bolometric correction for the nucleus. We conclude that the most

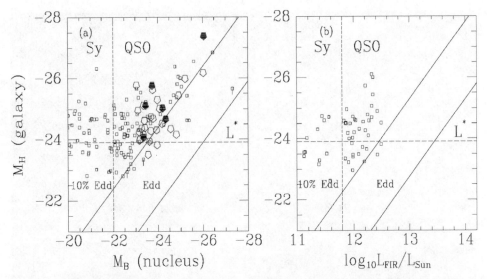

FIGURE 10. Like Fig. 7, but now including ULIRGS. For the ULIRGS, the integrated far-IR luminosity is taken as a proxy for nuclear power. If the ULIRGS are powered by quasar nuclei that follow the same black hole/spheroid relation as nonactive galaxies, they appear to radiate at similar Eddington fractions to the quasars. From McLeod, Rieke, & Storrie-Lombardi (1999).

luminous nearby quasars are radiating at $\sim$10–20% of the Eddington limit, as shown on Fig. 7. Recently, several groups have explored how continuum and emission-line-based virial mass estimators for quasar black holes can be used to refine these numbers (McLure & Jarvis 2002; Bechtold et al. 2003).

The large Eddington fractions imply a small duty cycle for activity over the Hubble time. Otherwise, the accretion process would produce higher-mass black holes than we infer today. Taken together with quasar and galaxy demographics in the local Universe, this all hangs together as a self-consistent picture in which *all galaxies go through a quasar phase* for a small fraction of their lives, with the galaxy and the black hole related in a fundamental way. The interesting question now becomes "How does the black hole know how big its host is?" In other words, by what mechanisms have black holes and galaxies grown up together over the history of the Universe to give the relation seen today?

One possible answer to this question has been presented in the form of a semi-analytical model of hierarchical galaxy formation by Kauffmann & Haehnelt (2000). In these models, the parameters chosen to fit observations such as the quasar luminosity function lead naturally to an observed luminosity/host-mass limit like the one we have seen. The model implies that the massive hosts of today's highest-luminosity quasars formed relatively recently, but that at earlier epochs, luminous quasars will be found in progressively lower-luminosity hosts. This is illustrated in Fig. 9.

We already can guess that this model is incomplete, and that the black hole/spheroid relation for quasars will break down at high redshift, where there are quasars whose large black hole masses would imply too-massive galaxies at early times.

5.3. *The role of Interactions and Fueling*

Before *HST*, ground-based observers had already determined that many quasar hosts are currently undergoing mergers or strong collisions. This meshed nicely with the Sanders et al. (1988) merger idea based on spectral energy distributions of ultraluminous infrared galaxies (ULIRGs). In this scenario, a quasar is formed following the merger of two gas-rich spirals, which go through a dusty phase until the central engine is finally revealed

FIGURE 11. Sample $z = 4$ quasar field obtained in four hours of exposure in the K-band with the PANIC camera on the Magellan I 6.5 m telescope. The FWHM is only $\sim 0\rlap{.}''3$ thanks to excellent seeing. Note the abundance of stars that can be used to measure the PSF. Images like these obtained in several colors will also be used to characterize the environments of quasars at high redshift. From Bechtold & McLeod (2005).

when the dust is blown away. If this were true, then we might expect the ULIRGs to follow a luminosity/host-mass limit like that of the quasars. Fig. 10 shows that this might be the case. Studies of red 2MASS quasar hosts might shed more light on this subject (Marble et al. 2003; Hutchings et al. 2003).

However, we also know, at least in the case of Seyferts and some low-luminosity quasars, that many hosts are decidedly not interacting or even disturbed. Thus at the lower end of the quasar luminosity range, strong interactions cannot be a requirement for nuclear activity. Given the new evidence that the more luminous quasars live in massive early-type galaxies, we can speculate that mergers are likely to precede the formation of the highest-power quasars, but only because these objects are destined to become giant ellipticals. According to the black hole/spheroid relation, their black holes pre-existed, and grew, in the progenitor galaxies.

That leaves open the important question of how activity is fueled through the lifetime of a galaxy. *HST* observations of Seyfert nuclei on small spatial scales have the potential

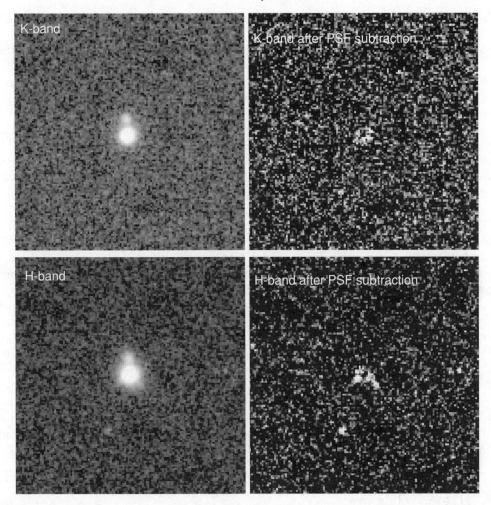

FIGURE 12. Example of fits to a $z = 4$ quasar from an image like the one in Fig. 11. In the images on the right a point source has been subtracted from the quasar, and a deVaucouleurs model has been subtracted from the galaxy above it. There is possible evidence for a host galaxy in the residuals. From Bechtold & McLeod (2005).

to help us understand the mechanisms at work. However, the galaxies are not giving up their secrets easily. As shown in Martini et al. (2003), both nonactive and active galaxies exhibit similar nuclear dust structure, and there is no universal fueling mechanism apparent.

5.4. *Beyond the back yard*

To test the models of the galaxy formation and black hole growth and fueling, we will need to characterize the hosts of quasars at earlier cosmic times. Two regimes of interest are the times around $z = 2$ when quasars were in their heyday, and $z = 4$, where they were extremely young. This will prove challenging for the same old reasons of contrast and resolution that affected pre-*HST* observations at low redshift.

At $z = 4$, the observed K band is sampling the rest-frame B band, and we once again find ourselves in the unhappy position where the CCD observers once were with low-z objects: observing where the contrast is poor (recall Fig. 1). For now, though, we must live with this situation, because there are not yet mid-IR instruments with high enough angular resolution to be of use.

In addition, at $z = 4$, because the galaxies are so distant, we are once again in the position of observing at low effective resolution despite our high angular resolution. Fortunately, by comparing the results from our NICMOS and ground-based near-IR images of the same low-z quasars, we have come to understand the limitations of our analysis techniques for objects at high redshift.

The quest for high-z hosts is underway with *HST*, and has already seen some successes, e.g., Kukula et al. (2001) at $z = 1$–2, Hutchings et al. (2002) at $z = 2$, and Ridgway et al. 2002 at $z = 2$–3. The CASTLES (CfA-Arizona Space Telescope LEns Survey) project has also been combining the mirrors of *HST* with the lenses of nature to probe hosts of gravitationally lensed quasars. See Peng et al. (2004) for very nice demonstration of the technique.

The hunt is also on with ground-based adaptive optics systems, e.g., Hutchings et al. (1999), Lacy et al. (2002). AO of course has a tremendous aperture advantage over *HST*, but because PSF characterization and stability are huge issues in this business, the full promise of AO might not be realized for some time, especially for the most luminous quasars.

Ground-based observations on large telescopes, without AO but with naturally good seeing, are also detecting high-z hosts. The larger field sizes of the non-AO cameras ensure that each frame has plenty of stars that can be used to characterize the PSF, as shown in Fig. 11 from Bechtold & McLeod (2005). These images will also be used to characterize the quasar environments at $z = 4$. Fig. 12 shows residual host flux from a $z = 4$ quasar.

By combining all of these observing techniques for high-z hosts, we can hope in the coming years to measure the host properties for a range of quasar types and luminosities, and to test galaxy formation models using the quasars and their black hole/spheroid relation as probes.

6. $H(\text{o})ST$ Galaxies

For a conference entitled *Essential Science in* Hubble*'s Final Years*, it is fitting that I point out again that quasar host galaxies were intended all along to be part of *Hubble*'s essential science. In fact, the two things are so closely intertwined that the name of one is embedded in the other: certainly, there is more work to be done on quasar $H(\text{o})ST$s!

The questions we will now ask of *Hubble*, some not yet conceived, will be different from, and more fundamentally interesting than, the original ones because of what we have learned from *Hubble* already. Keeping with the tradition of Astro 101, I offer here a choose-your-own-ending story:

"In 2009, *HST* observations of...

- ...large samples of nearby hosts/hosts discovered by the Sloan Digital Sky Survey/hosts of red quasars
  [Despite the strides we've made with *HST*, we've only observed a small number of nearby quasars!]

- ...fuzz around $z = 4$ quasars
  [The revived NICMOS is an obvious choice for this.]

- ...lensed quasars at high redshift
  [Magnification is a good thing!]

- ...elemental abundances of high-z quasars
  [Does the fact that the highest redshift quasars are already metal-rich tell us something about the dark matter potential in which they sit?]

- ...the nuclei of nearby dwarf galaxies
  [Can we understand the feeding mechanisms at work in our local analogs of the small black holes that all galaxies once had?]

- ...insert your own favorite here

helped us uncover physical processes by which stars and black holes grow together starting from seed masses of ____$M_\odot$ inside dark matter halos to become the galaxies we see around us today."

The author thanks the organizing committee for the invitation.

REFERENCES

ABELL, G. 1984. In *Realm of the Universe*. 3rd ed., p. 429. Saunders College Publishing.

BAHCALL, J. N., KIRHAKOS, S., & SCHNEIDER, D. P. 1994 *ApJ*, **435**, L11.

BAHCALL, J. N., KIRHAKOS, S., SAXE, D. H., & SCHNEIDER, D. P. 1997 *ApJ*, **479**, 642.

BECHTOLD, J., ET AL. 2003 *ApJ*, **588**, 119.

BECHTOLD, J. & MCLEOD, K. K. 2005 In preparation.

BOROSON, T. A. & OKE, J. B. 1984 *ApJ*, **281**, 535

BOYCE, P. J., ET AL. 1998 *MNRAS*, **298**, 121

DUNLOP, J. S., TAYLOR, G. L., HUGHES, D. H., & ROBSON, E. I. 1993 *MNRAS*, **264**, 455.

DUNLOP, J. S., MCLURE, R. J., KUKULA, M. J., BAUM, S. A., O'DEA, C. P., & HUGHES, D. H. 2003 *MNRAS*, **340**, 1095.

DUNLOP, J. S. 2004. In *Coevolution of Black Holes and Galaxies*, (ed. L. C. Ho). Carnegie Observatories Astrophysics Series, Vol. 1, p. 342. Cambridge University Press.

FRAKNOI, A., MORRISON, D., & WOLF, S. 2004. In *Voyages to the Stars and Galaxies*, 3rd ed., p. 407. Brooks/Cole.

HST WEBSITE *http://hubble.nasa.gov/overview/conception-part2.html*, accessed 8 July 2004.

HUGHES, D. H., KUKULA, M. J., DUNLOP, J. S., & BOROSON, T. 2000 *MNRAS*, **316**, 204.

HUTCHINGS, J. B., CAMPBELL, B., PRITCHET, C., & CRAMPTON, D. 1981 *ApJ*, **247**, 743.

HUTCHINGS, J. B., CRAMPTON, D., MORRIS, S. L., DURAND, D., & STEINBRING, E. 1999 *AJ*, **117**, 1109.

HUTCHINGS, J. B., FRENETTE, D., HANISCH, R., MO, J., DUMONT, P. J., REDDING, D. C., & NEFF, S. G. 2002 *AJ*, **123**, 2936.

HUTCHINGS, J. B., MADDOX, N., CUTRI, R. M., & NELSON, B. O. 2003 *AJ*, **126**, 63.

HUTCHINGS, J. B., JANSON, T., & NEFF, S. G. 1989 *ApJ*, **342**, 660.

HUTCHINGS, J. B. & MORRIS, S. C. 1995 *AJ*, **109**, 1541.

KAUFMANN, W. J. III. 1994. In *Universe*, 4th. ed., p. 511. W. H. Freeman and Co.

KAUFFMANN, G. & HAEHNELT, M. 2000. *MNRAS*, **311**, 576.

KRISTIAN, J. 1973 *ApJ*, **179**, L61.

KUKULA, M. J., DUNLOP, J. S., MCLURE, R. J., MILLER, L., PERCIVAL, W. J., BAUM, S. A., & O'DEA, C. P. 2001 *MNRAS*, **326**, 1533.

LACY, M., GATES, E. L., RIDGWAY, S. E., DE VRIES, W., CANALIZO, G., LLOYD, J. P., & GRAHAM, J. R. 2002 *AJ*, **124**, 3023.

MAGORRIAN, J., ET AL. 1998 *AJ*, **115**, 2285.

MARBLE, A. R. ET AL. 2003 *ApJ*, **590**, 707.

MARTINI, P., REGAN, M. W., MULCHAEY, J. S., & POGGE, R. W. 2003 *ApJ*, **589**, 774.

MCLEOD, K. K. & MCLEOD, B. A. 2001 *ApJ*, **546**, 782.

MCLEOD, K. K. & RIEKE, G. H. 1994a *ApJ*, **420**, 58.

MCLEOD, K. K. & RIEKE, G. H. 1994b *ApJ*, **431**, 137.

MCLEOD, K. K. & RIEKE, G. H. 1995 *ApJ*, **454**, L77.

MCLEOD, K. K., RIEKE, G. H., & STORRIE-LOMBARDI, L. J. 1999. *ApJ*, **511**, L67.

McLURE, R. J., KUKULA, M. J., DUNLOP, J. S., BAUM, S. A., O'DEA, C. P., & HUGHES, D. H. 1999 *MNRAS*, **308**, 377.

McLURE, R. J., DUNLOP, J. S., & KUKULA, M. J. 1999 *MNRAS*, **318**, 693.

McLURE, R. J. & JARVIS, M. J. 2002 *MNRAS*, **337**, 109.

NOLAN, L. A., DUNLOP, J. S., KUKULA, M. J., HUGHES, D. H., BOROSON, T., & JIMENEZ, R. 2001 *MNRAS*, **323**, 308.

PENG, C. Y. ET AL. 2004. In *Coevolution of Black Holes and Galaxies*, (ed. L. C. Ho). Carnegie Observatories Astrophysics Series, Vol. 1, p. 49.
(http://www.ociw.edu/ociw/symposia/series/symposium1/proceedings.html)

RIDGWAY, S., HECKMAN, T., CALZETTI, D., & LEHNERT, M. 2002 *New Astronomy Reviews*, **46**, 175.

SANDERS, D. B., SOIFER, B. T., ELIAS, J. H., MADORE, B. F., MATTHEWS, K., NEUGEBAUER, G., & SCOVILLE, N. Z. 1988 *ApJ*, **325**, 74.

SMITH, E. P., HECKMAN, T. M., BOTHUN, G. D., ROMANISHIN, W., & BALICK, B. 1986 *ApJ*, **306**, 64.

VERON-CETTY, M.-P. & WOLTJER, L. 1990 *A&A*, **236**, 69.

WYCKOFF, S., GEHREN, T., & WEHINGER, P. A. 1981 *ApJ*, **247**, 750.

Reverberation mapping of active galactic nuclei

By B. M. PETERSON[1] AND K. HORNE[2]

[1]Department of Astronomy, The Ohio State University, 140 West 18th Avenue, Columbus, OH, USA

[2]School of Physics and Astronomy, University of St. Andrews, St. Andrews KY16 9SS, Scotland

Reverberation mapping is a proven technique that is used to measure the size of the broad emission-line region and central black hole mass in active galactic nuclei. More ambitious reverberation mapping programs that are well within the capabilities of the *Hubble Space Telescope* could allow us to determine the nature and flow of line-emitting gas in active nuclei and to assess accurately the systematic uncertainties in reverberation-based black hole mass measurements.

1. Introduction: The inner structure of AGNs

There is now general consensus that the long-standing paradigm for active galactic nuclei (AGNs) is basically correct, i.e., that AGNs are fundamentally powered by gravitational accretion onto supermassive collapsed objects. Details of the inner structure of AGNs, however, remain sketchy, although both emission lines and absorption lines reveal the presence of large-scale gas flows on scales of hundreds to thousands of gravitational radii. The accretion disk produces a time-variable high-energy continuum that ionizes and heats this nuclear gas, and the broad emission-line fluxes respond to the changes in the illuminating flux from the continuum source. The geometry and kinematics of the broad-line region (BLR), and fundamentally, its role in the accretion process, are not understood. Immediate prospects for understanding this key element of AGN structure do not seem especially promising with the realization that the angular size of the nuclear regions projects to only microarcsecond scales even in the case of the nearest AGNs. Unfortunately, there is only very limited information about the BLR from the emission-line profiles alone, since many simple kinematic models are highly degenerate. Nevertheless, it has been possible to draw a few basic interferences about the nature of the BLR:

(*a*) *There is strong evidence for a disk component in at least some AGNs.* In particular, there is a relatively small subset of AGNs whose spectra show double-peaked Balmer-line profiles. Double-peaked profiles are generally associated with rotating Keplerian disks.

(*b*) *There is strong evidence for an outflowing component in many AGNs.* Some emission lines have strong blueward asymmetries, suggesting that we preferentially observe outflowing material on the nearer side of an AGN. Slightly blueshifted (relative to the systemic redshift of the host galaxy) absorption features are quite common in AGNs, and there is a good deal of evidence that this absorption, seen primarily in ultraviolet and X-ray spectra, arises on scales similar to that of the broad emission lines.

(*c*) *There is strong evidence that gravitational acceleration by the central source is important.* As discussed below, a physical scale for the size of the line-emitting region can be obtained by the process of reverberation mapping. The derived scale length for each line is different, with lines that are characteristic of high-ionization gas arising closer to the central source than lines that are more characteristic of low-ionization gas, thus demonstrating ionization stratification within the BLR. Moreover, the higher ionization lines are broader, and indeed the relationship between size and velocity dispersion of the line-emitting region shows a virial-like relationship, i.e., $r \propto \Delta V^{-2}$, where r is the characteristic scale for a line which has Doppler width ΔV.

The conclusion that gravity is important leads us directly to an estimate of the black hole mass, which we take to be

$$M_{\rm BH} = \frac{fr\Delta V^2}{G},\tag{1.1}$$

where G is the gravitational constant and f is a scaling factor of order unity that depends on the presently unknown geometry and kinematics of the BLR.

In this brief introduction, we already see the two major reasons that understanding the BLR is of critical importance to understanding the entire quasar phenomenon: (1) we need to understand how the accretion/outflow processes work in AGNs and (2) we need to understand the geometry and kinematics of the BLR to assess possible systematic uncertainties in AGN black-hole mass measurements.

2. Reverberation mapping basics

Simply put, the idea behind reverberation mapping is to learn about the structure and kinematics of the BLR by observing the detailed response of the broad emission lines to changes in the continuum. The basic assumptions needed are few and straightforward, and can largely be justified after the fact:

(*a*) *The continuum originates in a single central source.* The size of the accretion disk in a typical bright Seyfert galaxy is expected to be of order 10^{13}–10^{14} cm, or about a factor of 100 or so smaller than the BLR turns out to be. It is worth noting that we do not necessarily have to assume that the continuum is emitted isotropically.

(*b*) *Light-travel time $\tau_{\rm LT} = r/c$ is the most important time scale.* The other potentially important time scales include:

- The recombination time scale $\tau_{\rm rec} = (n_e \alpha_{\rm B})^{-1}$, which is the time for emission-line gas to re-establish photoionization equilibrium in response to a change in the continuum brightness. For typical BLR densities, $n_e \approx 10^{10}$ cm^{-3}, $\tau_{\rm rec} \approx 0.1$ hr, i.e., virtually instantaneous relative to the light-travel timescales of days to weeks for luminous Seyfert galaxies.
- The dynamical time scale for the BLR gas, $\tau_{\rm dyn} \approx r/\Delta V$. For typical luminous Seyferts, this works out to be of order 3–5 years. Reverberation experiments must be kept short relative to the dynamical timescale to avoid smearing the light travel-time effects.

(*c*) *There is a simple, though not necessarily linear, relationship between the observed continuum and the ionizing continuum.* In particular, the observed continuum must vary in phase with the ionizing continuum, which is what is driving the line variations. This is probably the most fragile of these assumptions since there is some evidence that long-wavelength continuum variations follow those at shorter wavelengths, but the timescales involved are still significantly shorter than the timescales for emission-line response.

Given these assumptions, a linearized response model can be written as

$$\Delta L(V,t) = \int \Psi(V,\tau)\Delta C(t-\tau)\, d\tau,\tag{2.1}$$

where $\Delta C(t)$ is the continuum light curve relative to its mean value $\overline{C}$, i.e., $\Delta C(t) = C(t) - \overline{C}$, and, $\Delta L(V,t)$ is the emission-line light curve as a function of line-of-sight Doppler velocity V relative to its mean value $\overline{L}(V)$. The function $\Psi(V,\tau)$ is the "velocity-delay map," i.e., the BLR responsivity mapped into line-of-sight velocity/time-delay space. It is also sometimes referred to as the "transfer function" and eq. (1.1) is called

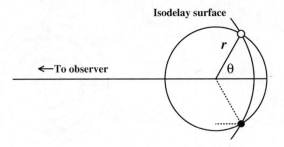

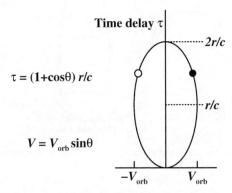

FIGURE 1. *Upper diagram:* In this simple, illustrative model, the line-emitting clouds are taken to be on a circular orbit of radius r around the central black hole. The observer is to the left. In response to an instantaneous continuum outburst, the clouds seen by the distant observer at a time delay τ after detection of the continuum outburst will be those that lie along an "isodelay surface," for which the time delay relative to the continuum signal will be $\tau = (1 + \cos\theta)r/c$, the length of the dotted path shown. *Lower diagram:* The circular orbit is mapped into the line-of-sight velocity/time-delay plane.

the transfer equation. Inspection of this formula shows that the velocity-delay map is essentially the observed response to a delta-function continuum outburst. This makes it easy to construct model velocity-delay maps from first principles.

Consider first what an observer at the central source would see in response to a delta-function (instantaneous) outburst. Photons from the outburst will travel out to some distance r where they will be intercepted and absorbed by BLR clouds and produce emission-line photons in response. Some of the emission-line photons will travel back to the central source, reaching it after a time delay $\tau = 2r/c$. Thus, a spherical surface at distance r defines an "isodelay surface," since all emission-line photons produced on this surface are observed to have the same time delay relative to the continuum outburst. For an observer at any other location, the isodelay surface would be the locus of points for which the travel from the common initial point (the continuum source) to the observer is constant. It is obvious that such a locus is an ellipsoid. When the observer is moved to infinity, the isodelay surface becomes a paraboloid. We show a typical isodelay surface for this geometry in the top panel of Figure 1.

We can now construct a simple velocity delay map. Consider the trivial case of BLR that is comprised of an edge-on (inclination 90°) ring of clouds in a circular Keplerian orbit, as shown on the top panel of Figure 1; in the lower panel, we map the points from

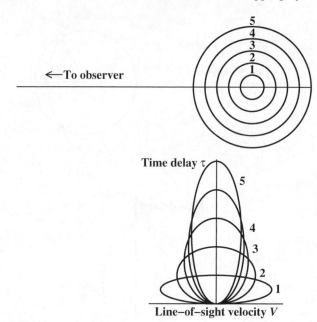

FIGURE 2. This diagram is similar to Figure 1. Here we show how circular Keplerian orbits of different radii map into the velocity-time delay plane. Inner orbits have a larger velocity range ($V \propto r^{-1/2}$) and shorter range of time delay ($\tau_{\max} = 2r/c$), resulting in tapering of the map in velocity with increasing time delay, a general feature of gravitationally dominated systems.

polar coordinates in configuration space to points in velocity-time delay space. Points (r, θ) in configuration space map into line-of-sight velocity/time-delay space (V, τ) according to $V = -V_{\mathrm{orb}} \sin\theta$, where V_{orb} is the orbital speed, and $\tau = (1 + \cos\theta)r/c$. Inspection of Figure 1 shows that a circular Keplerian orbit projects to an ellipse in velocity-time delay space. Generalization to radially extended geometries is simple: a disk is a system of rings of different radii and a spherical shell is a system of rings of different inclinations. Figure 2 shows a system of circular Keplerian orbits, i.e., $V_{\mathrm{orb}}(r) \propto r^{-1/2}$, and how these project into velocity-delay space. A key feature of Keplerian systems is the "taper" in the velocity-delay map with increasing time delay.

In Figure 3, we show two complete velocity-delay maps for radially extended systems: in one case a Keplerian disk, and in the other, a spherical system of clouds in circular Keplerian orbits of random inclination. In both examples, the velocity-delay map is shown in the upper left panel in grayscale. The lower left panel shows the result of integrating the velocity-delay map over time delay, thus yielding the emission-line profile for the system. The upper right panel shows the result of integrating over velocity, yielding the total time response of the line; this is referred to as the "delay map" or the "one-dimensional transfer function." Inspection of Figure 3 shows that these two velocity-delay maps are superficially similar; both show clearly the tapering with time delay that is characteristic of Keplerian systems and have double-peaked line profiles. However, it is also clear that they can be easily distinguished from one another. This, of course, is the key: the goal of reverberation mapping is to use the observables, namely the continuum light curve $C(t)$ and the emission-line light curve $L(V, t)$, and invert eq. (2.1) to recover the velocity-delay map $\Psi(V, \tau)$. Equation (2.1) represents a fairly common type of problem that arises in many applications in physics and engineering. Indeed, the velocity-delay map is the Green's function for the system. Solution of eq. (2.1) by Fourier transforms immediately suggests itself, but real reverberation data are far too sparsely sampled and usually too noisy for this method to be effective. Other methods have to be employed,

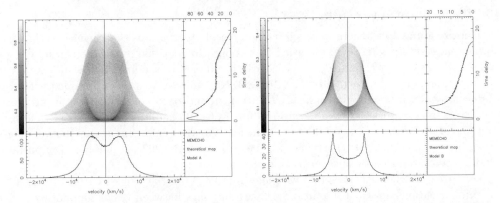

FIGURE 3. Theoretical velocity-delay maps $\Psi(V, \tau)$ shown in grayscale for a spherical distribution of line-emitting clouds in circular Keplerian orbits of random inclination (left) and an inclined Keplerian disk of line-emitting clouds (right). Projections in velocity and time-delay show the line profile (below) and delay map (right). From Horne et al. (2004).

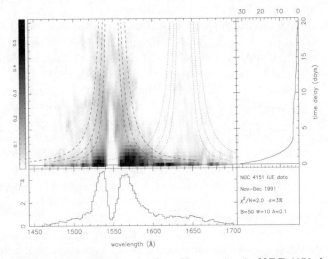

FIGURE 4. A velocity-delay map for the C IV–He II region in NGC 4151, based on data obtained with the *International Ultraviolet Explorer*. From Ulrich & Horne (1996).

such as reconstruction by the maximum entropy method (Horne 1994). Unfortunately, even the best reverberation data obtained to date have not been up to the task of yielding a high-fidelity velocity-delay map. Existing velocity-delay maps are noisy and ambiguous. Figure 4 shows the result of an attempt to recover a velocity-delay map for the C IV–He II spectral region in NGC 4151 (Ulrich & Horne 1996). The Keplerian taper of the map is seen, but other possible structure is only hinted at, as it is in other attempts to recover a velocity-delay map from real data (e.g., Wanders et al. 1995; Done & Krolik 1996; Kollatschny 2003). It must be pointed out, however, that in no case to date has the recovery of the velocity-delay map been a design goal for an experiment. Previous reverberation-mapping experiments have had the more modest goal of recovering only the mean response time of emission lines, from which one can still draw considerable information. By integrating eq. (2.1) over velocity and then convolving it with the continuum light curve, we find that under reasonable conditions, cross-correlation of the continuum and emission-line light curves yields the mean response time, or "lag," for the emission lines.

3. Reverberation results

Prior to about 1988, there were a large number of observations that suggested that the broad emission lines in Seyferts varied in response to continuum variations, and did so on surprisingly short time scales. These early results led to the first highly successful reverberation campaign, carried out in 1988–89, combining UV spectra obtained with the *International Ultraviolet Explorer* (*IUE*) with ground-based optical observations from numerous observatories. The program ran for over eight months and achieved time resolution of a few days in several continuum and emission-line time series (Clavel et al. 1991; Peterson et al. 1991; Dietrich et al. 1993; Maoz et al. 1993). A number of important results were produced by this project, including:

(*a*) From the shortest measured wavelength (1350 Å) to the longest (5100 Å), the continuum variations appear to be in phase, with any lags between bands amounting to no more than a couple of days.

(*b*) The highest ionization emission lines respond most rapidly to continuum variations (e.g., ∼2 days for He II λ1640 and ∼10 days for Lyα and C IV λ1549) and the lower ionization lines respond less rapidly (e.g., ∼20 days for Hβ and nearly 30 days for C III] λ1909). The BLR thus shows radial ionization stratification.

Optical spectroscopic monitoring of NGC 5548 continued for a total of 13 years, and during the fifth year of the program (1993), concurrent high-time resolution (daily observations) were made for about 60 days with *IUE* and for 39 days with the Faint Object Spectrograph on the *Hubble Space Telescope* (Korista et al. 1995). Over time, it became clear that the Hβ emission-line lag is a *dynamic* quantity; it varies with time and is dependent on the current mean continuum luminosity (Peterson et al. 2002). In other words, there is much more nuclear gas on scales of thousands of gravitational radii than previously thought: at any given time, most of the emission in any particular line arises primarily in that gas for which the physical conditions optimally produce that particular emission line (cf. the "locally optimized cloud" model of Baldwin et al. 1995).

Peterson et al. (2004) recently completed a comprehensive reanalysis of 117 independent reverberation mapping data sets on 37 AGNs, measuring emission-line lags, line widths, and black hole masses for all but two of these sources. Calibration of the reverberation-based mass scale, as embodied in the scaling factor f in eq. (1.1), is set by assuming that AGNs follow the same relationship between black hole mass and the host-galaxy bulge velocity dispersion (the $M_{BH} - \sigma_*$ relationship) seen in quiescent galaxies (Onken et al. 2004). The range of measured masses runs from $\sim 2 \times 10^6 \, M_\odot$ for the narrow-line Seyfert 1 galaxy NGC 4051 to $\sim 1.3 \times 10^9 \, M_\odot$ for the quasar PG 1426+015. The statistical errors in the mass measurements (due to uncertainties in lag and line-width measurement) are typically only about 30%. However, the systematic errors, due to scatter in the $M_{BH} - \sigma_*$ relationship, amount to about a factor of three; this systematic uncertainty can decrease only by understanding the geometry and kinematics of the BLR. Figure 5 shows a current version of the mass-luminosity relationship for AGNs, based on these reverberation-based black hole masses.

4. The future: What will it take to map the broad-line region?

While we still do not have a velocity-delay map in hand, we certainly know how to get one. More than a dozen years of experience in reverberation mapping have led to a reasonably good understanding of the timescales for response of various lines as a function of luminosity and of how the continuum itself varies with time. On the basis of this information, we have carried out extensive simulations to determine the observational

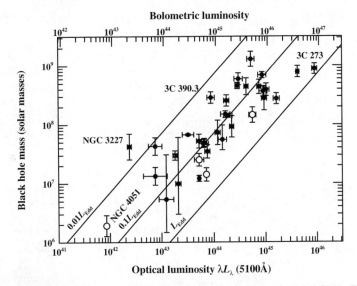

FIGURE 5. Black hole mass vs. luminosity for 35 reverberation-mapped AGNs. The luminosity scale on the lower x-axis is $\log \lambda L_\lambda$ in units of ergs s^{-1}. The upper x-axis shows the bolometric luminosity, assuming that $L_{\mathrm{bol}} \approx 9\lambda L_\lambda$. The diagonal lines show the Eddington limit L_{Edd}, $0.1L_{\mathrm{Edd}}$, and $0.01L_{\mathrm{Edd}}$. The open circles represent narrow-line Seyfert 1 galaxies. From Peterson et al. (2004).

requirements to obtain high-fidelity velocity-delay maps for emission lines in moderate luminosity Seyfert galaxies (quantities that follow are based specifically on NGC 5548, by far the AGN best studied by reverberation). A sample numerical simulation is shown in Figure 6. As described more completely by Horne et al. (2004), the principal requirements are:

(*a*) *High time resolution* (less than 0.2–1 day, depending on the emission line). The interval between observations translates directly into the resolution in the time-delay axis.

(*b*) *Long duration* (several months). A rule of thumb in time series analysis is that the duration of the experiment should exceed the maximum timescale to be probed by at least a factor of three. The lag for Hβ in NGC 5548 is typically around 20 days, so the longest timescale to be probed is $2r/c$. The duration should thus be at least $\sim$120 days to map the Hβ-emitting region. However, since C IV seems to respond twice as fast as Hβ, the C IV-emitting region might be mapped in as little as $\sim$60 days. A more important consideration, however, is detection in the time series a strong continuum signal, such as a change in sign of the derivative of the light curve. This produces a similarly strong emission-line response. We find that $\sim$200 days of observations are required *to be certain* that such an event occurs, and to observe its consequences in the emission lines.

(*c*) *Moderate spectral resolution* ($\leqslant$600 km s^{-1}). While higher spectral resolution is always desirable, AGN emission lines show little additional structure at resolution better than several hundred kilometers per second. Higher resolution does, however, make it in principle possible to detect a gravitational redshift (e.g., Kollatschny 2004), providing an independent and complementary measure of the black hole mass.

(*d*) *High homogeneity and signal-to-noise ratios* ($S/N \approx 100$). Both continuum and emission-line flux variations are small on short time scales, typically no more than a few percent on diurnal timescales. Excellent *relative* flux calibration and signal-to-noise ratios are necessary to make use of the high time resolution.

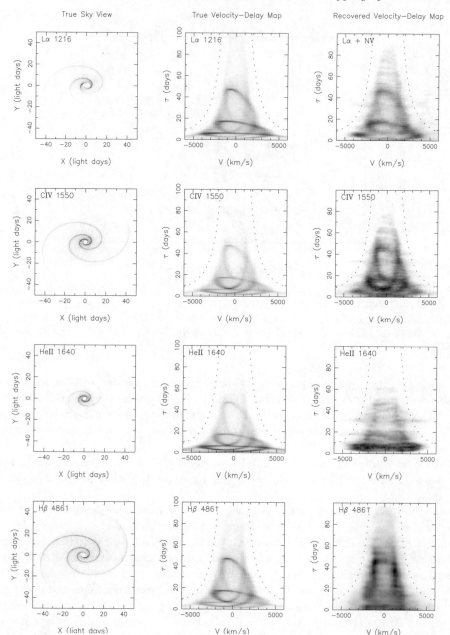

FIGURE 6. Numerical simulations of velocity-delay map recovery. An arbitrary but complex geometry was chosen to show that a complicated BLR structure can be recovered. The left column shows the BLR geometry in four lines, Lyα λ1215, C IV λ1549, He II λ1640, and Hβ λ4861, from top to bottom. The middle column shows the BLR model mapped into velocity-delay space. The right column shows the recovered velocity-delay map recovered from simulated data. From Horne et al. (2004).

The final point makes it clear that this will be difficult to do with ground-based observations: it is hard to maintain such high levels of homogeneity and thus accurate relative flux calibration with a variable point-spread function, such as one dominated by the effects of atmospheric seeing. Space-based observations are much more likely to succeed, and moreover, the ultraviolet part of the spectrum gives access to the important strong lines Lyα, N v λ1240, Si iv λ1400, C iv λ1549, and He ii λ1640, all of which vary strongly.

We have carried out a series of simulations specifically assuming observation of NGC 5548 with the Space Telescope Imaging Spectrograph (STIS) on *HST*. We assume that the typical BLR response times are as observed during the first major monitoring campaign in 1989, as these values appear to be typical. For practical reasons, we also assume that we can obtain only one observation per day (mostly on account of restrictions against the use of the STIS UV detectors during orbits when *HST* passes through the South Atlantic Anomaly), which we must complete in one *HST* orbit, using the typical time that NGC 5548 is observable per orbit. Furthermore, we included in the simulations nominal spacecraft- and instrument-safing events of typical duration (generally a few days) and frequency. We also considered the effects of early termination of the experiment due, for example, to a more catastrophic failure. We carried out 10 individual simulations, with each one using a different continuum model; all of the continuum models were "conservative" in the sense that the continuum activity is weaker and less pronounced than it is usually observed to be.

Given our assumptions, we find that all 10 of the simulated experiments succeed within 200 days. For the most favorable continuum behavior, success can be achieved in as little as $\sim$60 days—though this is rare—or more commonly in around $\sim$150 days. We also find that these results are robust against occasional random data losses.

These generally conservative assumptions argue strongly that velocity-delay maps for all the strong UV lines in NGC 5548 could be obtained in a 200-orbit *HST* program based on one orbit per day. There are clearly elements of risk associated with the program, the most obvious being early termination on account of a systems failure or a major safing event that would end the time series prematurely. This risk is somewhat mitigated by the conservatism of our simulations; it is possible that the experiment could succeed in much less time. Indeed, if we define "success" as obtaining a velocity-delay map that is stable for 50 days, we find that the probability of success is as high as $\sim$90% in 150 days.

Finally, one might also ask about the scientific risk: for example, what if the velocity-delay map, though of high fidelity, cannot be interpreted? In other words, what if the velocity-delay map is a "mess" and has no discernible structure? First of all, this is not a likely outcome since long-term monitoring shows persistent features in emission-line profiles that imply there is some order or symmetry to the BLR. Moreover, even if the BLR turns out to be a "mess" we'll still have learned an important fact about the BLR structure, namely that it is basically chaotic. But the bottom line is that right now we have *nearly complete ignorance* about the BLR structure. We cannot even assess to *any* level of confidence how many velocity-delay maps of AGNs we will need to solve the problem until we have obtained at least *one* velocity-delay map of at least *one* emission line. Until we have that, our knowledge of the role of the BLR in AGN fueling and outflows will remain based on theoretical speculation alone, historically a very dangerous situation for astrophysicists.

5. Summary

We have argued that reverberation mapping provides a unique probe of the inner structure of AGNs. The reverberation technique has been very successful in determining the BLR sizes and black hole masses in 35 AGNs. The masses obtained are accurate to about a factor of three, based on the observed scatter in the AGN $M_{\mathrm{BH}} - \sigma_*$ relationship. The accuracy of these masses is fundamentally limited by unknown geometry and kinematics of the BLR. We have also argued that it is possible to obtain complete, high-fidelity velocity-delay maps of the strong ultraviolet lines in relatively nearby, moderately luminous AGNs with *HST*, and we specifically argue that this can be done with high confidence of success for NGC 5548 with one *HST* orbit per day for a period of no longer than 200 days.

We are grateful for support by the National Science Foundation (grant AST-0205964) and PPARC.

REFERENCES

BALDWIN, J., FERLAND, G., KORISTA, K., & VERNER, D. 1995 *ApJ* **455**, L119.

CLAVEL, J., ET AL. 1991 *ApJ* **366**, 64.

DIETRICH, M., ET AL. 1993 *ApJ* **408**, 416.

DONE, C. & KROLIK, J. H. 1996 *ApJ* **463**, 144.

HORNE, K. 1994. In *Reverberation Mapping of the Broad-Line Region in Active Galactic Nuclei* (eds. P. M. Gondhalekar, K. Horne, & B. M. Peterson). Astronomical Society of the Pacific Conference Series, vol. 69, p. 23. ASP.

HORNE, K., PETERSON, B. M., COLLIER, S., & NETZER, H. 2004 *PASP* **116**, 465.

KOLLATSCHNY, W. 2003 *A&A* **407**, 461.

KOLLATSCHNY, W. 2004 *A&A* **412**, L61.

KORISTA, K. T., ET AL. 1995 *ApJS* **97**, 285.

MAOZ, D., NETZER, H., PETERSON, B. M., BECHTOLD, J., BERTRAM, R., BOCHKAREV, N. G., CARONE, T. E., DIETRICH, M., FILIPPENKO, A. V., KOLLATSCHNY, W., KORISTA, K. T., SHAPOVALOVA, A. I., SHIELDS, J. C., SMITH, P. S., THIELE, U., & WAGNER, R. M. 1993 *ApJ* **404**, 576.

ONKEN, C. A. FERRARESE, L., MERRITT, D., PETERSON, B. M., POGGE, R. W., VESTERGAARD, M., & WANDEL, A. 2004 *ApJ*, **615**, 645.

PETERSON, B. M., ET AL. 1991 *ApJ* **368**, 119.

PETERSON, B. M., ET AL. 2002 *ApJ* **581**, 197.

PETERSON, B. M., FERRARESE, L., GILBERT, K. M., KASPI, S., MALKAN, M. A., MAOZ, D., MERRITT, D., NETZER, H., ONKEN, C. A., POGGE, R. W., VESTERGAARD, M., & WANDEL, A. 2004 *ApJ* **613**, 682.

ULRICH, M.-H. & HORNE, K. 1996 *MNRAS* **283**, 748.

WANDERS, I., GOAD, M. R., KORISTA, K. T., PETERSON, B. M., HORNE, K., FERLAND, G. J., KORATKAR, A. P., POGGE, R. W., & SHIELDS, J. C. 1995 *ApJ* **453**, L87.

Feedback at high redshift

By ALICE E. SHAPLEY

[1]Department of Astronomy, 601 Campbell Hall, UC Berkeley, Berkeley, CA, 94720, USA

We examine the process of feedback in star-forming galaxies at $2 \leqslant z \leqslant 3$. Large-scale outflows of interstellar material are observed in starburst galaxies in the nearby universe, and have long been invoked as a means to address important shortcomings in current models of galaxy formation. At $z \sim 3$, superwinds appear to be a generic feature of color-selected star-forming galaxies with spectroscopic information, and may explain both the apparent lack of neutral hydrogen near star-forming galaxies, and also the strong cross-correlation between galaxies and C IV metal-absorption systems. Another type of star-formation feedback is the leakage of hydrogen-ionizing radiation from galaxies, which may also have a profound effect on the physical state of the intergalactic medium (IGM), especially as the number density of QSOs drops off at $z > 2.5$. Between $z = 3$ and $z = 2$, there is strong evolution in the number density of H I absorption systems in the Lyα forest. Therefore, it is also of interest to trace how the effect of galactic superwinds on the IGM evolves from $z = 3$ to $z = 2$. We show preliminary results that many properties of superwinds are similar in star-forming galaxies at $z \sim 2$, and direct evidence that enriched gas reaches radii of at least $\sim$100 kpc. Finally, we discuss future directions for the study of outflows in the high-redshift universe. Specifically, we highlight the unique combination of existing deep *HST*/ACS imaging in the GOODS-N field with high signal-to-noise rest-frame UV spectra. Using the morphological information provided by the *HST*/ACS will enable us to probe a complementary, spatial dimension of feedback at high redshift, which has been unexplored until now.

1. Introduction: The importance of feedback

The term star-formation feedback refers to the collective effect of multiple overlapping supernova explosions stemming from an episode of star formation. The associated mechanical energy input leads to a shock-heated expanding superbubble with a characteristic temperature of $T \sim 10^8$ K. According to the above picture described in Heckman et al. (1990), the superbubble expands in the direction of the steepest pressure gradient (i.e., perpendicular to a galaxy disk), breaks out into the surrounding environment, and enters the so-called "superwind" phase. Ambient cold and warm interstellar gas is swept up into a shell during the superbubble phase, and is also entrained in the expanding superwind phase. Superwinds are observed in local starburst galaxies with star-formation surface densities $\Sigma_* > 0.1$ $M_\odot$ yr^{-1} kpc^{-2} (Heckman 2002), a classic example of which is the nearby star-forming galaxy M82. As described in this proceedings, superwinds also appear to be a generic feature of star-forming galaxies at high redshift (Pettini et al. 2001; Adelberger et al. 2003).

Characterizing star-formation feedback may have profound implications for our understanding of how galaxies form and how the physical state of the intergalactic medium (IGM) evolves. Indeed, large-scale outflows of hot, enriched gas may explain the degree of enrichment and possible entropy floor in the intracluster medium (Mushotzky & Loewenstein 1997; Ponman et al. 1999). Additionally, two fundamental issues in models of galaxy formation are the "overcooling" (White & Frenk 1991) and "angular momentum" problems (Navarro & White 1994). In the absence of heating associated with supernova feedback, gas cools too efficiently in the dense environments of simulated high-redshift halos and too many baryons are converted into stars, relative to what is observed. Also, during simulations of disk galaxy assembly, baryons lose angular momentum to the surrounding dark matter halo, and therefore model disk galaxies are much smaller than observed disks with the same rotational velocity. An accurate description of the process of supernova feedback during galaxy formation has long been invoked as a means to resolve both the over-cooling and angular momentum problems. However, robust observational constraints on feedback at high redshift are only starting to be assembled. In this proceedings, we summarize observations of feedback at $z \sim 3$ (§2) and $z \sim 2$ (§3). In §4, we describe the unique way in which *Hubble Space Telescope* (*HST*) observations can be combined with rest-frame UV spectra in order to provide insight into the nature of large-scale galaxy outflows. In §5, we conclude with a brief summary.

2. Observations of feedback at $z \sim 3$

2.1. *Superwinds*

Our understanding of the importance and ubiquity of star-formation induced outflows in high-redshift UV-color-selected Lyman Break galaxies (LBGs; Steidel et al. 1996) has progressed considerably in the last few years. These outflows, or superwinds, have been extensively observed in local starburst galaxies over the preceding decade (Heckman et al. 2000, and references therein). When the rest-frame UV spectra of LBGs were first collected, one of the remarkable properties which was immediately apparent was the large width of the strong low-ionization interstellar absorption lines, FWHM = 400–700 km s^{-1} (Steidel et al. 1996; Lowenthal et al. 1997). One proposed explanation for these large widths was the impact of non-gravitational processes—such as interstellar shocks and supernova winds—rather than the effects of gravitational virial dynamics. Franx et al. (1997) first noted the blueshift of interstellar absorption lines relative to redshifted Lyα emission in the spectrum of a $z = 4.92$ gravitationally lensed arc, and compared this redshift offset with observations and models of galactic-scale outflows in local starburst galaxies. The evidence for outflows was again presented by Steidel et al. (1998), who found that the average redshift offset between Lyα and interstellar absorption lines for a sample of galaxies in the SSA22 field was $\Delta z_{\mathrm{Ly\alpha - abs}} = 0.008 \pm 0.004$, corresponding to $\Delta v = 600 \pm 300$ km s^{-1}. Neither redshifted Lyα emission nor blueshifted interstellar absorption provide accurate estimates of the true systemic redshifts of the galaxies in most cases, as Pettini et al. (2001) demonstrated for a sample of 17 LBGs with near-infrared spectroscopic measurements (see Figure 1). In this sample, the rest-frame optical emission lines such as [O III] and Hβ, which are produced in H II regions and should therefore be fairly reliable tracers of the systemic velocity of a galaxy, yielded redshifts which fell somewhere in between the Lyα emission and interstellar absorption measurements most of the time (these near-IR spectroscopic results were preceded by the earlier related work of Pettini et al. 1998, which contained a much smaller sample of objects).

Roughly concurrent with the observations of rest-frame optical nebular lines, a detailed analysis of the velocity field associated with interstellar gas in the gravitationally lensed $z = 2.73$ LBG, MS1512-cB58, also revealed the signatures of a large-scale outflow (Pettini et al. 2000, 2002). These high-resolution rest-frame UV spectroscopic observations showed that neutral and ionized interstellar gas was blueshifted by as much as ~ 600 km s^{-1} relative to the stellar systemic velocity, while the peak of the Lyα emission profile was redshifted by ~ 400–500 km s^{-1} relative to the stars, with Lyα emission extending redwards to $\sim +1000$ km s^{-1}. Based on the measurement of the outflow neutral H I column density from the damped Lyα profile in cB58, and based on an assumption of a spherical geometry and plausible radius for the location of the absorbing gas, it was also shown that the mass outflow rate in cB58 is comparable to the star-formation

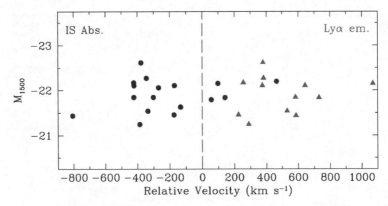

FIGURE 1. Velocity offsets relative to nebular emission line redshifts. The circles and triangles correspond, respectively, to the values of Δv_{IS} and $\Delta v_{\mathrm{Ly}\alpha}$. We see the indication of large-scale motions with respect to the systemic redshift defined by the [O III], [O II], and Hβ nebular emission lines. From Pettini et al. (2001).

rate (Pettini et al. 2000). Therefore, not only in the sample of LBGs with near-IR observations of rest-frame optical emission lines, but also in the best-studied high-redshift star-forming galaxy with the most detailed and high-quality data, there is evidence for the large-scale outflow of interstellar material.

While there is no hope of collecting data of comparable quality to the cB58 spectra for individual unlensed LBGs due to their faint optical magnitudes ($\mathcal{R}_{\mathrm{AB}} = 24$–25.5), almost 1000 spectroscopically confirmed $z \sim 3$ galaxies have been assembled over the past eight years. By dividing the LBG spectroscopic database into subsamples according to specific criteria and creating high S/N composite spectra of each subsample, we hope to understand how the LBG spectroscopic properties depend in a systematic way on other galaxy properties. Specifically, we would like to characterize the nature of both star-forming regions and the large-scale outflows of interstellar gas caused by the kinetic energy input from frequent supernova explosions. Figure 2 shows a composite spectrum which is the average of the entire spectroscopic sample of 811 LBGs (Shapley et al. 2003). Rest-frame UV spectra of LBGs are dominated by the emission from O and B stars with masses higher than 10 $M_\odot$ and $T \geqslant 25000$ K. Marked in Figure 2 are spectroscopic features that trace the photospheres and winds of massive stars, neutral and ionized gas associated with large-scale outflows, and ionized gas in H II regions where star formation is taking place. The velocity field of this average spectrum supports the general picture outlined above, which was based on a smaller sample of galaxies with rest-frame optical nebular emission-line redshifts, and cB58. On this total composite spectrum, both low- and high-ionization interstellar absorption lines are blue-shifted by $\sim$170 km s^{-1} with respect to the stellar photospheric absorption features (which define the average galaxy rest frame). At the same time, Lyα is redshifted by $\sim$350 km s^{-1} with respect to the stellar systemic redshift.

By grouping the database of LBG spectra according to galaxy parameters such as Lyα equivalent width, UV spectral slope, and interstellar kinematics, the major trends in LBG spectra least compromised by selection effects are isolated. We find that LBGs with stronger Lyα emission have bluer UV continua, weaker low-ionization interstellar absorption lines, smaller kinematic offsets between Lyα and the interstellar absorption lines, and lower star-formation rates. There is a decoupling between the dependence of low- and high-ionization outflow features on other spectral properties. Most of the above trends can be explained in terms of the properties of the large-scale outflows seen in

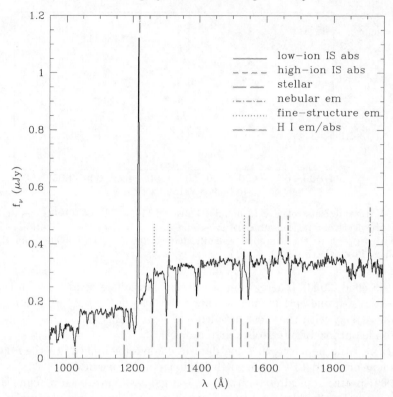

FIGURE 2. A composite rest-frame UV spectrum constructed from 811 individual LBG spectra. Dominated by the emission from massive O and B stars, the overall shape of the UV continuum is modified shortward of Lyα by a decrement due to inter-galactic H I absorption. Several different sets of UV features are marked: stellar photospheric and wind; interstellar low- and high-ionization absorption; nebular emission from H II regions; Si II* fine-structure emission, whose origin is ambiguous; and emission and absorption due to interstellar H I (Lyα and Lyβ). There are numerous weak features which are not marked, as well as several features bluewards of Lyα which only become visible by averaging over many sightlines through the IGM. From Shapley et al. (2003).

LBGs. According to this scenario, the appearance of LBG spectra is determined by a combination of the covering fraction of outflowing neutral gas which contains dust, and the range of velocities over which this gas is absorbing (Shapley et al. 2003). Higher sensitivity and spectral resolution observations are still required for a full understanding of the covering fraction and velocity dispersion of the outflowing neutral gas in LBGs and its relationship to the escape fraction of Lyman Continuum radiation in galaxies at $z \sim 3$.

Finally, based on simple physical arguments, given their typical star-formation rates and physical sizes (Shapley et al. 2001; Giavalisco et al. 1996), LBGs easily satisfy and exceed the necessary criteria for driving a superwind: $\Sigma_* \geqslant 0.1\ M_\odot\,\mathrm{yr}^{-1}\,\mathrm{kpc}^{-2}$ (Heckman 2002).

Most recently, an entirely complementary study underscores the importance of outflows in LBGs. Comparing the large-scale distributions of LBGs, intergalactic neutral hydrogen, and metals in the same cosmic volumes, Adelberger et al. (2003) find evidence that superwinds from LBGs clear out large ($r \sim 0.5$ comoving Mpc) regions in the IGM, and enrich the galaxies' surroundings with heavy elements (see Figure 3). This claim stems from the apparent deficit of neutral hydrogen within $r \sim 0.5$ comoving

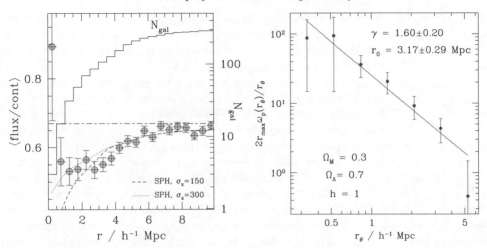

FIGURE 3. The large-scale distribution of LBGs relative to neutral intergalactic hydrogen and metals. Left: Mean intergalactic Lyα transmissivity as a function of comoving distance from LBGs. LBGs appear to be associated with intergalactic H I underdensities on the smallest spatial scales, as indicated by the increase in the Lyα forest transmissivity near LBGs. Right: The projected cross-correlation function of 217 C IV systems and 318 Lyman Break galaxies. The LBG-C IV cross-correlation function is similar to the LBG autocorrelation function, and it is inferred that LBGs reside in regions of enhanced C IV abundance relative to the intergalactic mean. From Adelberger et al. (2003).

Mpc of LBGs (unlikely to be caused by escaping Lyman Continuum radiation), and the strong cross-correlation between the locations of LBGs and C IV absorption systems. The newest, complementary evidence for superwinds emphasizes the significance of feedback, not only in the evolution of galaxies (if they are losing material at the same rate as they are forming stars), but also in the enrichment and physical conditions in the IGM.

2.2. *Lyman Continuum emission*

Recent observational results from the Sloan Digital Sky Survey (Fan et al. 2002; Becker et al. 2001) and the *Wilkinson Microwave Anisotropy Probe* (*WMAP*; Spergel et al. 2003), when combined, indicate that the Universe may have had a complex history of reionization at $z > 6$. The question of the origin of the UV radiation field that re-ionized the universe and maintained the ionization at $z < 6$ has been revisited often; several different groups have probed different redshifts. In particular, the relative contributions of QSOs and star-forming galaxies to the meta-galactic radiation field has important implications for the spectral energy distribution of the ionizing background. This, in turn, affects the physical conditions in the IGM, and therefore the inferences drawn from the hydrogen and metal absorption lines in the Lyα forest. As we move towards higher redshift, by $z \sim 3$, it appears that the known population of QSOs may have a difficult time accounting for the measured ionizing background (Hunt et al. 2004; Scott et al. 2000), and that the contribution of galaxies to the ionizing background may overtake, and actually dominate, the QSO contribution at $z > 3$. One critical parameter in determining the galaxy contribution to the ionizing background is $f_{\rm esc}$: the escape fraction of hydrogen-ionizing Lyman Continuum photons from galaxies.

For relatively nearby galaxies, observations in the vacuum far-UV are required, and the best existing data (*Hopkins Ultraviolet Telescope* spectra) have set only upper limits of $< 5\%$ on the fraction of ionizing photons which escape the galaxies in which they are produced (Leitherer et al. 1995). At intermediate redshifts ($z = 1.1$–1.4), Malkan et al.

(2003) use *Hubble Space Telescope*/Space Telescope Imaging Spectrograph (*HST*/STIS) UV imaging to find an even more stringent upper limit of < 1% for a sample of 11 star-forming galaxies, concluding that galaxies can contribute no more than 20%, and possibly no more than 2%, to the ionizing background at $z \sim 1.2$.

At $z \sim 3$, there has been much controversy regarding the contributions of galaxies to the ionizing background. Steidel et al. (2001) created a composite spectrum of 29 LBGs drawn from the high-redshift end of the LBG selection function. This subsample of galaxies had $\langle z \rangle = 3.4 \pm 0.09$ in order to place the Lyman limit at rest-frame 912 Å in a region where the LRIS spectrograph had sensitivity (since this experiment was conducted prior to the advent of the UV-sensitive LRIS-B spectrograph). Towards the high-redshift tail of the LBG selection function, only the bluest quartile of LBGs (in terms of the intrinsic UV slope) satisfy the color-selection criteria; furthermore, the galaxies going into this composite spectrum showed stronger than average Lyα emission lines, and weaker interstellar absorption lines (possibly indicating a smaller than average covering fraction of outflowing interstellar gas). Figure 4 shows the composite spectrum from these objects, which appears to exhibit significant Lyman Continuum emission (emission shortward of 912 Å in the rest frame), and an escape fraction of $\sim$10%. This constitutes the first direct detection of ionizing flux from galaxies at any redshift. *If* the Lyman Continuum emission is real, and *if* the composite spectrum is representative of the LBG population as a whole, then the contribution of galaxies to the metagalactic ionizing radiation field at $z \sim 3.4$ apparently exceeds that of QSOs by a factor of $\sim$5, and can easily produce the bulk of the metagalactic radiation field measured from QSO proximity-effect experiments (e.g., Scott et al. 2000).

In contradiction to the apparent detection of Lyman Continuum emission, Giallongo et al. (2002) derive an upper limit for the escape fraction in two bright LBGs that is four times lower than that detected by Steidel et al. (2001). One of these two galaxies is Q0000−D6, which apparently has stronger than average Lyα emission and only a 60% covering fraction of gas capable of producing significant low-ionization metal absorption lines! Furthermore, using *HST*'s Wide Field Planetary Camera 2 (WFPC2) *UBVI* photometry of 27 spectroscopically confirmed galaxies at $1.9 < z < 3.5$ in the Hubble Deep Field (HDF), Fernández-Soto et al. (2003) compare theoretical galaxy spectral energy distributions with the observed colors. Considering models with different amounts of neutral hydrogen opacity (i.e., different f_{esc}), Fernández-Soto et al. show that the average f_{esc} for their sample must be < 4%, and that star-forming galaxies at $z \sim 3$ are highly opaque to ionizing photons.

Clearly, there is a need for more sensitive rest-frame UV spectra for a sample of representative LBGs (as opposed to bluer than average) with spectral coverage in the Lyman Continuum region. Such an experiment has now been conducted using the rest-frame UV sensitive LRIS-B spectrograph (see Steidel et al. 2004, appendix, for a full description of the LRIS-B spectrograph). There now exist spectra with 18-plus hours of exposure time for a sample of $\sim$15 objects, including sensitive coverage of the Lyman Continuum region (Shapley 2003). These observations will provide a more robust detection of Lyman Continuum emission, or else place a much tighter constraint on f_{esc}. There is already preliminary evidence for a detection of Lyman Continuum emission in an individual object, SSA22a−D3, at $z = 3.07$. This object has a range of multi-wavelength information, including *HST* Near-Infrared Camera and Multi-Object Spectrometer (NICMOS) *H*-band imaging, that reveals its double morphology in detail. An accurate determination of f_{esc} from star-forming galaxies becomes more pressing as samples of galaxies are assembled at redshifts coeval with the end of reionization. For example, using the *i*-dropout technique to find galaxies at $z \sim 6$ in the Hubble Ultra Deep Field (HUDF), Bunker et al. (2004)

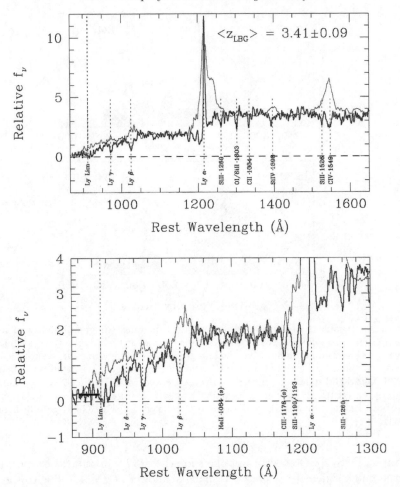

FIGURE 4. Lyman Continuum emission. Composite spectrum of Lyman Break galaxies at $\langle z \rangle = 3.40$ (dark histogram), together with a composite QSO spectrum drawn from QSOs over the same range of redshifts (light histogram), as described in the text. The spectrum has been boxcar smoothed by one resolution element. The position of the rest frame Lyman limit is indicated with a vertical dotted line. Note the average residual flux in the Lyman Continuum, evaluated from 880–910 Å in the rest frame, indicated with the dark horizontal line segment in the bottom panel. Positions of some notable interstellar and stellar absorption features are indicated. Stellar features with no possible interstellar contributions are indicated with (s). From Steidel et al. (2001).

claim that even if $f_{esc} = 1$ (which is very unlikely), the star-formation rate density associated with the i-dropout sample is insufficient to reionize the Universe. Constraining f_{esc} in galaxies is therefore a necessary ingredient to testing models of reionization.

3. Observations of feedback at $z \sim 2$

The evolution in the number density of H I absorption systems in the Lyα forest is governed by the balance between the ionization background (ionization rate) and the Hubble expansion (recombination rate). At $1.5 < z < 4$, the number density of absorption systems decreases as a strong function of decreasing redshift, $\frac{dn}{dz} \propto (1+z)^{2.47}$ (Kim et al. 2002). As shown by Weymann et al. (1998), using a sample of 63 *HST* Faint Object

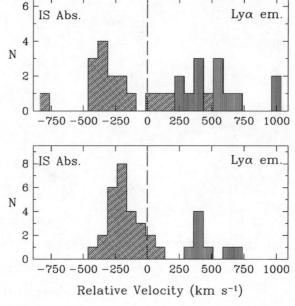

FIGURE 5. Velocity offsets of the interstellar absorption lines (diagonally-hatched histogram) and, when present, Lyα emission (vertically-hatched histogram) relative to the systemic redshifts defined by the nebular emission lines (vertical long-dash line). The top panel shows the results for $z \sim 3$ LBGs presented by Pettini et al. (2001); the bottom panel shows results for a sub-sample of 27 $z \sim 2$ BX and BM galaxies which have been observed with NIRSPEC at Hα and also have high quality Keck I/LRIS-B spectra. The velocity offsets seen in $z \sim 2$ BX and BM galaxies are similar, in both magnitude and distribution, to those typical of LBGs at $z \sim 3$. From Steidel et al. (2004).

Spectrograph (FOS) QSO spectra obtained for the *HST* Quasar Absorption Line Key Project, the number density evolution is much shallower at $z \leqslant 1.5$, due to the decrease in ionizing radiation density from QSOs. Our focus here is the redshift range $2 \leqslant z \leqslant 3$, where the number density of absorption systems is evolving strongly. We would also like to examine how the relationship between galaxies and IGM evolves over this period, in terms of the nature of large-scale star-formation-induced outflows from galaxies, and their effects on the surrounding intergalactic environments.

In order to address this question, we turn to the sample of $1.4 \leqslant z \leqslant 2.5$ UV-selected galaxies assembled by Steidel et al. (2004)—more than 700 of which currently have spectroscopic identifications. This survey was conducted in seven separate fields, five of which contain at least one bright, background QSO at $z \geqslant 2.5$ with a high-quality Lyα forest spectrum. The Lyα forest spectra of these background QSOs probe the the H I and metal content in the same cosmic volumes occupied by the numerous foreground UV-selected galaxies. This aspect of the $z \sim 2$ survey is entirely analogous to the observations presented by Adelberger et al. (2003). However, one relative advantage of studying the galaxy/IGM connection at $z \sim 2$ is that, down to the same limiting apparent magnitude, the galaxy surface density is $\sim$4 times higher than at $z \sim 3$; the number of suitable QSO sightlines increases from $z \sim 3.5$ to $z \sim 2.5$ as well, since the overall QSO number density is higher.

An important feature of the spectra of UV-selected $z \sim 2$ galaxies is that the same systematic offset is observed between Lyα emission and interstellar absorption lines, as is found in the spectra of $z \sim 3$ LBGs. Furthermore, as shown in Figure 5, when compared with the nebular systemic redshifts, $z \sim 2$ and $z \sim 3$ UV spectroscopic emission and absorption features exhibit similar distributions of velocity offsets (Steidel et al. 2004). Therefore, $z \sim 2$ star-forming galaxies appear to show the same kinematic signature of large-scale interstellar outflows as do $z \sim 3$ LBGs.

Though still a work in progress, it appears that the effects of galaxy winds on their nearby IGM environments are complex, both in terms of the neutral hydrogen content and the distribution of heavy elements. Furthermore, there appears to be a direct association between galaxies and shock-heated gas traced by broad absorption features from high-ionization transitions such as C IV, N V, and O VI (Simcoe et al. 2002)—a "smoking gun" demonstrating that metals tracing outflows reach more than 100 kpc from star-forming galaxies. Future work will include characterizing the IGM environments of galaxies as a function of star-formation rate, stellar mass, and star-formation history, all of which information will be available for a statistical sample of galaxies near QSO sightlines. With such information, it will be possible to describe important aspects of the evolution of galaxy feedback from $z \sim 3$ to $z \sim 2$.

4. Future directions with *Hubble*

Up until now, there have only been detailed *HST* morphological studies for a small number of star-forming high-redshift galaxies (Giavalisco et al. 1996; Lowenthal et al. 1997), with each study arriving at somewhat different conclusions about the morphologies of LBGs. Furthermore, the morphological properties of the LBGs were considered in isolation, without any regard for how morphology and spectroscopic properties were related. Now, however, the Great Observatories Origins Deep Survey (GOODS) has provided a uniform and deep Advanced Camera for Surveys (ACS) imaging dataset (Giavalisco et al. 2004), with 240 spectroscopically-confirmed UV-selected galaxies at $z = 1.4$–3.5 contained in the GOODS-N field. With this dataset, it will be possible to trace the evolution of the morphological properties of a large sample of star-forming galaxies as a function of redshift during the epoch when large disk and elliptical galaxies were assembled. It will also be possible to study in detail how the outflow properties, as traced by features in the rest-frame UV spectra, are related to what we can observe of the galaxy morphologies.

Figure 6 shows the Keck I/LRIS-B rest-frame UV (un-flux-calibrated) spectra and GOODS-N/ACS *Bviz* images for HDF-BMZ1148 ($z = 2.04$), HDF-BX1564 ($z = 2.22$), and HDF-BX1567 ($z = 2.23$). The ACS images are $4''$ on a side and represent the sum of 3, 2.5, 2.5, and 5 orbits in B, v, i, and z, respectively. These three examples demonstrate the range of spectroscopic and morphological properties present in the UV-selected star-forming sample of Steidel et al. (2004): Lyα properties ranging from strong emission (BMZ1148) to absorption (BX1564) to even damped absorption (BX1567); low-ionization interstellar absorption lines with equivalent widths varying by more than a factor of two; and C IV profiles with different relative contributions from P-Cygni massive stellar winds and interstellar absorption. It is also possible to measure the large-scale outflow velocity fields as traced by the kinematic offsets between interstellar absorption, Lyα emission, and stellar (or H II region) features.

The high-resolution ACS images also indicate a range of morphological properties, from simple compact regions of emission, to extended, elongated configurations, to complex and non-uniform brightness distributions that either indicate merger events, or else simply H II regions lit up against lower-surface-brightness material that is beyond the limit of detection, even in the deep GOODS images (or both). Furthermore, the detailed morphological information will enable us to understand how the spectroscopic

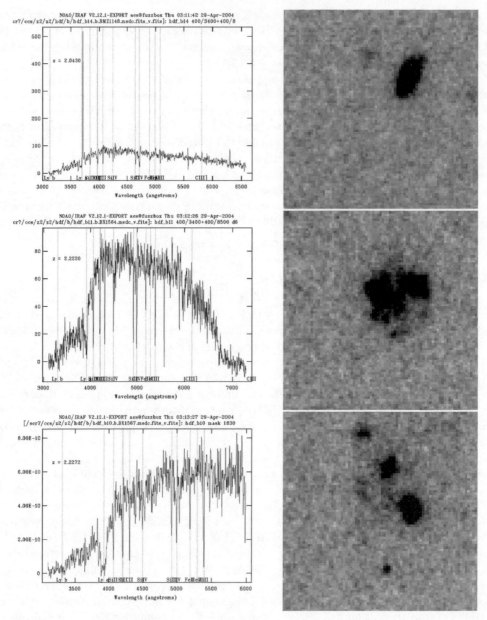

FIGURE 6. Spectroscopic and morphological properties of $z \sim 2$ star-forming galaxies in the GOODS-N field. This panel shows the Keck I/LRIS-B rest-frame UV (un-flux-calibrated) spectra and GOODS/ACS $Bviz$ images for HDF-BMZ1148 (top), which has $z = 2.05$; HDF-BX1564 (middle), which has $z = 2.22$; and HDF-BX1567 (bottom), which has $z = 2.23$. The ACS images are $4''$ on a side and represent the sum of 3, 2.5, 2.5, and 5 orbits in B, v, i, and z, respectively. These three examples demonstrate the range of spectroscopic and morphological properties present in the UV-selected star-forming sample of Steidel et al. (2004).

observations (in terms of slit position angle and width) sampled the light distributions, and may suggest future, complementary spectroscopic observations that sample the galaxy morphologies at different spatial locations or position angles. Such experiments will only be feasible for fairly extended morphologies, like that of BX1567, which has knots of emission extending more than $2''$ in one direction.

We would like to understand whether or not the morphological information can teach us anything about the outflow properties we observe: Does the presence of a disk configuration relate to the properties of the rest-frame UV spectra? Can we learn anything about the orientation or geometric configuration of the outflow that we are observing, based on the morphologies? Does the presence of a merger event affect what we observe in spectra probing the outflow? Can we obtain spatially distinct probes of an outflow using different slit positions for objects with extended morphologies, and therefore gain insight into the spatial distribution of absorption gas? Obtaining spatial information will be very difficult, due to the small sizes (typical half-light radii are $< 0.5''$) of these high-redshift star-forming galaxies, but is also crucial for understanding the nature of the outflows. Using close pairs of high redshift galaxies will also grant insight into the spatial distribution of outflowing interstellar gas. In such pairs, the spectrum of the high-redshift member of the pair skewers the outflowing interstellar gas of the lower-redshift pair-member at a different impact parameter from that of the lower-redshift galaxy spectrum itself. Determining the physical extent and covering fraction of the gas is crucial for determining quantities such as mass outflow rates.

5. Conclusions

We have presented observational evidence for the effects of supernova feedback and its effect on the IGM at $z \sim 2$–3. We have also discussed the importance of constraining the amount of radiative feedback from galaxies at high redshift. There is much work to be done in the future, including the determination of the detailed properties of the galaxies sustaining large-scale outflows. These properties consist of morphologies, ages, masses, and outflow geometry. Ultimately, the desired physical quantities are the amount of mass and energy, and number of heavy elements being transported from the galaxies. The combination of deep HST/ACS morphological information with rest-frame UV spectra probing the large-scale outflows of $z \sim 2$–3 galaxies will provide unique insight into the nature of star-formation feedback during an epoch when star formation, stellar mass assembly, and AGN activity were at their most vigorous.

REFERENCES

ADELBERGER, K. L., STEIDEL, C. C., SHAPLEY, A. E., & PETTINI, M. 2003 *ApJ* **584**, 45.
BECKER, R. H., FAN, X., WHITE, R. L., STRAUSS, M. A., NARAYANAN, V. K., LUPTON, R. H., GUNN, J. E., ANNIS, J., BAHCALL, N. A., BRINKMANN, J., CONNOLLY, A. J., CSABAI, I., CZARAPATA, P. C., DOI, M., HECKMAN, T. M., HENNESSY, G. S., IVEZIĆ, Ž., KNAPP, G. R., LAMB, D. Q., McKAY, T. A., MUNN, J. A., NASH, T., NICHOL, R., PIER, J. R., RICHARDS, G. T., SCHNEIDER, D. P., STOUGHTON, C., SZALAY, A. S., THAKAR, A. R., & YORK, D. G. 2001 *AJ* **122**, 2850.
BUNKER, A. J., STANWAY, E. R., ELLIS, R. S., & McMAHON, R. G. 2004 *MNRAS* **355**, 374.
FAN, X., NARAYANAN, V. K., STRAUSS, M. A., WHITE, R. L., BECKER, R. H., PENTERICCI, L., & RIX, H. 2002 *AJ* **123**, 1247.
FERNÁNDEZ-SOTO, A., LANZETTA, K. M., & CHEN, H.-W. 2003 *MNRAS* **342**, 1215.
FRANX, M., ILLINGWORTH, G. D., KELSON, D. D., VAN DOKKUM, P. G., & TRAN, K. 1997 *ApJ* **486**, L75.
GIALLONGO, E., CRISTIANI, S., D'ODORICO, S., & FONTANA, A. 2002 *ApJ* **568**, L9.

GIAVALISCO, M., FERGUSON, H. C., KOEKEMOER, A. M., DICKINSON, M., ALEXANDER, D. M., BAUER, F. E., BERGERON, J., BIAGETTI, C., BRANDT, W. N., CASERTANO, S., CESARSKY, C., CHATZICHRISTOU, E., CONSELICE, C., CRISTIANI, S., DA COSTA, L., DAHLEN, T., DE MELLO, D., EISENHARDT, P., ERBEN, T., FALL, S. M., FASSNACHT, C., FOSBURY, R., FRUCHTER, A., GARDNER, J. P., GROGIN, N., HOOK, R. N., HORNSCHEMEIER, A. E., IDZI, R., JOGEE, S., KRETCHMER, C., LAIDLER, V., LEE, K. S., LIVIO, M., LUCAS, R., MADAU, P., MOBASHER, B., MOUSTAKAS, L. A., NONINO, M., PADOVANI, P., PAPOVICH, C., PARK, Y., RAVINDRANATH, S., RENZINI, A., RICHARDSON, M., RIESS, A., ROSATI, P., SCHIRMER, M., SCHREIER, E., SOMERVILLE, R. S., SPINRAD, H., STERN, D., STIAVELLI, M., STROLGER, L., URRY, C. M., VANDAME, B., WILLIAMS, R., & WOLF, C. 2004 *ApJ* **600**, L93.

GIAVALISCO, M., STEIDEL, C. C., & MACCHETTO, F. D. 1996 *ApJ* **470**, 189.

HECKMAN, T. M. 2002. In *Extragalactic Gas at Low Redshift* (eds. J. S. Mulchaey & J. Stocke). ASP Conf. Ser. 254, p. 292. ASP.

HECKMAN, T. M., ARMUS, L., & MILEY, G. K. 1990 *ApJS* **74**, 833.

HECKMAN, T. M., LEHNERT, M. D., STRICKLAND, D. K., & ARMUS, L. 2000 *ApJS* **129**, 493.

HUNT, M. P., STEIDEL, C. C., ADELBERGER, K. L., & SHAPLEY, A. E. 2004 *ApJ* **605**, 625.

KIM, T.-S., CARSWELL, R. F., CRISTIANI, S., D'ODORICO, S., & GIALLONGO, E. 2002 *MNRAS* **335**, 555.

LEITHERER, C., FERGUSON, H. C., HECKMAN, T. M., & LOWENTHAL, J. D. 1995 *ApJ* **454**, L19.

LOWENTHAL, J. D., KOO, D. C., GUZMAN, R., GALLEGO, J., PHILLIPS, A. C., FABER, S. M., VOGT, N. P., ILLINGWORTH, G. D., & GRONWALL, C. 1997 *ApJ* **481**, 673.

MALKAN, M., WEBB, W., & KONOPACKY, Q. 2003 *ApJ* **598**, 878.

MUSHOTZKY, R. F. & LOEWENSTEIN, M. 1997 *ApJ* **481**, L63.

NAVARRO, J. F. & WHITE, S. D. M. 1994 *MNRAS* **267**, 401.

PETTINI, M., KELLOGG, M., STEIDEL, C. C., DICKINSON, M., ADELBERGER, K. L., & GIAVALISCO, M. 1998 *ApJ* **508**, 539.

PETTINI, M., RIX, S. A., STEIDEL, C. C., ADELBERGER, K. L., HUNT, M. P., & SHAPLEY, A. E. 2002 *ApJ* **569**, 742.

PETTINI, M., SHAPLEY, A. E., STEIDEL, C. C., CUBY, J., DICKINSON, M., MOORWOOD, A. F. M., ADELBERGER, K. L., & GIAVALISCO, M. 2001 *ApJ* **554**, 981.

PETTINI, M., STEIDEL, C. C., ADELBERGER, K. L., DICKINSON, M., & GIAVALISCO, M. 2000 *ApJ* **528**, 96.

PONMAN, T. J., CANNON, D. B., & NAVARRO, J. F. 1999 *Nature* **397**, 135.

SCOTT, J., BECHTOLD, J., DOBRZYCKI, A., & KULKARNI, V. P. 2000 *ApJS* **130**, 67.

SHAPLEY, A. E. 2003 *Ph.D. Thesis*.

SHAPLEY, A. E., STEIDEL, C. C., ADELBERGER, K. L., DICKINSON, M., GIAVALISCO, M., & PETTINI, M. 2001 *ApJ* **562**, 95.

SHAPLEY, A. E., STEIDEL, C. C., PETTINI, M., & ADELBERGER, K. L. 2003 *ApJ* **588**, 65.

SIMCOE, R. A., SARGENT, W. L. W., & RAUCH, M. 2002 *ApJ* **578**, 737.

SPERGEL, D. N., VERDE, L., PEIRIS, H. V., KOMATSU, E., NOLTA, M. R., BENNETT, C. L., HALPERN, M., HINSHAW, G., JAROSIK, N., KOGUT, A., LIMON, M., MEYER, S. S., PAGE, L., TUCKER, G. S., WEILAND, J. L., WOLLACK, E., & WRIGHT, E. L. 2003 *ApJS* **148**, 175.

STEIDEL, C. C., ADELBERGER, K. L., DICKINSON, M., GIAVALISCO, M., PETTINI, M., & KELLOGG, M. 1998 *ApJ* **492**, 428.

STEIDEL, C. C., GIAVALISCO, M., PETTINI, M., DICKINSON, M., & ADELBERGER, K. L. 1996 *ApJ* **462**, L17.

STEIDEL, C. C., PETTINI, M., & ADELBERGER, K. L. 2001 *ApJ* **546**, 665.

STEIDEL, C. C., SHAPLEY, A. E., PETTINI, M., ADELBERGER, K. L., ERB, D. K., REDDY, N. A., & HUNT, M. P. 2004 *ApJ* **604**, 534.

WEYMANN, R. J., JANNUZI, B. T., LU, L., BAHCALL, J. N., BERGERON, J., BOKSENBERG, A., HARTIG, G. F., KIRHAKOS, S., SARGENT, W. L. W., SAVAGE, B. D., SCHNEIDER, D. P., TURNSHEK, D. A., & WOLFE, A. M. 1998 *ApJ* **506**, 1.

WHITE, S. D. M. & FRENK, C. S. 1991 *ApJ* **379**, 52.

The baryon content of the local intergalactic medium

By JOHN T. STOCKE,
J. MICHAEL SHULL, AND STEVEN V. PENTON

Center for Astrophysics & Space Astronomy, and Dept. of Astrophysical & Planetary Sciences,
University of Colorado, Boulder, CO 80309-0389, USA

In this review, we describe our surveys of low column density (Lyα) absorbers ($N_{\rm HI} = 10^{12.5-16}$ cm^{-2}), which show that the warm photoionized IGM contains $\sim$30% of all baryons at $z \leq 0.1$. This fraction is consistent with cosmological hydrodynamical simulations, which also predict that an additional 20–40% of the baryons reside in much hotter 10^{5-7} K gas, the warm-hot IGM (WHIM). The observed line density of Lyα absorbers, $d\mathcal{N}/dz \approx 170$ for $N_{\rm HI} \geq 10^{12.8}$ cm^{-2}, is dominated by low-$N_{\rm HI}$ systems that exhibit slower redshift evolution than those with $N_{\rm HI} \geq 10^{14}$ cm^{-2}. HST/FUSE surveys of O VI absorbers, together with recent detections of O VII with $Chandra$ and $XMM/Newton$, suggest that 10–40% of all baryons could reside in the WHIM, depending on its assumed abundance (O/H $\approx$ 10% solar). We also review the relationship between the various types of Lyα absorbers and galaxies. At the highest column densities, $N_{\rm HI} \geq 10^{20.3}$ cm^{-2}, the damped Lyα (DLA) systems are often identified with gas-rich disks of galaxies over a large range in luminosities (0.03–1 L^*) and morphologies. Lyman-limit systems ($N_{\rm HI} = 10^{17.3-20.3}$ cm^{-2}) appear to be associated with bound bright ($\geq$ 0.1–0.3 L^*) galaxy halos. The Lyα absorbers with $N_{\rm HI} = 10^{13-17}$ cm^{-2} are associated with filaments of large-scale structure in the galaxy distribution, although some may arise in unbound winds from dwarf galaxies. Our discovery that $\sim$20% of low-z Lyα absorbers reside in galaxy voids suggests that a substantial fraction of baryons may be entirely unrelated to galaxies. In the future, HST can play a crucial role in a precise accounting of the local baryons and the distribution of heavy elements in the IGM. These studies will be especially effective if NASA finds a way to install the Cosmic Origins Spectrograph (COS) on $Hubble$, allowing an order-of-magnitude improvement in throughput and a comparable increase in our ability to study the IGM.

1. Introduction

In its first year, the $Hubble$ $Space$ $Telescope$ (HST) discovered that a majority of all baryons in the current universe are not in galaxies, but instead remain in the intergalactic medium (IGM). In subsequent years, the UV spectrographs aboard HST and the Far $Ultraviolet$ $Spectroscopic$ $Explorer$ ($FUSE$) have continued these investigations of the multiphase IGM, using sensitive UV tracers of diffuse gas: the Lyman series of H I (Lyα at 1215.67 Å, Lyβ at 1025.72 Å, etc.) and the O VI doublet (1031.926, 1037.617 Å). These HST and $FUSE$ studies have led to a preliminary "baryon census" of the "warm" (photoionized) and "warm-hot" (collisionally ionized) IGM. With spectrographs aboard the $Chandra$ and $XMM/Newton$ X-ray telescopes, astronomers are beginning to search for even more highly ionized gas through resonance absorption lines of O VII, O VIII, N VII, and Ne IX.

Unlike virtually all other astronomical objects, the Lyα absorption systems were first discovered at great distances ($z \geq 2$) owing to their cosmological redshifts and the near-UV atmospheric cutoff. Only with the advent of HST have nearby examples been found. The first low-z Lyα absorbers were seen in the spectrum of 3C 273 at $z_{\rm abs} < 0.158$ (Bahcall et al. 1991; Morris et al. 1991). While the number of absorbers was significantly less than the line density at high-z, the "local Lyα forest" contains far more absorbers than expected from extrapolating the ground-based data (Bahcall et al. 1993 and subsequent QSO Absorption-Line Key Project papers by Jannuzi et al. 1998 and Weymann

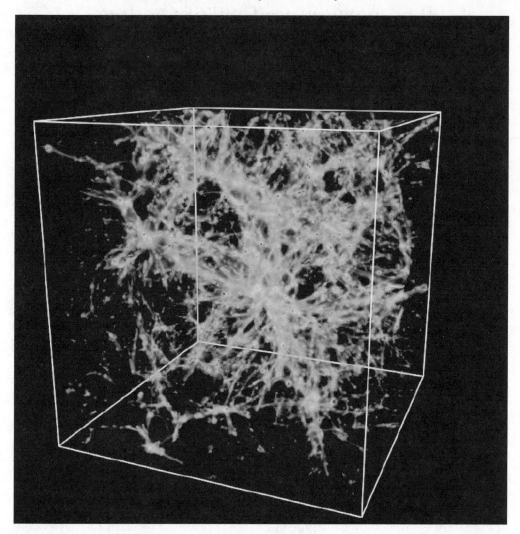

FIGURE 1. The distribution of baryons in the "cosmic web" of dark matter (Cen & Ostriker 1999). In this simulation the Lyα, O VI, and other absorption lines are produced in the denser filaments, representing a fluctuating distribution of gas structured by the dark-matter gravitational potentials.

et al. 1998). Although the Lyα absorbers at $z \geq 2$ are sufficiently abundant to account for nearly all the baryons (Rauch et al. 1997; Schaye 2001), their substantial numbers at $z \leq 0.1$ imply that $\sim$30% of all baryons remain in these photoionized clouds locally (Penton, Stocke, & Shull 2000a, Penton, Shull, & Stocke 2000b, 2004, hereafter denoted Papers I, II, and IV).

Numerical simulations (Fig. 1) of the evolving IGM (Cen & Ostriker 1999; Davé et al. 1999, 2001) explain not only the general features of the Lyα number density evolution, but also many detailed properties, including distributions in column density ($N_{\rm HI}$) and doppler b-value (Papers II and IV; Davé & Tripp 2001; Ricotti, Gnedin, & Shull 2000), and their relationship to galaxies (Davé et al. 1999; Impey, Petry, & Flint 1999; Penton, Stocke, & Shull 2002, hereafter denoted Paper III). Any accounting of the present-day distribution of baryons must include an accurate census of these absorbers and the associated mass, inferred from their ionized fractions, column densities, and physical extents.

Moderate-resolution UV spectroscopy of bright quasars, QSOs, blazars, and Seyfert galaxies has provided a substantial database of low-z Lyα absorbers. At the same time, several ground-based galaxy surveys (Morris et al. 1993; Lanzetta et al. 1995; Chen et al. 1998; Tripp, Lu, & Savage 1998; Rao & Turnshek 1998, 2000; Impey, Petry, & Flint 1999; Nestor et al. 2002; Bowen et al. 2002; Paper III; Bowen & Pettini 2003; Stocke et al. 2005, hereafter denoted Paper V) have probed the relationship between Lyα absorbers and galaxies, filaments of galaxies, and voids. Using nearby examples of the Lyα phenomenon, these authors sought to identify the galaxy types responsible for the absorption and thus assist in interpreting the wealth of information (number densities, metallicities, ionization states, line widths) of Lyα absorption systems at high-z. These efforts have been somewhat successful, although the results in most column-density regimes remain controversial (see conference proceedings edited by Mulchaey & Stocke 2002).

In this review, we describe the various *HST* QSO absorption line surveys that have been undertaken (§2), review our present knowledge of the baryon content of the IGM (§3), and describe the emerging, but still controversial, evidence for the relationship between the various column densities of Lyα absorbers and galaxies (§4). The last information has come largely from studying low-z absorbers discovered with *HST*. We conclude (§5) with a brief prospectus on low-z IGM studies facilitated by the Cosmic Origins Spectrograph (COS), a new instrument that may be installed on *HST* in the coming years.

2. *HST* surveys the low-z Lyα absorbers

HST has conducted several important surveys of the IGM with its UV spectrographs (FOS, GHRS, STIS), which provided basic data for studying the bulk of local baryons. Owing to its modest spectral resolution (200–300 $\mathrm{km\,s^{-1}}$), the Faint Object Spectrograph (FOS)—used for the initial QSO Absorption-Line Key Project (Bahcall et al. 1993)—primarily detected high column density Lyα absorbers with equivalent widths $W_\lambda \geq 240$ mÅ. The Key Project provided examples of the various types of Lyα absorbers: damped Lyα (DLA) absorbers, Lyman-limit/strong Mg II absorbers, weak high-ionization (C IV) and low-ionization (Mg II) metal-line absorbers, and Lyα-only absorbers (Bahcall et al. 1996; Jannuzi et al. 1998). Even though the broad UV wavelength coverage (G130H, G190H, G270H gratings) of the Key Project spectra allowed the discovery of many Lyα absorbers at $z \leq 1.6$, the detection efficiency of low redshift ($z \leq 0.2$) absorbers was reduced by lower than expected far-UV sensitivity of the FOS digicon. The FOS Key Project survey firmly established the column density distribution, $f(N_{\mathrm{HI}})$, for high-N_{HI} absorbers and $d\mathcal{N}/dz$, the number density of absorbers per unit redshift. Above limiting Lyα equivalent width, $W_\lambda = 240$ mÅ ($N_{\mathrm{HI}} \geq 10^{14}$ cm^{-2}), Weymann et al. (1998) found $d\mathcal{N}/dz = 32.7 \pm 4.2$ over a substantial redshift pathlength ($\Delta z \approx 30$). As we will discuss later (§3.1), the Lyα line density increases substantially to lower columns, reaching $d\mathcal{N}/dz \approx 170$ for $N_{\mathrm{HI}} \geq 10^{12.8}$ cm^{-2} (Paper IV).

The absorber number density (Weymann et al. 1998) shows a dramatic break from rapid evolution ($z \geq 1.5$) to almost no evolution ($z \leq 1.5$). This observation was explained by cosmological hydrodynamical simulations (Davé et al. 1999), in which the rapid evolution in $d\mathcal{N}/dz$ is controlled by the response of photoionized gas to evolving QSO populations in a hierarchical distribution of large-scale structure. At high redshift, the number of Lyα absorbers decreased rapidly with time, owing to a nearly constant extragalactic ionizing flux from QSOs in an expanding universe. (The cosmological expansion decreases recombinations, thus decreasing the neutral hydrogen column density.) At $z \leq 2$, the ionizing flux began a rapid decline with the rapidly decreasing QSO

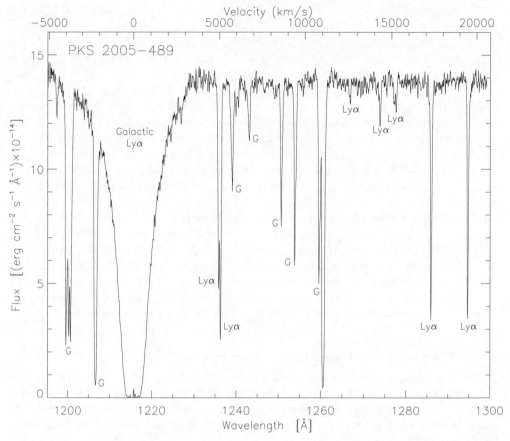

FIGURE 2. An *HST*/STIS medium-resolution (20 km s^{-1}) spectrum (Paper IV) of the bright BL Lac Object PKS 2005−489 illustrates the highest-quality data obtained for this project. The deep, broad absorption at 1216 Å is the damped Lyα absorption due to the Milky Way. Other Galactic interstellar metal lines (S II, Si II, Mg II, N I, N V and Si III) are marked with a "G." The weakest Lyα absorbers are at 1266.740 Å (12,594 km s^{-1}, $W_\lambda = 22 \pm 13$ mÅ) and 1277.572 Å (15,265 km s^{-1}, $W_\lambda = 27 \pm 12$ mÅ) with column densities $N_{\mathrm{HI}}= 10^{12.6-12.7}$ cm^{-2}). The heliocentric velocity scale along the top is for the Lyα absorbers only.

numbers and/or luminosities, thereby decreasing the ionized fraction in the Lyα absorbers. As a consequence of reduced photoionization, along with large-scale structure formation, the $d\mathcal{N}/dz$ evolution slowed rapidly. Although the Key Project discovered only one DLA (Jannuzi et al. 1998), Rao & Turnshek (1998) used the FOS to increase the number of low-z DLAs by targeting strong Mg II absorbers found by ground-based telescopes. Bowen et al. (1996) used archival FOS spectra to search for strong Lyα and metal-line absorption from known foreground galaxies. See Bowen, Pettini, & Blades (2002) for a STIS extension of this successful project.

The Goddard High-Resolution Spectrograph (GHRS) was used in the co-discovery (Morris et al. 1991) of the Lyα absorbers in the 3C 273 sightline. However, few QSOs were sufficiently bright to observe in moderate-length integrations. Tripp, Lu, & Savage (1998) and Impey, Petry, & Flint (1999) observed a few UV-bright AGN targets with the low-resolution far-UV GHRS grating, obtaining observations only slightly more sensitive than the FOS Key Project spectra. In the mid-1990s, the Colorado group conducted a moderate-resolution ($\sim$20 km s^{-1}) survey of 15 sightlines with the GHRS/G160M

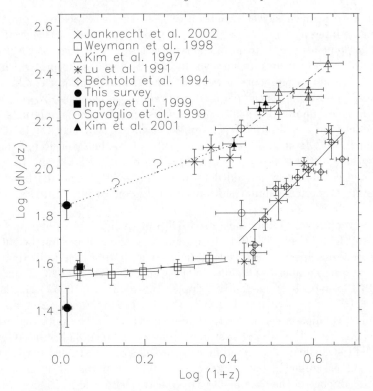

FIGURE 3. Comparison of the Lyα absorber evolutionary plot, dN/dz versus $\log(1+z)$, for two N_{HI} ranges. The lower distribution corresponds to $W_\lambda \geq 240$ mÅ ($N_{\mathrm{HI}} \geq 10^{14}$ cm^{-2} for an assumed $b = 25$ km s^{-1}). The upper distribution corresponds to absorbers in the range $N_{\mathrm{HI}} = 10^{13.1-14.0}$ cm^{-2}. The two $z \approx 0$ points (**solid circles**) are taken from our survey (Paper IV) for each of the two N_{HI} ranges specified above. Solid lines are taken from Weymann et al. (1998) and have evolutionary slopes $[dN/dz \propto (1+z)^\gamma]$ with $\gamma = 0.16$ at $\log(1+z) < 0.4$ and $\gamma = 1.85$ at $\log(1+z) > 0.4$. The complete evolutionary picture for low-N_{HI} absorbers is unavailable, owing to the lack of data at intermediate redshifts, as indicated by question marks (see Paper IV for details).

first-order grating (Papers I and II), followed by a comparable survey of 15 additional targets with the STIS first-order, medium-resolution grating G140M (Paper IV). In these surveys, the key strategy was to select only the brightest targets (see Fig. 2), while observing only a modest pathlength per target ($cz \approx 10,000$ km s^{-1} with GHRS and 20,000 km s^{-1} with STIS). The number density of Lyα absorbers rises steeply with decreasing equivalent width. After subtracting intrinsic absorbers (near the redshift of the QSO) we found 187 intervening low-N_{HI} Lyα absorbers at $\geq 4\sigma$ significance over a total unobscured pathlength $\Delta z = 1.157$. A careful analysis, accounting for S/N sensitivity bias and overlap with Galactic interstellar lines, gives a line density $d\mathcal{N}/dz \approx 170$ for $N_{\mathrm{HI}} \geq 10^{12.8}$ cm^{-2}, or one absorber every $\sim$25 h_{70}^{-1} Mpc. Our more sensitive GHRS/STIS survey provides an important extension of the Key Project to lower-N_{HI} absorbers. As shown in Figure 2, the very best spectra in our survey reach equivalent-width limits of $W_\lambda = 10$–20 mÅ (4σ), corresponding to $N_{\mathrm{HI}} \approx 10^{12.3-12.5}$ cm^{-2}, comparable to the best spectra obtained with Keck/HIRES or VLT/UVES.

The cosmic evolution of the Lyα forest absorbers shown in Figure 3 represents the historical record of a large fraction of the baryons during the emergence of the modern universe at $z \leq 2$. The $d\mathcal{N}/dz$ evolution of the high-N_{HI} absorbers exhibits a substantial break in slope to a much slower evolution at $z \leq 1.5$, as expected from the physical effects acting on intergalactic absorbers in an expanding universe (Davé et al. 1999). The low-N_{HI} data point at $z \approx 0$ from our survey is lower than that of the Key Project. We suspect this results from our survey's higher spectral resolution, which affects the counting and N_{HI}-bin assignments of weak absorbers. The break in the $d\mathcal{N}/dz$ evolution of the higher-N_{HI} absorbers is therefore not as dramatic as the Key Project data suggest. A more gradual decline in slope is also more consistent with the simulations in a ΛCDM Universe model (Davé et al. 1999). The physical effects that account for the break in the high-column $d\mathcal{N}/dz$ should create a similar break in the evolution of the lower-N_{HI} absorbers, although this is not yet evident in Figure 3. New STIS spectra obtained by B. Jannuzi and collaborators should determine whether the expected break is present, and whether the numerical simulations have used the correct physics to account for these absorbers.

Recently, several investigators have used STIS in its medium-resolution ($\sim$7 km s^{-1}) echelle modes (E140M and E230M) to obtain long integrations on bright AGN targets. Many of these spectra have been analyzed for interesting individual absorbers, including a growing number of redshifted O VI absorbers: H1821+643 (Tripp et al. 2000, 2001); 3C 273 (Sembach et al. 2001; Tripp et al. 2002); PG 0953+41 (Savage et al. 2002); PKS 2155−304 (Shull et al. 2003); PG 1116+215 (Sembach et al. 2004); and PG 1211+143 (Tumlinson et al. 2004). These spectra have the potential to be *HST*'s best resource for low-z Lyα absorber studies. As of July 2004, 35 bright AGN have been observed in STIS-echelle mode, providing in some cases the most sensitive detection limits for Lyα, as well as significant pathlengths at $z \geq 0.11$ for potential discovery of O VI absorbers. In the absence of the Cosmic Origins Spectrograph, the STIS medium-resolution echelle gratings continue to provide high-quality QSO spectra for IGM studies. However, obtaining the S/N needed to make sound inferences about the IGM absorbers requires increasingly long integration times on fainter targets.

3. Baryon content of the low-z Lyα forest

At high redshift ($z = 2$–4), the number density of Lyα absorbers (after substantial ionization corrections for photoionization equilibrium) is so large that estimates of their total baryonic masses suggest that the entire baryon content of the universe ($\Omega_b h_{70}^2 = 0.046 \pm 0.002$ as measured by the Cosmic Microwave Background anisotropies or cosmic [D/H] ratio) can be accommodated within the warm ($\sim$10^4 K) photoionized IGM (Rauch et al. 1997; Schaye 2001). This result is consistent with numerical simulations of large-scale structure formation (Cen & Ostriker 1999; Davé et al. 1999, 2001), which also predict that by $z \approx 0$ the baryons are approximately evenly divided among several components as follows:

• **Stars and gas in or near galaxies; 30% predicted**. Salucci & Persic (1999) and Fukugita (2004) account for only $\sim$6% in this reservoir.

• **Very hot (10^{7-8} K) intracluster and intragroup gas; $\leq$10% predicted**. Fukugita (2004) estimates only $\sim$5% of the baryons in this component, but the gas in groups may not be fully accounted for.

• **Warm (10^4 K) photoionized gas; 30–40% predicted in Lyα forest absorbers**. In our Paper IV, we find 29 ± 4% of the baryons in local Lyα absorbers.

• **Warm-hot intergalactic medium (WHIM) at 10^{5-7} K; 20–40% predicted**. A series of observations with *HST* and *FUSE* (Tripp, Savage, & Jenkins 2000; Savage et al. 2002) estimate that 5–10% of the baryons have been detected (at 10% assumed O/H metallicity) through O VI absorption. Recent *Chandra* observations of O VII absorption

(Fang et al. 2002; Nicastro et al. 2004) suggest even larger percentages, although with enormous statistical uncertainties and a few unconfirmed observations.

Thus, we submit that the current-epoch baryon census remains an unsolved problem, with large and uncertain corrections for ionization state (H I, O VI, and O VII) and an overall uncertainty over the appropriate metallicity (for O VI, O VII, O VIII in WHIM). These problems can be addressed by *HST* and *FUSE* in the coming years, and by future high-throughput X-ray spectroscopic missions such as *Constellation-X* or *Xeus*.

3.1. *Photoionized Lyα absorbers*

In Paper II, we described a method for estimating the baryon content of the local Lyα absorbers, based upon the observed column density distribution. In the context of optically-thin, photoionized clouds, the assumptions of this simplified model are: (1) spherically symmetric absorbers; (2) an isothermal density profile; (3) absorber sizes of $100h_{70}^{-1}$ kpc, based on QSO pairs experiments conducted at somewhat higher redshifts (Dinshaw et al. 1997; but see also Rosenberg et al. 2003); and (4) a value and slope for the extragalactic ionizing flux (Shull et al. 1999) using space densities and ionizing spectra of local Seyfert galaxies (Telfer et al. 2002). However, numerical simulations show that Lyα absorbers are not at all spherical, especially at lower column densities. Cen & Simcoe (1997) used numerical simulations at $z = 2$–4 to show that lower column density ($N_{\mathrm{HI}} = 10^{13-14}$ cm^{-2}) absorbers have large physical sizes, consistent with or somewhat larger than $100h_{70}^{-1}$ kpc (equivalent spherical radius). We have integrated this simplified model over our observed column density distribution, from $12.5 \leq \log N_{\mathrm{HI}} \leq 16$, the range where the above assumptions are most valid. In Paper II we obtained a baryon fraction $\geq 20\%$, with specific dependences on measurable parameters as follows:

$$\Omega_{\mathrm{Ly}\alpha} = (0.008 \pm 0.001)\left[N_{14}\, J_{-23}\, b_{100}\left(\frac{4.8}{\alpha_s + 3} \right) \right]^{1/2} h_{70}^{-1} \,. \tag{3.1}$$

Here, N_{14} is the characteristic H I column density in the Lyα forest in units of 10^{14} cm^{-2}, J_{-23} is the extragalactic ionizing specific intensity at 1 ryd in units of 10^{-23} erg cm^{-2} s^{-1} Hz^{-1} sr^{-1}, b_{100} is the characteristic absorber radius in units of $100h_{70}^{-1}$ kpc, and $\alpha_s \approx 1.8$ is the mean power-law spectral index ($F_\nu \propto \nu^{-\alpha_s}$) for the metagalactic ionizing radiation, which presumably comes from QSOs (Shull et al. 1999; Telfer et al. 2002).

Schaye (2001) developed a methodology for estimating the baryon content that makes somewhat different assumptions. As with the method of our Paper II, this model assumes photoionization of optically-thin absorbers, but it also assumes gravitationally-bound clouds whose observed column densities are equal to their characteristic column densities over a Jeans length. This estimate has the following specific dependences:

$$\Omega_{\mathrm{Ly}\alpha} = (2.2 \times 10^{-9})h_{100}^{-1}\, \Gamma_{-12}^{1/3}\, T_4^{0.59} \int N_{\mathrm{HI}}^{1/3} f(N_{\mathrm{HI}}, z)dN_{\mathrm{HI}} \,, \tag{3.2}$$

where Γ_{-12} is the H I photoionization rate in units of 10^{-12} s^{-1}, T_4 is the IGM temperature in units of 10^4 K, and $f(N_{\mathrm{HI}}, z) \propto N_{\mathrm{HI}}^{-\beta}$ is the H I column density distribution. In this model, the absorber baryon content depends on the minimum column density of H I absorbers as $N_{\mathrm{min}}^{-(\beta-4/3)}$ or $N_{\mathrm{min}}^{-0.32\pm0.07}$ for $\beta = -1.65 \pm 0.07$ (see below). The above expression assumes that the Lyα absorbers contain the universal ratio of baryons to dark matter, with no bias.

Here we employ both of these methods to estimate the baryon content of the local Lyα absorbers, based on our latest column density distribution from Paper IV, which fitted power-law slopes of $\beta = -1.65 \pm 0.07$ and -1.33 ± 0.30 on the lower and upper sides, respectively, of a break at $N_{\mathrm{HI}} = 10^{14.5}$ cm^{-2}. From our enlarged sample, we are confident that the $\beta = -1.65$ slope extends down to at least log $N_{HI} = 12.5$, so that we can reliably integrate this distribution from log $N_{HI} = 12.5$ to 16.0. We recognize that, at or near our adopted lower limit (which corresponds approximately to an overdensity $\delta \approx 3$ at the current epoch), some of the assumptions of both methods may break down. Also, at $N_{HI} \geq 10^{17}$ cm^{-2}, absorbers become optically thick in the Lyman continuum and can hide additional gas mass. Therefore, neither method can hope to derive an extremely accurate measurement for the baryon content of the Lyα forest, even if the column density distribution and other absorber properties (size, ionizing flux, temperature) are known to some precision.

We quote the resulting baryon densities as percentages of the total baryon density from measurements of the Cosmic Microwave Background anisotropy ($\Omega_b h_{70}^2 = 0.046$, Spergel et al. 2003). A slightly smaller total baryon density has been obtained by measurement of primordial D/H nucleosynthesis ($\Omega_b h_{70}^2 = 0.041$, Burles & Tytler 1998). The baryon fraction in the IGM is dominated by the lower column density absorbers (22% out of 29% total baryons in Lyα absorbers at $N_{HI} \leq 10^{14.5}$ cm^{-2}), owing to the steep slope of the column density distribution. The Schaye (2001) prescription yields a divergent baryon fraction at low N_{HI}, owing to an inverse dependence of size on column density, which must break down well above the limit employed here. On the other hand, at $N_{HI} \geq 10^{14.5}$ cm^{-2}, the 100 kpc size assumed in Paper II probably becomes too large for many absorbers at higher column densities (Cen & Simcoe 1997; Tripp et al. 2002). Thus—the most conservative—and we believe most accurate approach is to use our Paper II method at low columns and the Schaye (2001) method at high columns. This technique yields values of $22 \pm 2\%$ below $N_{HI} = 10^{14.5}$ cm^{-2}, and $7 \pm 3\%$ above that column density, for a total baryon fraction in the low-z photoionized IGM of $29 \pm 4\%$ (Paper IV). Because the baryon fraction is dominated by the lowest column density absorbers, our result is relatively insensitive to the assumed b-values for individual absorbers. Here, we have used $b = 25$ km s^{-1}, which is comparable to the median $b = 22$ km s^{-1} found by Davé & Tripp (2001) using medium-resolution echelle spectra.

In order to make further progress in the baryon census of warm IGM gas, it is important to: (1) estimate absorber sizes and shapes (and thus equivalent cloud radii) as a function of column density. In the absence of observations of many QSO pairs, this is best done using numerical simulations (the Cen & Simcoe [1997] analysis, but for simulations at $z \approx 0$); (2) estimate the percentage of Lyα absorbers that are collisionally ionized rather than photoionized. This will require observations of O VI with *FUSE* or *HST* to address the "double-counting" issue; see §3.2; (3) observe a much longer pathlength for Lyα absorbers, since we find that the bias introduced by cosmic variance is still significant for the observed redshift pathlength, $\Delta z = 1.157$ (Paper IV); (4) determine more accurate H I column densities from *FUSE* observations of higher Lyman lines and a curve-of-growth analysis (e.g., Shull et al. 2000; Sembach et al. 2001) for saturated absorbers; and (5) observe a few very bright targets over very long integration times (≥ 30 orbits) to push the detection limits below $N_{HI} \leq 10^{12}$ cm^{-2}; very weak absorbers can still contribute to the baryon budget.

3.2. *The Warm-Hot IGM*

The search for the WHIM gas has now begun, using sensitive UV resonance absorption lines: primarily the O VI lithium-like ($2s$–$2p$) doublet (1031.926, 1037.617 Å) but also the Li-like doublet of Ne VIII (770.409, 780.324 Å)—see Savage et al. (2004). The O VI lines can be observed with *HST*/STIS for IGM absorbers with $z \geq 0.11$, while the *FUSE* spectrographs can observe O VI essentially down to $z = 0$. The K-shell X-ray absorption lines of O VII (21.602 Å), O VIII (18.97 Å), N VII (24.782 Å), Ne IX (13.447 Å), and a few other highly ionized species can (barely) be observed with *Chandra* and *XMM/Newton*, primarily toward X-ray flaring blazars. As mentioned in §2, there now exists a substantial number (35 to our knowledge) of STIS medium-resolution echelle spectra of QSOs. However, fewer than half of these targets have sufficient redshift ($z \geq 0.11$) and continuum signal-to-noise ratio (S/N ≥ 10 per resolution element) to facilitate a sensitive search for O VI absorption. Thus far, only a modest pathlength has been searched for O VI absorbers with *HST* (Tripp, Savage, & Jenkins 2000; Savage et al. 2002). These spectra yield measured baryon fractions of 5–10% (assuming 10% O/H metallicity), consistent with expectations from simulations (Gnedin & Ostriker 1997; Cen et al. 2001; Fang & Bryan 2001). We caution that these estimates involve large uncertainties, including untested assumptions of cloud sizes and shapes, metallicities, ionization equilibrium, and multi-phase structure.

The multiphase nature of IGM absorbers leads directly to the classic census problem of "double counting." In the IGM context, this means that the photoionized H I and O VI absorbers could count the same baryons twice. Recent surveys down to 50 mÅ equivalent width give H I line densities approximately ten times those of the O VI absorbers: $d\mathcal{N}_{\rm HI}/dz \approx 112 \pm 9$ (Paper IV) and $d\mathcal{N}_{\rm OVI}/dz \approx 14^{+9}_{-6}$ (Savage et al. 2002). Many IGM absorbers (Shull et al. 2003) as well as Galactic high velocity clouds (Collins et al. 2004) clearly contain gas at vastly different temperatures, in order to explain the observed range of ionization stages. Deriving the metallicities of such complex systems requires subtle ionization modeling: collisional vs. photoionization, time-dependent ionization, range of temperatures, inhomogeneities in density and velocity.

Theoretical considerations of the shocked IGM (Davé et al. 1999) suggest that collisional ionization dominates the stronger O VI systems, although some weak O VI absorbers have been modeled successfully using photoionization equilibrium (Tripp et al. 2001). The problem with these photoionization models for high ions is that they require a large photoionization parameter ($U \propto J_0/n_H$) to produce sufficient O VI. At fixed ionizing radiation intensity, J_0, this requires very low hydrogen densities, $n_H \approx (10^{-6}$ cm$^{-3})n_{-6}$. As a result, photoionization models with high-U and low-n_H are physically implausible, since they produce a low neutral fraction,

$$f_{\rm HI} \equiv \frac{n_{\rm HI}}{n_H} \approx (3 \times 10^{-5})\, n_{-6}\, T_4^{-0.845}\, \Gamma_{-14}^{-1} \;, \qquad (3.3)$$

and unrealistically large absorption pathlengths,

$$L_{\rm abs} \equiv \frac{N_{\rm HI}}{n_H\, f_{\rm HI}} \approx (1\ {\rm Mpc})\, \Gamma_{-14}\, T_4^{0.845}\, n_{-6}^{-2}\, N_{14} \;. \qquad (3.4)$$

Here, we have scaled the photoionization parameters to an H I photoionization rate Γ_{-14} at $z \approx 0$ in units of 10^{-14} s^{-1} and a temperature of $(10^4$ K$)T_4$ and adopted an absorber with $N_{\rm HI} = (10^{14}$ cm$^{-2})N_{14}$. Thus, some of the photoionized baryons in the O VI absorbers may have been accounted for already by the H I calculation in the previous section. However, the hotter, collisionally ionized O VI should represent a distinct thermal phase from the H I.

A substantial survey of O VI absorbers may reveal systematics among the O VI absorbers that will allow us to separate the collisionally-ionized absorbers from the photoionized absorbers and thus unravel the double-counting problem. Ongoing STIS echelle spectroscopy by Tripp, Savage, Howk and collaborators will go a long way towards answering these questions and better quantifying the "O VI forest" statistics. Equally important to this effort is ongoing *FUSE* spectroscopy of lower-redshift O VI with complementary Lyα data from *Hubble*. To date, we have found 114 Lyα absorbers ($W_\lambda \geq 80$ mÅ) with acceptable O VI data, and roughly half (57 systems) are detected in O VI at $\geq 3\sigma$ significance (Danforth & Shull 2004, in preparation). This combined *HST/FUSE* database can address important questions concerning the extent to which metals are spread away from galaxies into the IGM. While much data exist and more is scheduled to be obtained, these are still early days.

If the study of the O VI absorbers is still in its infancy, the detection of the remainder of the WHIM gas is not yet "out of the womb." Simulations predict that this hotter WHIM (10^{6-7} K) contains $\geq$20–40% of the total local baryons. Nicastro (2003) claims that WHIM gas at $T \approx 10^{5.8}$ K has been detected at $z = 0$ along several sightlines, through absorption lines of O VI with *FUSE* and of C VI, O VII, O VIII, N VII, and Ne IX with *Chandra* and/or *XMM/Newton*. However, it is possible that this gas is not intergalactic at all, but instead resides in smaller physical regions in the halo of the Milky Way. The interpretation of these absorptions remains controversial (e.g., Sembach 2002; Shull 2003; Fox et al. 2004; Collins et al. 2004). While two groups have claimed WHIM detections at $z > 0$ in the sightlines of PKS 2155−304 (Fang et al. 2002) and H1821+643 with *Chandra* (Mathur et al. 2003), neither of these results has been confirmed. The PKS 2155−304 detection has been refuted by *XMM/Newton* observations (Rasmussen et al. 2003; Cagnoni et al. 2004).

On the other hand, two new WHIM detections are claimed (Nicastro et al. 2004) toward Mrk 421, a blazar at $z = 0.03$. The data were obtained when the object flared in X-rays, and the claimed absorbers lie at redshifts $z = 0.011$ (where a Lyα system was identified by Shull, Stocke, & Penton 1996) and $z = 0.027$ (where no Lyα absorber is seen to a 4σ limit of 30 mÅ). The $z = 0.027$ absorber is detected in three lines (N VI, N VII, O VII), at approximately 3σ significance each, while the $z = 0.011$ absorber is seen in only a single line (O VII) at 3.8σ significance. The ion ratios are consistent with a WHIM filament at $T \approx 10^7$ K. Over the pathlength $\Delta z \approx 0.03$, these $N = 1$–2 detections yield a line density of O VII absorbers, $N/(\Delta z)$, of 2–4 times the O VI estimate, but with enormous uncertainties. The formal baryon fraction in the O VII-bearing gas could be anywhere from 10–40%. A possible one-line WHIM detection in the *Chandra* spectrum of 3C 120 (McKernan et al. 2003) yields a similar O VII line density and baryon fraction. Obviously, the statistics of these studies are poor, and the model assumptions substantial. Sadly, the prospects of further observations with *Chandra* are remote; when Mrk 421 was observed it was the brightest extragalactic source in the sky. Because the brightest extragalactic soft X-ray sources are all blazars, and the velocity resolution of *Chandra's* Low Energy Transmission Grating is only ~1000 km s^{-1}, one worries that the $z = 0.027$ absorber is within 900 km s^{-1} of the blazar redshift ($z = 0.03$). Thus, it might be outflowing gas along a jet and not intervening WHIM at all. Taking this observational program to the level of *FUSE* (or even *HST*) will have to await the *Constellation-X* grating spectrographs.

4. The relationship between Lyα absorbers and galaxies

While the baryon census and its evolution are ample reasons for studying the local Lyα forest in detail, it is also only at low redshift that the locations of Lyα absorbers and galaxies can be compared accurately. This allows the relationship between these "clouds" and galaxies to be determined to some degree of certainty. However, the degree to which absorbers correlate with individual galaxies has been controversial. Lanzetta et al. (1995) and Chen et al. (1998, 2001) argue that the Lyα absorbers are the very extended halos of individual galaxies, while others (Morris et al. 1993; Stocke et al. 1995; Impey, Petry, & Flint 1999; Davé et al. 1999; Papers III and V) argue that the absorbers are related to galaxies only through their common association with large-scale gaseous filaments, arising from overdensities in the high-redshift universe. Much of the difference between these results primarily reflects the column density range of the Lyα absorbers studied in each case; at some level both sides to this argument are correct (see conference papers in Mulchaey & Stocke 2002). In this section, we review the current evidence for the relationship between galaxies and Lyα absorbers as a function of H$_I$ column density.

4.1. The Damped Lyα absorbers (DLAs)

At the highest column densities ($N_{\rm HI} \geq 10^{20.3}$ cm^{-2}) Lyα exhibits broad damping wings that allow an accurate determination of H$_I$ column density. This has proved valuable for studying the evolution of the chemical abundance of galaxies with cosmic time, since even quite weak metal lines are detectable in these systems with high-quality spectra (e.g., Pettini et al. 1999; Prochaska & Wolfe 2000, 2002). DLAs have also been used to measure the evolution of the amount of neutral gas in galaxies (Boissier et al. 2003). However, clues about the types of galaxies probed by the DLAs are largely circumstantial at high redshift. Thick disks of massive galaxies are suggested by the kinematics (Wolfe & Prochaska 2000), but are hardly demanded by the evidence. Surveys of DLAs with *HST* (e.g., Rao & Turnshek 2000; Turnshek et al. 2002) and their subsequent imaging with large-aperture ground-based telescopes (Nestor et al. 2002; Rao et al. 2003) have shown convincingly that no single galaxy population dominates the DLAs. Surprisingly, there is a large contribution to DLAs made by dwarf and low surface brightness (LSB) galaxies (see Bowen, Tripp, & Jenkins 2001 for an excellent example), in contrast to the Wolfe & Prochaska (2000) kinematic analysis. Until recently, an apparent contradiction existed between the cross-sectional areas on the sky covered by high column density H$_I$ at $z = 0$ as measured either by DLAs or by 21 cm emission maps (Rao & Briggs 1993). However, new "blind" 21 cm surveys (Zwaan et al. 2001; Rosenberg & Schneider 2003) show that the 21 cm cross-sectional area now agrees with the DLA measurement at low-z.

These new surveys also show that the galaxy populations detected by these two different methods now agree. Rosenberg & Schneider (2003) find that DLA galaxies span a wide range of total H$_I$ masses (two-thirds are between $M_{\rm HI} = 10^{8.5-9.7}$ $M_{\odot}$) and luminosities 0.01–1 L^*. Thus, the chemical abundance data from DLAs at high-z must be interpreted with these new *HST* + ground-based results in mind; that is, the chemical abundance evolution refers to H$_I$ disks and amorphous H$_I$ clumps in a wide variety of galaxies. This certainly explains why the spread in DLA metallicities at any one redshift is so broad.

4.2. Lyman Limit absorbers

In the range $N_{\rm HI} = 10^{17.3-20.3}$ cm^{-2}, the Lyman Limit System (LLS) absorbers constitute a substantial number ($dN/dz \approx$ 1–3 per unit redshift) of intervening systems in the spectra of high-z QSOs. As shown by Steidel (1995, 1998), strong ($W_\lambda \geq 0.3$ Å) Mg$_{II}$ absorbers at $z = 0.2$–1 are invariably also LLSs, although a few DLAs are also present in such samples. Physically, this can be understood because H$_I$ and Mg$_{II}$ have similar ionization potentials, so that once hydrogen becomes optically thin, Mg$_{II}$ rapidly ionizes to Mg$_{III}$. In an important pre-*HST* result, Steidel (1995, 1998) found that of 58 LLSs in his sample, 55 could be identified with nearby ($\leq 50h_{70}^{-1}$ kpc offsets), bright ($L > 0.1$–0.3 L^*) galaxies at the same redshift (see also Bergeron & Boissé 1991). These results have been strengthened by obtaining ground-based spectra of the underlying stellar kinematics of the associated galaxies and *HST* imaging and UV spectra of the C$_{IV}$ absorptions in these systems (Churchill et al. 2000; Steidel et al. 2002). These results suggest that the LLSs are almost exclusively the bound gaseous halos of luminous galaxies, and that these halos share the kinematics of the underlying stellar disk.

Recently, *HST* observations of the quasar 3C 232 probed the halo of the modest starburst galaxy NGC 3067. The spectra show a complex LLS with three absorption components at small radial velocity differences ($\Delta v = -260$, -130 and $+170$ km s^{-1}) from the nucleus of NGC 3067, in ions ranging from Na I and Ca II detected optically (Stocke et al. 1991) to Mg I, Mg II, Fe II and Mn II in the near-UV (Tumlinson et al. 1999) to Si IV and C IV in the far-UV. Surprisingly, all these species share the same three velocities (Keeney et al. 2005), which are gravitationally bound to NGC 3067; two of the systems, including the strongest at $N_{\mathrm{HI}} = 10^{20.0}$ cm^{-2}, are infalling. The 3C 232 sightline has an offset of $8h_{70}^{-1}$ kpc from the nucleus of NGC 3067; the sightline clearly penetrates extra-planar H I, seen in 21 cm emission with the VLA (Carilli & van Gorkom 1992). The H I column densities, kinematics, metallicity, spin temperature, and inferred size of these clouds are similar to high velocity clouds in the Milky Way. Thus, despite a modest starburst ($\geq 0.6\ M_\odot$ yr^{-1}) in this 0.5 L^* galaxy, even the high ionization gas does not appear to be escaping. In summary, both the detailed *HST* and ground-based observations of this nearby LLS and the cumulative data from the Steidel LLS sample are consistent with LLSs being the bound, gaseous halos of luminous galaxies.

4.3. *Weak Metal-line systems*

Both the *HST* QSO Absorption Line Key Project and our GHRS/STIS Lyα survey of Papers I–IV have found large numbers of intermediate column density ($N_{\mathrm{HI}} = 10^{13-16}$ cm^{-2}) Lyα absorbers at low redshift. Available *HST* or *FUSE* spectra are not sensitive enough to detect metals (C II/III/IV, Si II/III/IV, O VI) in these absorbers, except in a few cases at $N_{\mathrm{HI}} = 10^{15-16}$ cm^{-2}. For example, metallicities have been measured to 1–10% of solar values in strong Lyα absorbers in the sightlines toward 3C 273 (Sembach et al. 2001; Tripp et al. 2002), PKS 2155-304 (Shull et al. 1998, 2003), and PG 1211+143 (Tumlinson et al. 2004). In an ongoing survey, Danforth & Shull (2004, in preparation) have found that ∼50% of all Lyα absorbers contain both H I and O VI in the N_{HI} range for which high-quality *FUSE* spectra exist. The "multiphase ratio," N(H I)/N(O VI), varies substantially, from ∼ 0.1 to greater than 100 (Shull 2003; Danforth & Shull 2004), probably indicating a wide range in shock velocities, and possibly O/H abundances, in these multiphase systems.

Given these statistics, it is conceivable that a large fraction of $z \approx 0$ Lyα absorbers at $N_{\mathrm{HI}} = 10^{13-17}$ cm^{-2} contain metals. The O VI non-detections are either at lower ionization (lower shock velocity) than allows significant columns of O VI, or at slightly lower metallicity than the *FUSE* sensitivity limits can detect. Since most or all Lyα absorbers at $z = 2$–4 in this column density range contain C IV and/or O VI absorption lines at metallicities $[Z] = -1$ to -3 (e.g., Schaye et al. 2000), it is not surprising that many low column density, low-z absorbers would contain some metals. Thus, some relationship with nearby galaxies is expected. Figure 4 shows the galaxies surrounding the sightline to PKS 2155–304, coincident with a cluster of strong Lyα absorbers, in which C IV or Si III have not been detected to a limit of ∼3% solar metallicity (there are two weak O VI absorbers detected by *FUSE*).

Lanzetta et al. (1995) and Chen et al. (1998, 2001) obtained deep images and multi-object spectroscopy for a large number of galaxies in the fields of Key Project absorbers. Lanzetta et al. (2002) summarizes the evidence for a ∼30% success rate in matching redshifts of absorbers and luminous galaxies within $240h_{70}^{-1}$ kpc of the sightline. They then extrapolate to the conclusion that all FOS-discovered Lyα absorbers can be associated with very extended galaxy halos. Chen et al. (1998) found that, with the exception of a weak dependence on galaxy luminosity, there is no nearest-galaxy property that correlates with absorber properties. On the other hand, using Lyα absorbers at somewhat lower

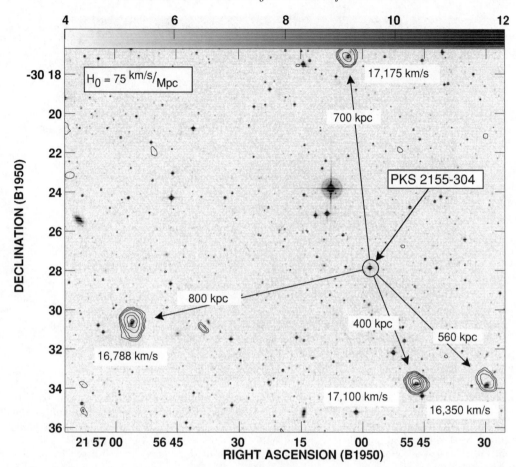

FIGURE 4. Optical and VLA field toward PKS 2155−304 (Shull et al. 1998, 2003) showing galaxies located at velocities (16,350–17,175 km s^{-1}) similar to a cluster of 7 Lyα absorbers. These H I galaxies are located at projected offsets of $(400–800)h_{75}^{-1}$ kpc, and the group has an overdensity $\delta \approx 10^2$. The multiphase gaseous medium has been detected in H I (Lyα, Lyβ, Lyγ), O VI, and perhaps O VII (the claimed X-ray detection by Fang et al. 2002 has not been confirmed).

column densities, Morris et al. (1993), Tripp, Lu, & Savage (1998), Impey, Petry, & Flint (1999) and our Papers III and V found no convincing statistical evidence that these GHRS- and STIS-discovered absorbers are associated with individual galaxies down to 0.1 L^* luminosities. However, over three-quarters of these weak absorbers are found in galaxy large-scale structure filaments (Paper III), with bright galaxies several hundred kpc away. Stocke (2002) summarizes the evidence in favor of this position. At the extremes of the column-density range are obvious examples that both of these positions are correct for some absorbers; at the high-N_{HI} end of this range, LLSs are so close to the nearby bright galaxy that an association of some sort seems inevitable.

On the other hand, an increasing percentage of lower column density absorbers are found in galaxy voids (Papers III and V), more than $3h_{70}^{-1}$ Mpc from the nearest bright or faint galaxy (McLin et al. 2001). At intermediate N_{HI}, the absorbers adjacent to gas-rich galaxies do not have velocities consistent with rotation curves of the nearby galaxy. This suggests a kinematic link to the nearby galaxy (Côté et al. 2002; Putman et al. 2004), in contrast to the Steidel et al. (2002) kinematic analysis of LLSs and their

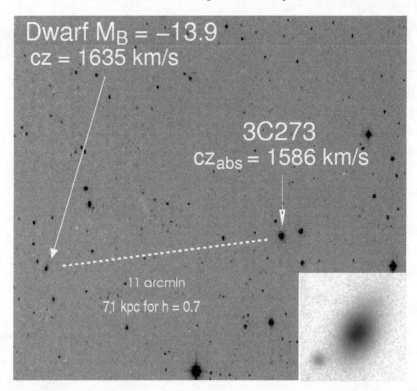

FIGURE 5. Digitized Sky Survey image of the region around 3C 273 (north up, east to left) showing the dwarf galaxy located $\sim 70 h_{70}^{-1}$ kpc away from the absorber on the sky and at the same recession velocity to within the errors (± 50 km s^{-1}); see Stocke et al. 2004. The R-band image of the dwarf galaxy taken with the ARC 3.5m telescope at Apache Point is shown as an insert at lower right. The surface brightness profile of this galaxy is well-fitted by a pure exponential disk.

associated galaxies that often shows a relationship between absorber metal-line velocities and the kinematics of the underlying stellar disk.

Recently, the discovery (Stocke et al. 2004) of a dwarf ($M_B = -14$, $L = 0.004\ L^*$) post-starburst galaxy $70 h_{70}^{-1}$ kpc from a weak-metal line absorber (Figure 5) with $N_{\rm HI} \approx 10^{16}$ cm^{-2} suggests a solution compatible with all the data presented above. Not only do the absorber and galaxy redshifts match to within their combined errors, but the absorber metallicity (6% solar) and the mean stellar metallicity of the galaxy ($\sim 10\%$ solar) approximately match. Further, the absorber has an overabundance of silicon to carbon indicative of recent supernova type II enrichment. The galaxy is a pure disk system whose optical spectrum has both strong Balmer and metal absorption lines, and no evidence for dust or gas (no emission lines and $M_{\rm HI} \leq 5 \times 10^6\ M_\odot$; van Gorkom et al. 1993). From ratios of Lick absorption-line indices, we estimate that the mean stellar age in this galaxy is 3.5 ± 1.5 Gyrs. Taken together, this information provides a consistent picture in which a massive ($\geq 0.3\ M_\odot$ yr^{-1}) starburst ~ 3 Gyrs ago created enough supernovae to blow away the remaining gas from this galaxy into the IGM. Because the dwarf is quite low mass, this wind can easily escape from the galaxy and move to $\sim 100 h_{70}^{-1}$ kpc at the required 20–30 km s^{-1} to create the metal-line absorber that we see between us and 3C 273.

Because this is the nearest weak metal-line system known, many other similar absorbers could arise from starburst winds produced by dwarf galaxies too faint to be detected by the above surveys (at $z \approx 0.1$, a galaxy like this dwarf would have $m_B \sim 24$). This explains why Lanzetta et al. (1995) found bright galaxies $(100\text{–}300)h_{70}^{-1}$ kpc away that do not have properties that correlate with absorber properties. These absorbers were produced not by the bright galaxy, but by a dwarf or dwarfs that accompany the bright galaxy. This also explains the correlation between absorber locations and large-scale structure in the samples of Paper III and Impey, Petry & Flint (1999). Because dwarf galaxies are so numerous $(1\text{–}3\ \mathrm{Mpc}^{-3})$, if each dwarf galaxy had at least one massive starburst that ejected most or all of the gas from the galaxy to a distance of $\sim 100h_{70}^{-1}$ kpc, this process would be sufficient to create several hundred weak metal-line absorbers per unit redshift. This number is comparable to the line density of all metal-line absorbers at $z = 2$. While this dwarf galaxy would have been much more luminous ($M_B \approx -16.5$) when it was "starbursting," the present-day absence of gas means that it will no longer form stars and will eventually fade to the approximate luminosity of a Local Group dwarf spheroidal. Thus, the dwarf in Figure 5 is the expected intermediate stage between the "faint blue galaxies" seen at $z = 0.5\text{–}1$ and the present-day dwarf spheroids (Babul & Rees 1992). That such "Cheshire Cat" galaxies might be responsible for a large portion of the Lyα forest was suggested by Salpeter (1993) and Charlton (1995).

However attractive this solution appears—based upon this one example—an equally plausible solution consistent with all available data is that large-scale filaments of galaxies are enriched with metals throughout the filament to approximately the same metallicity. In this picture, metal enrichment is not due to any one galaxy. Indeed, in some environments, this almost has to be the case. In the Virgo Cluster, studied through QSO absorption-line probes by Impey, Petry & Flint (1999), no nearby dwarf galaxies were found, despite a galaxy redshift survey complete to $M_B = -16$. In some cases (the $cz = 1100$ km s^{-1} H I + O VI absorber toward 3C 273, Tripp et al. 2002; and the 1666 km s^{-1} metal-line absorber toward RXJ 1230.8+0115, Rosenberg et al. 2003) weak metal-line absorbers have no nearby ($\leq 100h_{70}^{-1}$ kpc) galaxies, despite a considerable redshift survey effort. While our galaxy surveying continues in these directions, these two hypotheses may not be easy to discriminate. However, a statistical study of the distances that metal-enriched gas extends away from galaxies may help to select the better model.

4.4. *Lyα-only Absorbers*

Many of the lowest-N_{HI} absorbers could be metal-free, a hypothesis consistent with all available data at high and low redshift. Although *HST* does not have the sensitivity to address this question directly, there is considerable circumstantial evidence in favor of this hypothesis. While $\sim 80\%$ of all low column density Lyα absorbers are found in galaxy filaments, the remaining 20% and an increasing percentage of absorbers with decreasing column densities, are found $> 3h_{70}^{-1}$ Mpc from the nearest bright galaxy. A deep survey of the few nearest examples of "void absorbers" (McLin et al. 2002) found no faint galaxies ($M_B \leq -12$ to -13.5, depending on sightline) within several hundred kpc of these absorbers. Thus, there is no plausible nearby source for metals in these absorbers. In Paper IV, we presented a two-point correlation function (TPCF) in absorber velocity separation that showed excess power at $\Delta v \leq 250$ km s^{-1}, but only for the higher column density absorbers in our sample ($W_\lambda \geq 65$ mÅ). The lower column density absorbers (log $N_{\mathrm{HI}} \leq 13$) appear to be randomly distributed in space with no excess power at any Δv (Paper IV). Searches for metals in the highest column density "void absorbers" are capable of setting metallicity limits at $[Z] \approx -2$ (B. Keeney 2004, private communication), but some high-z absorbers contain metals at even lower metallicities. Therefore, even deeper *HST* observations of very bright targets are required to test this interesting hypothesis.

5. The future

When *HST* Servicing Mission 4 was still scheduled, the installation of the Cosmic Origins Spectrograph (COS) on *HST* was eagerly awaited by astronomers who study the local IGM. With 10–20 times the far-UV throughput of STIS at comparable (15 km s^{-1}) resolution, COS would revolutionize low-z Lyα absorption work by allowing us to obtain high-resolution, high signal-to-noise ratio spectra of 16th–17th magnitude QSOs in a few orbits. STIS studies of the IGM require ever-increasing numbers of orbits to address new science goals. Below is a sampling of the questions that either STIS or COS can address; those that truly require COS are noted with asterisks (*):

(*a*) **Baryons in the WHIM**. How many are there? Where are they relative to galaxies? What are the ionization processes that create the observed absorption-line signatures?

(*b*) **Galaxy halo sizes, shapes, metallicities and kinematics**.* A few of the brightest examples can be attempted with STIS, but to provide gaseous halo properties as a function of galaxy type and luminosity is clearly a COS project.

(*c*) **Low column density Lyα absorbers**.* How many baryons do these absorbing clouds contain? Do they contain metals, and at what metallicity? This is a particularly interesting question for those absorbers in voids, which are the most obvious locations to search for primordial hydrogen clouds, if they exist. Their metals provide an important "fossil record" of very early star formation.

(*d*) **Extent of metal transport away from galaxies**.* How far away from star formation sites do metals spread in the IGM? What types of galaxies are responsible for enriching the IGM in metals? Can details of nucleosynthesis be found in the IGM to understand early star formation history and evolution?

(*e*) **Accurate H I column densities**. *FUSE* observations of higher Lyman lines can be used to obtain accurate H I column densities by curve-of-growth techniques. These are also essential for accurate metallicities. Therefore, this goal is best carried out while *FUSE* and *HST* are both still operational.

(*f*) **Physical properties of IGM absorbers**. The column density distribution, b-value distribution (temperature), detailed relationship to galaxies and other properties are still affected by cosmic variance. Obtaining 10–20 high signal-to-noise STIS medium-resolution echelle spectra or a similar number of high-quality COS spectra of $z \approx 0.5$ QSOs would be sufficient to address concerns about cosmic variance for Lyα absorbers, but not for the rarer O VI absorbers. For O VI, *FUSE* spectroscopy of already detected Lyα absorption lines appears to be a viable route to obtaining a sufficient sample. One could also use *HST*/STIS data on QSO targets at $z \geq 0.11$–0.12 (to shift O VI λ1031.926 or λ1037.617 into the *HST*/STIS band). This project also means observing fainter targets for many more orbits.

(*g*) **Size of IGM absorbers in galaxy filaments and voids**.* QSO pairs available to perform these observations are too faint to observe adequately with STIS.

As nearly everyone has noticed over the past decade, the IGM has attracted considerable interest from the fields of cosmology, galaxy formation, and galactic chemical evolution. Even at low redshift, a substantial fraction (30–60%) of the baryons reside in the IGM, where they influence the mass infall and chemical history of gas in galaxies, including the Milky Way. If this "cosmic web" of multi-phase (photoionized, collisionally ionized) gas truly represents a significant baryon reservoir (Shull 2003; Sembach 2003), one of the best scientific legacies from *Hubble* (and *FUSE*) will be an accurate characterization of its features.

Financial support for the local Lyα forest work at the University of Colorado comes from grants provided through *HST* GO programs 6593, 8182, 8125, 8571, 9221, 9506, 9520, and 9778 from the *FUSE* Project (NASA contract NAS5-32985, grant NAG5-13004), and theoretical grants from NASA/LTSA (NAG5-7262) and NSF (AST02-06042). We thank Ray Weymann for the faint-galaxy redshift survey work and his inspirational leadership in this field. We thank Mark Giroux for ionization modeling and insightful comments. We thank our other collaborators, Brian Keeney, Jessica Rosenberg, Jason Tumlinson, John Hibbard, Charles Danforth, and Jacqueline van Gorkom for valuable contributions and discussions, and F. Nicastro and B. Keeney for permission to quote their results before publication.

REFERENCES

BABUL, A. & REES, M. J. 1992 *MNRAS* **255**, 346.

BAHCALL, J. N., ET AL. 1991 *ApJ* **377**, L5.

BAHCALL, J. N., ET AL. 1993 *ApJS* **87**, 1.

BAHCALL, J. N., ET AL. 1996 *ApJ* **457**, 19.

BECHTOLD, J., ET AL. 1994 *ApJ* **437**, L38.

BERGERON, J. & BOISSÉ, P. 1991 *A&A* **243**, 344.

BOISSIER, S., PÉROUX, C., & PETTINI, M. 2003 *MNRAS* **338**, 131.

BOWEN, D. V., BLADES, J. C., & PETTINI, M. 1996, *ApJ* **464**, 141.

BOWEN, D. V. & PETTINI, M. 2003. In *The IGM/Galaxy Connection* (eds. J. L. Rosenberg & M. E. Putman). p. 133. Kluwer.

BOWEN, D. V., PETTINI, M., & BLADES, J. C. 2002 *ApJ* **580**, 169.

BOWEN, D. V., TRIPP, T. M., & JENKINS, E. B. 2001 *AJ* **121**, 1456.

BURLES, S. & TYTLER, D. 1998 *Space Sci. Rev.* **84**, 65.

CAGNONI, I., NICASTRO, F., MARASCHI, L., TREVES, A., & TAVECCHIO, 2004, *ApJ* **603**, 449.

CARILLI, C. L. & VAN GORKOM, J. H. 1992 *ApJ* **399**, 373.

CEN, R. & OSTRIKER, J. P. 1999 *ApJ* **519**, L109.

CEN, R. & SIMCOE, R. A. 1997 *ApJ* **483**, 8.

CEN, R., TRIPP, T. M., OSTRIKER, J. P., & JENKINS, E. B. 2001, *ApJ* **559**, L5.

CHARLTON, J. C. 1995. In *QSO Absorption Lines* (ed. G. Meylan). p. 405. Springer.

CHEN, H.-W., ET AL. 1998 *ApJ* **498**, 77.

CHEN, H.-W., ET AL. 2001 *ApJ* **559**, 654.

CHURCHILL, C. W., ET AL. 2000 *ApJ* **543**, 577.

COLLINS, J. A., SHULL, J. M., & GIROUX, M. L. 2004 *ApJ* **605**, 216.

CÔTÉ, S., ET AL. 2002. In *The Dynamics, Structure, & History of Galaxies* (ed. G. S. Da Costa & H. Jerjen). Vol. 254, 353. ASP.

DAVÉ, R., HERNQUIST, L., KATZ, N., & WEINBERG, D. H. 1999 *ApJ* **511**, 521.

DAVÉ, R., ET AL. 2001, *ApJ* **552**, 473.

DAVÉ, R. & TRIPP, T. M. 2001 *ApJ* **553**, 528.

DINSHAW, N., WEYMANN, R. J., IMPEY, C. D., FOLTZ, C. B., MORRIS, S. L., & AKE, T. 1997 *ApJ* **491**, 45.

FANG, T.-T. & BRYAN, G. L. 2001 *ApJ* **561**, L31.

FANG, T.-T., CANIZARES, C., ET AL. 2002 *ApJ* **572**, L127.

FOX, A. J., ET AL. 2004 *ApJ* **602**, 738.

FUKUGITA, M. 2004. In *Dark Matter in Galaxies* (eds. S. D. Ryder, D. J. Pisano, M. A. Walker, & K. C. Freeman). IAU Symposium 220, p. 227. ASP.

GNEDIN, N. Y. & OSTRIKER, J. P. 1997 *ApJ* **486**, 581.

IMPEY, C. D., PETRY, C. E., & FLINT, K. P. 1999 *ApJ* **524**, 536.

JANKNECHT, E., BAADE, R., & REIMERS, D. 2002 *A&A* **391**, 11.

JANNUZI, B. T., ET AL. 1998 *ApJS* **118**, 1.

KIM, T.-S., CRISTIANI, S., & D'ODORICO, S. 2001 *A&A* **373**, 757.

KEENEY, B. A., STOCKE, J. T., MOMJIAN, E., CARILLI, C. L., & TUMLINSON, J. 2005, in preparation

LANZETTA, K. M., ET AL. 2002. In *Extragalactic Gas at Low Redshift* (eds. J. Mulchaey & J. Stocke). ASP Conf. Proc. vol. 254, p. 78. ASP.

LANZETTA, K. M., BOWEN, D. V., TYTLER, D., & WEBB, J. K. 1995, *ApJ* **442**, 538.

MATHUR, S., WEINBERG, D. H., & CHEN, X. 2003. In *The IGM/Galaxy Connection* (eds. J. L. Rosenberg & M. E. Putman). p. 103. Kluwer.

McKERNAN, B., YAQOOB, T., MUSHOTZKY, R., GEORGE, I. M., & TURNER, T. J. 2003 *ApJ*, **598**, L83.

McLIN, K. M., STOCKE, J. T., WEYMANN, R. J., PENTON, S. V., & SHULL, J. M. 2002 *ApJ* **574**, L115.

MORRIS, S. L., WEYMANN, R. J., SAVAGE, B. D., & GILLILAND, R. L. 1991 *ApJ* **377**, L21.

MORRIS, S. L., ET AL. 1993 *ApJ* **419**, 524.

MULCHAEY, J. S. & STOCKE, J. T., EDS. 2002 *Extragalactic Gas at Low Redshift*. ASP Conf. Proc. vol. 254.

NESTOR, D. B., RAO, S. M., ET AL. 2002. In *Extragalactic Gas at Low Redshift* (eds. J. Mulchaey & J. Stocke. ASP Conf. Ser. vol. 254, p. 34. ASP.

NICASTRO, F. 2003. In *The IGM/Galaxy Connection* (eds. J. L. Rosenberg & M. E. Putman). p. 277. Kluwer.

NICASTRO, F., ET AL. 2004 *ApJ*; submitted.

PENTON, S. V., STOCKE, J. T., & SHULL, J. M. 2000a *ApJS* **130**, 121 (Paper I),

PENTON, S. V., SHULL, J. M., & STOCKE, J. T. 2000b *ApJ* **544**, 150 (Paper II).

PENTON, S. V., STOCKE, J. T., & SHULL, J. M. 2002 *ApJ* **565**, 720 (Paper III).

PENTON, S. V., STOCKE, J. T., & SHULL, J. M. 2004 *ApJS* **152**, 29 (Paper IV).

PETTINI, M., ELLISON, S. L., STEIDEL, C. C., & BOWEN, D. V. 1999 *ApJ* **510**, 576.

PROCHASKA, J. X. & WOLFE, A. M. 2000 *ApJ* **533**, L5.

PROCHASKA, J. X. & WOLFE, A. M. 2002 *ApJ* **566**, 68.

PUTMAN, M. E., ROSENBERG, J. L, RYAN-WEBER, E. V., & STOCKE, J. T. 2004 *BAAS* **204**, 7902.

RAO, S. M. & BRIGGS, F. 1993 *ApJ* **419**, 515.

RAO, S. M. & TURNSHEK, D. A. 1998 *ApJ* **500**, L11.

RAO, S. M. & TURNSHEK, D. A. 2000 *ApJS* **130**, 1.

RAO, S. M., ET AL. 2003 *ApJ* **595**, 94.

RASMUSSEN, A., KAHN, S. M., & PAERELS, F. 2003. In *The IGM/Galaxy Connection* (eds. J. L. Rosenberg & M. E. Putman). p. 109. Kluwer.

RAUCH, M., ET AL. 1997 *ApJ* **489**, 7.

RICOTTI, M., GNEDIN, N. Y., & SHULL, J. M. 2000 *ApJ* **534**, 41.

ROSENBERG, J. L. & SCHNEIDER, S. E. 2003 *ApJ* **585**, 256.

ROSENBERG, J. L., GANGULY, R., GIROUX, M. L., & STOCKE, J. T. 2003 *ApJ* **591**, 677.

SALPETER, E. E. 1993 *AJ* **106**, 1265.

SALUCCI, P. & PERSIC, M. 1999 *MNRAS* **309**, 923.

SAVAGE, B. D., SEMBACH, K. R, TRIPP, T. M, & RICHTER, P. 2002 *ApJ* **564**, 631.

SAVAGE, B. D., LEHNER, N., WAKKER, B. P., SEMBACH, K. R., & TRIPP, T. M. 2004 *ApJ*; submitted.

SAVAGLIO, S., ET AL. 1999 *ApJ* **515**, L5.

SCHAYE, J. 2001 *ApJ* **559**, 507.

SCHAYE, J., RAUCH, M., SARGENT, W. L. W., & KIM, T-S. 2000 *ApJ* **541**, L1.

SEMBACH, K. R., HOWK, J. C., SAVAGE, B. D., SHULL, J. M., & OEGERLE, W. R. 2001 *ApJ* **561**, 573.

SEMBACH, K. R. 2002. In *Extragalactic Gas at Low Redshift* (eds. J. Mulchaey & J. Stocke). ASP Conf. Ser. vol. 254, p. 283. ASP.

SEMBACH, K. R., TRIPP, T. M., SAVAGE, B. D., & RICHTER, P. 2004 *ApJS* **155**, 351.

SHULL, J. M., STOCKE, J. T., & PENTON, S. 1996 *AJ* **111**, 72.

SHULL, J. M., ET AL. 1998 *AJ* **116**, 2094.

SHULL, J. M., ET AL. 1999 *AJ* **118**, 1450.

SHULL, J. M., ET AL. 2000 *ApJ* **538**, L13.

SHULL, J. M., ET AL. 2003 *ApJ* **594**, L107.

SHULL, J. M. 2003, In *The IGM/Galaxy Connection* (eds. J. L. Rosenberg & M. E. Putman). p. 1. Kluwer.

SPERGEL, D. N., ET AL. 2003 *ApJS* **148**, 175.

STEIDEL, C. C. 1995. In *QSO Absorption Lines* (ed. G. Meylan). p. 139. Springer.

STEIDEL, C. C. 1998. In *Galaxy Halos* (ed. D. Zaritsky). ASP Conf. Ser. vol. 136, p. 167. ASP.

STEIDEL, C. C., ET AL. 2002 *ApJ* **570**, 526.

STOCKE, J. T. 2002. In *Extragalactic Gas at Low Redshift* (eds. J. Mulchaey & J. Stocke). ASP Conf. Ser. vol. 254, p. 98. ASP.

STOCKE, J. T., CASE, J., DONAHUE, M., SHULL, J. M., & SNOW, T. P. 1991 *ApJ* **374**, 72.

STOCKE, J. T., SHULL, J. M., PENTON, S. V., DONAHUE, M., & CARILLI, C. 1995 *ApJ* **451**, 24.

STOCKE, J. T., KEENEY, B. A., MCLIN, K. M., ROSENBERG, J. L., WEYMANN, R. J., & GIROUX, M. L. 2004 *ApJ* **609**, 94.

STOCKE, J. T., PENTON, S. V., MCLIN, K. M., SHULL, J. M., & WEYMANN, R. J. 2005, in preparation.

TELFER, R., ZHENG, W., KRISS, G. A., & DAVIDSEN, A. F. 2002 *ApJ* **565**, 773.

TRIPP, T. M., LU, L., & SAVAGE, B. D. 1998 *ApJ* **508**, 200.

TRIPP, T. M., SAVAGE, B. D., & JENKINS, E. B. 2000 *ApJ* **534**, L1.

TRIPP, T. M., ET AL. 2001 *ApJ* **563**, 724.

TRIPP, T. M., ET AL. 2002 *ApJ* **575**, 697.

TUMLINSON, J., GIROUX, M. L., SHULL, J. M., & STOCKE, J. T. 1999 *AJ* **118**, 2148.

TUMLINSON, J., SHULL, J. M., GIROUX, M. L., & STOCKE, J. T. 2005 *ApJ* **620**, 95.

TURNSHEK, D. A., RAO, S. M., & NESTOR, D. B. 2002. In *Extragalactic Gas at Low Redshift* (eds. J. Mulchaey & J. Stocke). ASP Conf. Ser. vol. 254, p. 42. ASP.

VAN GORKOM, J. H., BAHCALL, J. N., JANNUZI, B. T., & SCHNEIDER, D. P. 1993 *AJ* **106**, 2213.

WEYMANN, R. J., ET AL. 1998 *ApJ* **506**, 1.

WOLFE, A. M. & PROCHASKA, J. X. 2003 *ApJ* **545**, 603.

ZWAAN, M. A., BRIGGS, F., & SPRAYBERRY, D. 2001 *MNRAS* **327**, 124.

Hot baryons in supercluster filaments

By ERIC D. MILLER, RENATO A. DUPKE,
AND JOEL N. BREGMAN

Department of Astronomy, University of Michigan, Ann Arbor, MI 48109, USA

Most of the baryons in the local universe are "missing" in that they are not in galaxies or in the previously detected gaseous phases. These missing baryons are predicted to be in a moderately hot phase, 10^5–10^7 K, largely in the form of giant cosmic filaments that connect the denser virialized clusters and groups of galaxies. Models show that the highest covering fraction of such filaments occurs in superclusters. To determine whether such filaments exist, we have begun a project to search for UV absorption against AGNs projected behind possible supercluster filaments. Using data from the *HST* and *FUSE* archives along with new observations, we have detected UV absorption within about 1300 km s^{-1} of seven supercluster sightlines out of a sample of eight. The likelihood of such detections being generated by chance is less than 10^{-4}.

1. Introduction

A census of baryons in the local universe indicates that the majority of this normal matter is undetected, or "missing." At high redshifts ($z \sim 3$), big-bang nucleosynthesis models and QSO absorption line observations indicate a baryon mass fraction of $\Omega_b \sim 0.04$ (e.g., Fukugita, Hogan & Peebles 1998). The stars and gas detected in local galaxies account for only 20% of this ($\Omega_b \sim 0.008$). The absence of a local Lyα forest indicates that these baryons are likely in a hot ($T > 10^5$ K), diffuse medium which has heretofore remained undetectable (e.g., Fukugita, Hogan & Peebles 1998; Cen & Ostriker 1999a; Davé et al. 2001). A reservoir of hot gas would be organized similarly to the collapsed structure, i.e., into a web of filaments like those seen in structure formation simulations (e.g., Evrard et al. 2002). Gas is shock heated as it collapses into structures, and at the nodes of this web are found the rich clusters of galaxies. These contain large concentrations of hot gas seen in X-ray emission ($T \sim 10^7$–10^8). Galaxy groups are found along the filaments connecting nodes, and while several groups are detected in diffuse X-ray emission ($T \sim 10^7$ K), the extent of this gaseous component is not well-constrained. The filaments themselves contain most of the volume of collapsed structure, and the temperature in these regions is thought to be lower than that of the denser clusters and groups, or $T \sim 10^5$–10^7 K (e.g., Cen et al. 1995). Recent results have shown possible absorption by this hot cosmic web of baryons against background point sources (Tripp & Savage 2000; Savage et al. 2002; Richter et al. 2004). We have begun a project to (1) identify lines connecting clusters within superclusters and (2) search for absorption against AGNs projected behind these likely filament locations. Results from three AGNs probing four superclusters have been given in detail by Bregman, Dupke & Miller (2004). Here we summarize that study and present preliminary results for four additional AGN/supercluster sightlines. This project makes use of both archival and proprietary *HST* data, using all generations of spectrographs, from FOS to GHRS to STIS. As we primarily search for Lyα $\lambda1216$ absorption at low redshift, this project requires space-based UV spectroscopy and would be impossible without the sensitivity and wavelength coverage of *HST*.

2. Identifying filaments and background AGNs

Filaments in emission are not visible to current instruments, so knowing where to look requires predicting their locations. Structure formation simulations indicate that filaments should connect clusters in approximately straight lines (e.g., Evrard et al. 2002), with the width of the filaments on the order of the virial radius of a typical cluster ($3h^{-1}$ Mpc; Cen & Ostriker 1999b). The supercluster catalog of Einasto et al. (1997), which contains 220 superclusters at $z < 0.12$, allows us to visually identify filaments connecting clusters in space and redshift and cross-correlate these locations with AGNs from the Véron-Cetty & Véron (2001) catalog. Observation with *HST* and *FUSE* necessitates selection of bright ($V < 16$) AGNs, which are typically nearby Seyfert galaxies at $z < 0.2$, and it requires low Galactic extinction (from Schlegel, Finkbeiner & Davis 1998). The background AGNs must be well separated in redshift from the intervening superclusters ($z_{AGN} \gg z_{SC}$), and these combined requirements preferentially select nearby superclusters. The larger solid angle subtended by these nearer systems also increases the likelihood of finding a suitable background AGN within $\sim$3 Mpc of the predicted filament line.

We have identified 11 supercluster sightlines probed by 10 AGNs (PHL 1811 lies behind two distinct superclusters), and these are listed in Table 1. Maps of four superclusters are shown in Figure 1 to demonstrate the predicted filament structure and projected locations of background AGNs. At the time of this writing, seven AGNs (probing eight supercluster sightlines) have been observed by *HST*, and four AGNs (probing five superclusters) have been observed by *FUSE*; these results include a combination of archived and proprietary data. The object Ton S180 fails our $z_{AGN} \gg z_{SC}$ criterion, and we treat it as a special case in the following discussion. Data for four of the AGNs have been discussed in the literature: PHL 1811, studied extensively by Jenkins et al. (2003); PG 1402+261, studied by Bechtold et al. (2002) and Wakker et al. (2003); H1821+643, observed by Tripp, Lu & Savage (1998) and Oegerle et al. (2000); and Ton S180, reported on by a number of authors (Shull et al. 2000; Turner et al. 2002; Wakker et al. 2003; Penton, Stocke & Shull 2004). We have incorporated their results and our own analysis of the data here. The remainder of the AGNs were observed by us for this project with the exception of KAZ 102, which was obtained from the *HST* archive.

3. Absorption line results

The *HST* and *FUSE* spectra together probe a number of high- and low-ionization species, including the strong lines of Lyα λ1216, Lyβ λ1026, O VI $\lambda\lambda$1032,1038, C II λ1335, and C III λ977. At the expected filament temperatures of about 3×10^5 K, the low-ionization metal lines will be absent, but the Ly series of H I will still be present. The relative strengths of the Ly and O VI absorption depend strongly on such parameters as temperature and metallicity. We expect individual filaments to have a velocity dispersion no greater than that of a typical cluster ($\sim$1300 km s^{-1}), centered on the redshift of the nearest clusters, so we have searched for absorption from all available species within $\pm$1300 km s^{-1} of this average velocity.

Absorption from Lyα is detected in seven of the eight superclusters observed with *HST*, and O VI absorption is seen in two of the five sightlines probed with *FUSE*. The Lyα and O VI λ1032 line strengths (or upper limits) are listed in Table 1. Line widths were measured by fitting a Gaussian to the line profile, and from this we obtained a maximum temperature corresponding to the maximum allowed thermal line broadening. As is shown in Table 1, these temperatures range from 10^5–10^6 K; however it must be

AGN	supercluster(s)	z_{AGN}	z_{SC}	HST	FUSE	W_{Ly} (mÅ)	W_{OVI} (mÅ)	T_{max} (10^6 K)
PHL 1811	Aquarius B	0.192	0.084	●	●	300	50	0.2
	Aquarius-Cetus	0.192	0.056	●	●	600	<250	1.0
PG 1402+261	Bootes	0.164	0.068	●	●	300	<100	1.6
Ton S180	Pisces-Cetus	0.062	0.060	●	●	100	150	0.1
KAZ 102	North Ecliptic Pole	0.136	0.087	●	⋯	400	⋯	6.4
H1821+643	North Ecliptic Pole	0.297	0.087	●	●	50	<150	0.4
WGAJ2153	Aquarius-Cetus	0.078	0.056	●	⋯	400	⋯	3.6
RXS J01004-5113	Phoenix	0.062	0.027	●	⋯	<150	⋯	⋯
RXS J01149-4224	Phoenix	0.124	0.027	○	○	⋯	⋯	⋯
HE0348-5353	Horologium-Reticulum	0.130	0.064	○	⋯	⋯	⋯	⋯
TEX 1601+160	Hercules	0.109	0.035	○	⋯	⋯	⋯	⋯
Ton 730	Bootes	0.087	0.065	○	○	⋯	⋯	⋯

TABLE 1. Summary of AGN sightlines and detected absorption by intervening supercluster filaments. Filled circles indicate that data have been obtained for the given instrument, while open circles indicate targets that have been approved. The last column shows the upper limit on the temperature of the absorbing medium, using the width of the narrowest absorption feature as an upper limit to thermal line broadening.

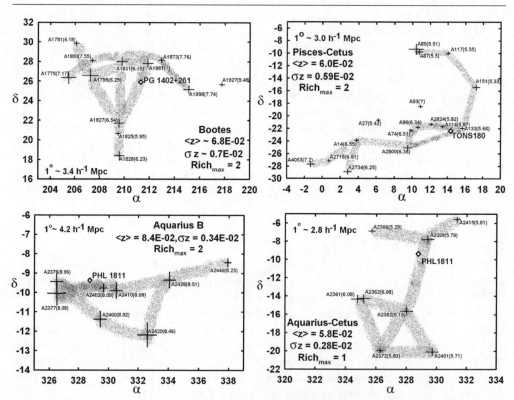

FIGURE 1. Spatial configuration of background AGNs (open diamonds) projected onto four superclusters. The crosses are galaxy clusters (many with Abell names), where the size of the cross indicates the relative richness. The linear grey stripes are connections between galaxy clusters that are possible filaments projected near the background AGN, based upon three-dimensional proximity and cluster richness; these filaments are 3 Mpc wide. We also indicate the redshift of the clusters in parenthesis (in units of 10^{-2}), the average redshift of the supercluster, and the standard deviation of cluster redshifts in the supercluster.

stressed that these are upper limits and it is possible the lines are composed of multiple, blended components. Two sample spectra are presented in Figures 2 and 3, showing Lyα absorption near the redshift of the Bootes supercluster (against PG 1402+261) and multiple Lyβ systems near the redshift of the Aquarius B supercluster (against PHL 1811).

4. Galaxy halo absorption?

It has been suggested that low-z Lyα absorbers are associated with galaxies, and specifically that all strong Lyα systems (with $W_\lambda \geqslant 240$ mÅ) arise from absorption within extended gaseous halos of galaxies near the line of sight (Lanzetta et al. 1995; Chen et al. 1998; Chen et al. 2001). Under this model, and given the W_λ measurements shown in Table 1, it is possible some of the observed absorption is due to intervening galaxies and not filaments.

One questionable case is the absorption against PHL 1811 near the redshift of the Aquarius B supercluster. Four Lyβ systems are seen by Jenkins et al. (2003) at redshifts of 0.0735, 0.0778, 0.0790 and 0.0809 (see Figure 3), with O VI observed at two of these redshifts, 0.0778 and 0.0809. Jenkins et al. (2003) identify an L^* galaxy at the redshift of the 0.0809 system (a Lyman limit system), projected 34 kpc away. The other three systems are shifted 580–2250 km s^{-1} in velocity and unlikely to be associated with gravitationally bound material in this galaxy. The authors suggest this material is due to tidal interactions, but the lack of tidal features in the LLS galaxy make this unlikely. The lack of galaxies at the redshifts of these systems leads us to suggest they are associated with the supercluster filament (in which the galaxy is embedded).

A second case is that of Ton S180, a Seyfert 1.2 galaxy at a similar redshift to (and possibly probing) the Pisces-Cetus supercluster. Three previously-unreported Lyα systems are seen and matched with O VI absorption at similar redshifts. However, it is impossible to determine the site of absorption, since it may be produced by the filament or by the AGN itself.

The remaining five absorption systems are not associated with halos of known galaxies. They have no apparent galaxies within 200 h^{-1} kpc, although this analysis is preliminary and based on shallow Digitized Sky Survey images. Further deep imaging around these sightlines is needed. Recent statistical studies suggest that the proposed absorber-galaxy link is valid only for strong Lyα absorbers ($W_\lambda \geqslant 400$ mÅ), and that most local absorption systems are associated with filaments (Penton, Stocke & Shull 2002, 2004).

5. Random, unassociated Lyα clouds?

It is possible that these absorbers are associated with neither filaments nor intervening galaxies, and by chance appear at the redshifts where we search. Without worrying about their physical nature, only that they are part of the ensemble of low-z Lyα systems well studied in the literature, we can derive the probably of detecting the number we have in the regions we have searched.

For a velocity range $\pm\Delta v$, the expected number of absorbers is $\mu = 2(\Delta v/c)(dN/dz)$, where dN/dz is the frequency of absorbers with redshift. This number has been estimated to be ∼28 for $W_\lambda(\text{Ly}\alpha) \geqslant 240$ mÅ and ∼15 for $W_\lambda(\text{Ly}\alpha) \geqslant 360$ mÅ (e.g., Dobrzycki et al. 2002). If we assume the absorbers are Poisson distributed in redshift, and use $dN(\text{Ly}\alpha)/dz = 28$ for a mean occurrence of $\mu = 0.24$ per supercluster redshift range, then there is a 21% chance of finding at least one absorber within a given 2600 km s^{-1} section of a pencil-beam spectrum. Excluding Ton S180 completely, and calling the W_λ

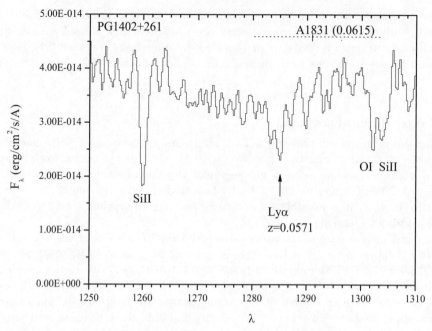

FIGURE 2. *HST*/FOS spectrum of PG 1402+261 showing absorption due to Galactic metal lines and a Lyα system near the Bootes supercluster redshift. The dotted line shows the expected region of absorption, within ± 1300 km s^{-1} of the supercluster redshift.

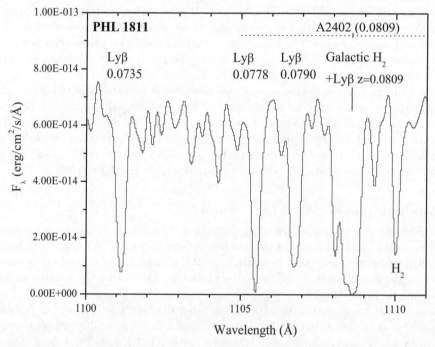

FIGURE 3. *FUSE* spectrum of PHL 1811 showing four Lyβ systems identified by Jenkins et al. (2003), three of which are near the redshift of the Aquarius B supercluster. Absorption from Galactic H$_2$ is also evident.

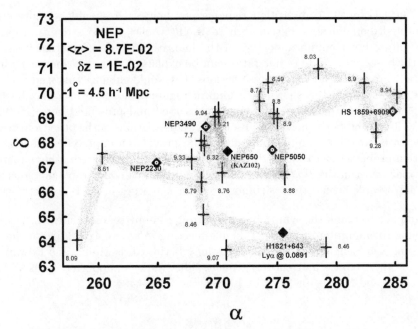

FIGURE 4. Map of the North Ecliptic Pole supercluster, with the locations of six background AGNs marked. Two of the AGNs (solid diamonds) show Lyα absorption and are discussed in this work. The remaining four (open diamonds) have not been observed.

< 240 mÅ system toward H 1821+643 a non-detection, this leads to a 0.6% probability of detecting Lyα absorption in five of seven trials (i.e., five of seven supercluster sightlines). If we include Ton S180 and H 1821+643, the probability drops to 0.01%.

The probability that these are unassociated clouds is likely much lower than this. We see three distinct absorption systems in both the Aquarius B and Pisces-Cetus superclusters, and as the dN/dz studies would count these as separate absorbers in their calculation, a more accurate figure is found by including these as separate components. This adds three additional absorbers (since the LLS is likely associated with a galaxy) and results in a probability of less than 10^{-5} that these are chance detections.

6. Conclusions and future work

We have detected absorption due to possible filaments along seven of eight supercluster sightlines. From a statistical analysis, it is likely that these absorbers are associated with the supercluster filaments, although it is unclear at this point whether they are produced by hot diffuse material in the filament or by some other mechanism. Galaxies should trace the filaments, and while we have argued against absorption due to galaxy halos, it is possible we are seeing absorption from galactic winds or ejecta. It can be argued that such material is really part of the diffuse filament gas, as it is gravitationally unbound from the galaxy and should eventually thermalize with a hot ambient medium, if present. Deep imaging of these sightlines will shed light on the presence of nearby galaxies, and datasets such as the Sloan Digital Sky Survey will provide redshifts for such systems and the clusters that lack redshifts.

The temperature of the absorbing gas is unknown at this time. The linewidths for most of the systems are consistent with temperatures up to 10^5–10^6 K, although if multiple velocity components are blended to produce the lines we see, this value will be much lower. The lack of O VI absorption remains puzzling, as it should be quite strong at the expected temperatures. Perhaps we are seeing cooler, denser streams of gas within a more diffuse hot medium; modeling of the ionization state and metallicity is needed before the observed Lyα/O VI line ratios can be explained. For the (uncertain) future of *HST*, there are several additional observations that would enhance this study. The North Ecliptic Pole supercluster, shown as a map in Figure 4, lies in front of six background AGNs. Two of those have been previously observed and presented here (H 1821+643 and KAZ 102), and both show Lyα absorption. Observations of the other four would not only increase our target sample, but they would provide an unprecedented study of the absorption properties across a single supercluster. Sightlines through other superclusters undoubtedly exist in the *HST* archive as well, but with no comparable missions planned for the foreseeable future, time is running out for new work in far-UV spectroscopy.

We would like to thank the symposium organizing committee, especially Mario Livio, for assembling such an interesting and valuable meeting. Also, we would like to thank B-G. Andersson, Ken Sembach, Bart Wakker, Jimmy Irwin, Chris Mullis, and Edward Lloyd-Davies for their advice and encouragement. This work is supported by NASA grants NAG5-10765 and NAG5-10806.

REFERENCES

BECHTOLD, J., DOBRZYCKI, A., WILDEN, B., MORITA, M., SCOTT, J., DOBRZYCKA, D., TRAN, K.-V, & ALDCROFT, T. L. 2002 *ApJS* **140**, 143.

BREGMAN, J. N., DUPKE, R. A., & MILLER, E. D. 2004 *ApJ*, in press.

CEN, R., KANG, H., OSTRIKER, J. P., & RYU, D. 1995 *ApJ* **451**, 436.

CEN, R. & OSTRIKER, J. P. 1999a *ApJ* **514**, 1.

CEN, R. & OSTRIKER, J. P. 1999b *ApJ* **519**, L109.

CHEN, H.-W., LANZETTA, K. M., WEBB, J. K., & BARCONS, X. 1998 *ApJ* **498**, 77.

CHEN, H.-W., LANZETTA, K. M., WEBB, J. K., & BARCONS, X. 2001 *ApJ* **559**, 654.

DAVÉ, R., CEN, R., OSTRIKER, J. P., BRYAN, G. L., HERNQUIST, L., KATZ, N., WEINBERG, D. H., NORMAN, M. L., & O'SHEA, B. 2001 *ApJ* **552**, 473.

DOBRZYCKI, A., BECHTOLD, J., SCOTT, J., & MORITA, M. 2002 *ApJ* **571**, 654.

EINASTO, M., TAGO, E., JAANISTE, J., EINASTO. J., & ANDERNACH, H. 1997 *A&AS* **123**, 119.

EVRARD, A. E., ET AL. 2002 *ApJ* **573**, 7.

FUKUGITA, M., HOGAN, C. J., & PEEBLES, P. J. E. 1998 *ApJ* **503**, 518.

JENKINS, E. B., BOWEN, D. V., TRIPP, T. M., SEMBACH, K. R., LEIGHLY, K. M., HALPERN, J. P., & LAUROESCH, J. T. 2003 *AJ* **125**, 2824.

LANZETTA, K. M., BOWEN, D. V., TYTLER, D., & WEBB, J. K. 1995 *ApJ* **442**, 538.

OEGERLE, W. R., ET AL. 2000 *ApJ* **538**, L23.

PENTON, S. V., STOCKE, J. T., & SHULL, J. M. 2002 *ApJ* **565**, 720.

PENTON, S. V., STOCKE, J. T., & SHULL, J. M. 2004 *ApJS* **152**, 29.

RICHTER, P., SAVAGE, B. D., TRIPP, T. M., & SEMBACH, K. R. 2004 *ApJS* **153**, 165.

SAVAGE, B. D., SEMBACH, K. R., TRIPP, T. M., & RICHTER, P. 2002 *ApJ* **564**, 631.

SCHLEGEL, D. J., FINKBEINER, D. P., & DAVIS, M. 1998 *ApJ* **500**, 525.

SHULL, J. M., ET AL. 2000 *ApJ* **538**, L13.

TRIPP, T. M., LU, L. , & SAVAGE, B. D. 1998 *ApJ* **508**, 200.

TRIPP, T. M. & SAVAGE, B. D. 2000 *ApJ* **542**, 42.

TURNER, T. J., ET AL. 2002 *ApJ* **568**, 120.

VÉRON-CETTY, M.-P. & VÉRON, P. 2001. *Catalogue of Quasars and Active Nuclei, 10th ed.* ESO.

WAKKER, B. P., ET AL. 2003 *ApJS* **146**, 1.

Galaxy assembly

By ERIC F. BELL

Max-Planck-Institut für Astronomie, Königstuhl 17, D-69117 Heidelberg, Germany;
bell@mpia.de

In a ΛCDM Universe, galaxies grow in mass both through star formation and through the addition of already-formed stars in galaxy mergers. Because of this partial decoupling of these two modes of galaxy growth, I discuss each separately in this biased and incomplete review of galaxy assembly—first giving an overview of the cosmic-averaged star formation history, and then moving on to discuss the importance of major mergers in shaping the properties of present-day massive galaxies. The cosmic-averaged star-formation rate, when integrated, is in reasonable agreement with the build-up of stellar mass density. Roughly 2/3 of all stellar mass is formed during an epoch of rapid star formation prior to $z \sim 1$, with the remaining 1/3 formed in the subsequent 9 Gyr during a period of rapidly-declining star-formation rate. The epoch of important star formation in massive galaxies is essentially over. In contrast, a significant fraction of massive galaxies undergo a major merger at $z \lesssim 1$, as evidenced by close-pair statistics, morphologically-disturbed galaxy counts, and the build-up of stellar mass in morphologically early-type galaxies. Each of these methods is highly uncertain; yet, taken together, it is not implausible that the massive galaxy population is strongly affected by late galaxy mergers, in excellent qualitative agreement with our understanding of galaxy evolution in a ΛCDM Universe.

1. Introduction

The last decade has witnessed amazing progress in our empirical and theoretical understanding of galaxy formation and evolution. This explosion in our understanding has been driven largely by technology: the profusion of 8–10-m class telescopes, the advent of wide-field multi-object spectrographs and imagers on large telescopes, servicing missions for the *Hubble Space Telescope* (*HST*), giving it higher resolution, higher sensitivity, larger field of view, and access to longer wavelengths; and the commissioning and/or launch of powerful observatories in the X-ray, ultraviolet, infrared, and sub-millimeter, are but a few of the important technological advances. This has led to a much-increased empirical understanding of broad phenomenologies: e.g., constraints on the overall shape of the cosmic history of star formation, the increased incidence of star-forming galaxies in less dense environments, the co-evolution of stellar bulges and the supermassive black holes that they host, and the increasing incidence of galaxy interactions at progressively higher redshifts, to name but a few. In turn, tension between these new observational constraints and models of galaxy formation and evolution have spurred on increasingly complex models, giving important (and sometimes predictive!) insight into the physical processes—star formation (SF), feedback, galaxy mergers, AGN activity—that are driving these phenomenologies.

In this review, I will give a grossly incomplete and biased overview of some hopefully interesting aspects of galaxy assembly. An important underpinning of this review is my (perhaps misguided) assumption that we live in a Universe whose broad properties are described reasonably well by the cold dark matter paradigm, with inclusion of a cosmological constant (ΛCDM): $\Omega_{\rm m} = 0.3$, $\Omega_{\Lambda} = 0.7$, and $H_0 = 70\,{\rm km\,s^{-1}\,Mpc^{-1}}$ following results from the *Wilkinson Microwave Anisotropy Probe* (Spergel et al. 2003) and the HST Key Project distance scale (Freedman et al. 2001). This model, while it has important and perplexing fine-tuning problems, seems to describe detection of 'cosmic jerk' using supernova type Ia (Riess et al. 2004), and the evolving clustering of the luminous and dark matter content of the Universe with truly impressive accuracy on a wide range of spatial scales (Seljak et al. 2004).

An important feature of CDM models in general is that galaxies are formed 'bottom-up'; that is, small dark matter halos form first, and halo growth continues to the present through a combination of essentially smooth accretion and mergers of dark matter halos (e.g., Peebles 1980). Thus, small halos were formed very early, whereas larger haloes should still be growing through mergers. To a greater or lesser extent, this has the important feature of decoupling the physics and timing of the formation of the stars in galaxies, and the assembly of the galaxies themselves from their progenitors through galaxy mergers and accretion (White & Rees 1978; White & Frenk 1991).

Accordingly, in this review I strive to explore the two issues separately. First I will explore the build-up of the stellar mass in the Universe, irrespective of how it is split up into individual galaxies. Then I will move on to describing some of the first efforts towards understanding how the galaxies themselves assembled into their present forms. This article will not touch on many important and interesting aspects of galaxy evolution; the co-evolution (or otherwise) of galaxy bulges and their supermassive black holes (see, e.g., Peterson, these proceedings; Ferrarese et al. 2001; Haehnelt 2004), the important influences of local environment on galaxy evolution (see, e.g., Bower & Balogh 2004), or the evolution of galaxy morphology (see, e.g., Franx, these proceedings). In this review, I adopt a ΛCDM cosmology and a Kroupa (2001) IMF; adoption of a Kennicutt (1983) IMF in this article would leave the results unchanged.

2. The assembly of stellar mass

In many ways, the empirical exploration of the assembly of stellar mass throughout cosmic history has been one of the defining features of the last decade of extragalactic astronomical effort. Yet, in spite of such effort, and the apparent simplicity of the goal, progress at times has been frustratingly slow. The difficulties are many: calibration of SF rate (SFR) and stellar mass indicators, the effects of dust on SFR estimates, relatively poor sensitivity for most SFR indicators, and field-to-field variations caused by large-scale structure. In this section, I will briefly summarize the progress made to date towards measuring the cosmic SF history (SFH) using two independent methods: exploration of the evolution of the cosmic-averaged SFR, and the evolution of the cosmic-averaged stellar mass density, both as a function of epoch. The two are intimately related—the cosmic SFH is the integral over the cosmic SFR—and, barring any strong variation in stellar initial mass functions (IMFs) as a function of galaxy properties and/or epoch, consistency between the two would be expected.

2.1. *The cosmic-averaged star formation rate density*

2.1.1. *Methodology*

One measures SFR by measuring galaxy luminosities in a passband or passbands which one hopes will reflect the total number of massive stars in a galaxy. Using stellar population synthesis and other models, one then attempts to convert this luminosity into a total SFR, assuming a given stellar IMF to allow conversion from the number of massive stars to the total mass in the newly-formed stellar population (see, e.g., Kennicutt 1998 for an excellent review).

In some cases, this calibration from total luminosity to SFR is relatively robust, because it relies on reasonably well-understood physics. For example, the total ultraviolet (UV) light from a galaxy is reasonably robustly translated into the number of massive O and B stars. The total infrared luminosity of a starbursting galaxy, if the galaxy is optically thick to the UV light, is a reasonable reflection of the bolometric output of this starburst. Balmer line emission, under weak assumptions about the physical conditions in H II regions, is a reasonable indicator of the number of very massive O stars. These quantities, in turn, can be converted into a total SFR, assuming a stellar IMF which does not depend on galaxy properties and epoch.

In other cases, the calibration from luminosity to SFR is much less direct: e.g., the GHz radio emission from star-forming galaxies is well-correlated with other indicators of SFR, such as IR or Balmer line luminosity, and is known to be dominated by synchrotron emission from cosmic-ray electrons spiraling in galactic magnetic fields for at least massive star-forming galaxies (see Condon 1992 for an excellent review; see also Bell 2003). Yet, there is no robust theoretical understanding of why the relationship between radio emission and SFR should show such modest scatter (see, e.g., Bressan, Silva, & Granato 2002; Niklas & Beck 1997; Lisenfeld, Völk, & Xu 1996 for models of the radio emission of star-forming galaxies).

On top of these interpretive challenges, there are other important difficulties. Dust extinguishes UV light very effectively; empirical dust corrections based on UV color and/or UV-optical properties calibrated on UV-bright starbursts in the local Universe (Calzetti, Kinney, & Storchi-Bergmann 1994; Calzetti 2001) do not apply to normal galaxies (Bell 2002; Kong et al. 2004), IR-bright starbursts (Goldader et al. 2002), or indeed even H II regions (Bell et al. 2002; Gordon et al. 2004). Estimates of dust reddening from Balmer line ratios may yield reasonably accurate Balmer line-derived SFRs for galaxies (Kennicutt 1983), yet are extremely challenging to measure at $z \gtrsim 0.4$. IR and radio facilities lack the sensitivity to probe to faint limits; only galaxies with SFRs in excess of $\sim 10 \ M_\odot \, \mathrm{yr}^{-1}$ are observable with current facilities at $z \gtrsim 0.5$ (e.g., Flores et al. 1999).

2.1.2. *Results*

There is a huge number of papers which have addressed the evolution of the cosmic SFR, which are too numerous to mention or discuss in any detail. A few particularly important examples are Lilly et al. (1996), Madau et al. (1996), Flores et al. (1999), Steidel et al. (1999), and Haarsma et al. (2000).

In Fig. 1, I show the general form of the cosmic SFR, as derived by Hopkins (2004) in a very nice compilation of cosmic SFR estimates, where he uses locally calibrated relationships between dust attenuation and SFR to correct for dust. His corrections are reasonably similar to those commonly used, but with the important advantages that they are uniformly derived, and account for the well-known SFR-dust correlation. The intention of showing this cosmic SFR is primarily to illustrate that despite the significant observational challenges and interpretive challenges, by using a variety of different observational methodologies, it has been possible to determine its broad shape to better than a factor of three over much of cosmic history.

There are two points which one should take away from the cosmic SFR, both of which have been known, at least at the qualitative level, since 1996 when the first cosmic SFRs were constructed (e.g., Lilly et al. 1996; Madau et al. 1996). First, it is abundantly clear that the cosmic SFR has dropped by a factor of nearly 10 since $z \sim 1.5$ (see, e.g., the interesting compilation from Hogg 2001). Second, there is no compelling evidence against a roughly constant cosmic SFR at redshifts higher than 1 (e.g., Steidel et al. 1999), with the exception of some recent explorations of UV-derived SFRs at $z \gtrsim 4$, which appear to be lower than those at $z \lesssim 4$ (e.g., Stanway et al. 2004).

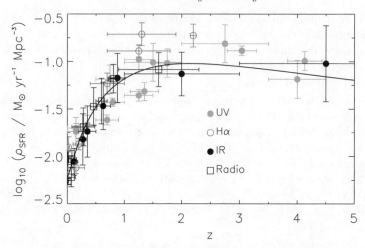

FIGURE 1. The evolution of the cosmic SFR density. SFRs assume a Kroupa (2001) IMF and $H_0 = 70\,\mathrm{km\,s^{-1}\,Mpc^{-1}}$. The data are taken from Hopkins (2004), and are corrected for dust assuming an SFR-dust correlation as found in the local Universe. The solid line shows an empirical fit to the cosmic SFR.

2.2. *The cosmic-averaged star-formation history*

A complementary way to understand the star-formation history of the Universe is to explore the build-up in stellar mass with cosmic time. This integral over the cosmic SFR contains the same information (under the assumption of a reasonably well-behaved stellar IMF), yet suffers from completely different systematic uncertainties, and is therefore an invaluable probe of the broad evolution of the stellar content of the Universe.

2.2.1. *Methodology*

Ideally, stellar masses would be estimated from spectroscopic data (e.g., velocity fields, rotation curves, and/or stellar velocity dispersions) coupled with multi-waveband photometry, under some set of assumptions about the dark matter content of the galaxy. Unfortunately, measurements of such quality are relatively uncommon, even in the local Universe.

In the absence of velocity data, one can attempt to estimate stellar masses using photometric data alone, with the aid of stellar-population synthesis models (e.g., Fioc & Rocca-Volmerange 1997; Bruzual & Charlot 2003). In these models, increasing the mean stellar age or the metallicity produces almost indistinguishable effects on their broadband optical colors, and indeed even in their spectra, with the exception of a few key absorption lines (e.g., Worthey 1994). Increasing dust extinction produces very similar effects, at least some of the time (e.g., Tully et al. 1998), although the relationship between reddening and extinction is rather sensitive to star/dust geometry (Witt & Gordon 2000). An increase in stellar population age, metallicity, or dust content reddens and dims the stellar population at a fixed stellar mass. Crucially, the relationship between reddening and dimming is similar for all three effects. While this makes it extremely challenging to measure the age, metallicity, or dust content of galaxies using optical broad-band colors alone, it does mean that one can invert the argument and use color and luminosity to rather robustly estimate stellar mass, almost independent of galaxy SFH, metallicity, or dust content (e.g., Bell & de Jong 2001).

The slope of the relationship between color and mass-to-light ratio (M/L) is passband dependent (steeper in the blue, shallower in the near infrared), but does not strongly

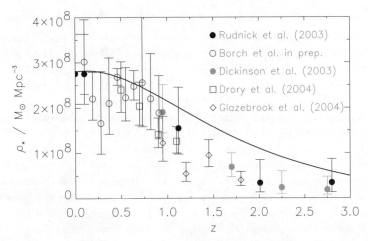

FIGURE 2. The evolution of the cosmic-averaged stellar mass density. Stellar masses assume a Kroupa (2001) IMF and $H_0 = 70\,\mathrm{km\,s^{-1}\,Mpc^{-1}}$. The data are taken from Rudnick et al. (2003; normalized to reproduce the $z = 0$ stellar mass density from Cole et al. 2001 & Bell et al. 2003), Dickinson et al. (2003), Drory et al. (2004), Glazebrook et al. (2004), and Borch et al.'s (in preparation) measurements from the COMBO-17 photometric redshift survey. The solid line shows the integral of the SFR from Fig. 1.

depend on stellar IMF. In contrast, the zero point of the color-M/L relation depends strongly on stellar IMF, especially to its shape at masses $\lesssim 1\ M_\odot$, where the bulk of the stellar mass resides but the contribution to the total luminosity is low. There are important sources of systematic error: dust does not always move galaxies along the same color-M/L relation as defined by age and metallicity (e.g., Witt & Gordon 2000), and most importantly, significant contributions from young stellar populations (either because the galaxy is truly young or because of recent starburst activity) can bias the stellar M/Ls at a given color towards lower values. Recently, a number of methodologies have been developed to address these limitations by inclusion of important variations in star formation history (e.g., Papovich, Dickinson, & Ferguson 2001), or by using spectral indices and colors to account for bursts and dust more explicitly (e.g., Kauffmann et al. 2003).

2.2.2. *Results*

The basic methodology has been applied in the last several years to a wide variety of galaxy surveys to explore the stellar mass density in the local Universe (e.g., Cole et al. 2001; Bell et al. 2003), its evolution out to $z \sim 1$ (e.g., Brinchmann & Ellis 2000; Drory et al. 2004; Borch et al., in preparation), and even out to $z \sim 3$ (e.g., Rudnick et al. 2003; Dickinson et al. 2003; Fontana et al. 2003; Glazebrook et al. 2004). In Fig. 2, I show a compilation of some of these stellar mass estimates, all transformed to a Kroupa (2001) IMF. The error bars in all cases give a rough idea of the uncertainties in stellar mass; the error bars for Drory et al. (2004), Glazebrook et al. (2004), and Borch et al. also include contributions from cosmic variance.

From Figs. 1 and 2, it is clear that the epoch $z \gtrsim 1$ is characterized by rapid star formation, with roughly two thirds of the stellar mass in the Universe being formed in the first 5 Gyr. In contrast, from $z \sim 1$ to the present day, one witnesses a striking decline in the cosmic SFR by a factor of roughly 10. Despite this strongly suppressed SFR, roughly one third of all stellar mass is formed in this interval, owing to the large amount of time between $z \sim 1$ and the present day.

2.3. *Are the cosmic star formation rates and histories consistent?*

It is interesting to test for consistency between the cosmic SFR and cosmic SFH—a lack of consistency between these two may give important insight into the form of the stellar IMF between ~ 1 $M_\odot$ (the mass range which optical/NIR light is most sensitive to) and $\gtrsim 5$ $M_\odot$ (the stellar mass range probed by SFR indicators), or the calibrations of or systematic errors in stellar mass and/or SFR determinations.

In order to integrate the cosmic SFR, I choose to roughly parameterize the form of the cosmic SFR following Cole et al. (2001). The cosmic SFR $\psi = (0.006 + 0.072z^{1.35})/[1 + (z/2)^{2.4}]$ $M_\odot$ yr^{-1} Mpc^{-3} provides a reasonable fit to the observations (Fig. 1). The data are insufficient to constrain whether or not the cosmic SFR declines towards high redshifts from a maximum at $z \sim 1.5$; I choose to impose a mild decrease towards high redshift, primarily because that matches the evolution in integrated stellar mass slightly better than a flat evolution. This cosmic SFR is then integrated using the PEGASE stellar population model assuming a Kroupa (2001) IMF and an initial formation redshift $z_f = 5$. The exact integration in PEGASE accounts explicitly for the recycling of some of the initial stellar bass back into the interstellar medium; for the Kroupa (2001) IMF this fraction is $\sim 1/2$, i.e., stellar mass in long-lived stars is one half of the stellar mass initially formed (for a Kennicutt 1983 IMF the fraction is similar). I show the expected cosmic SFH as the solid line in Fig. 2.

It is clear that the form of the cosmic SFR required in Fig. 1 reproduces rather well the cosmic SFH as presented in Fig. 2. There are some slight discrepancies; a cosmic SFR as flat as that shown in Fig. 1 appears to overpredict the amount of stellar mass that one sees at $z \sim 3$. This might indicate that a drop-off in cosmic SFR towards higher redshift is required, or may indicate that an IMF richer in high-mass stars is favored for high-redshift starbursts. Yet, it is important to remember that estimates of cosmic SFR and SFH are almost impossible to nail down with better than 30% accuracy at any redshift. At $z \gtrsim 1$ the constraints are substantially weaker still, owing to large uncertainties from large-scale structure, uncertainties in the faint-end slope of the stellar mass or SFR functions used to extrapolate to total SFRs or masses, and the difficulty in measuring SFRs and masses of intensely star-forming, dusty galaxies. Therefore, I would tend to downweight this disagreement at $z \gtrsim 1.5$ until better and substantially deeper data are available, focusing instead on the rather pleasing overall agreement between these two independent probes of the cosmic SFH.

3. The assembly of galaxies

Section 2 focused on the broadest possible picture of the assembly of stellar mass—the build-up of the stellar population, averaged over cosmologically-significant volumes. Yet, the physical processes contributing to this evolution will be much more strongly probed by studying the demographics of the galaxy populations as they evolve. This splitting of the cosmic SFR/SFH can happen in many ways: study of the evolution of the luminosity function, stellar-mass function, or SFR 'function,' study of galaxies split by morphological type or rest-frame color, or by identification of galaxies during particular phases of their evolution (e.g., interactions). There has been a great deal of activity over the last decade towards this goal: implicit in the exploration of Figs. 1 and 2 is the construction of SFR and stellar mass functions, and a number of studies have explored the evolution of the galaxy population split by morphological type (e.g., Brinchmann & Ellis 2000; Im et al. 2002), rest-frame color (e.g., Lilly et al. 1996; Wolf et al. 2003; Bell et al. 2004b), or focusing on the role of galaxy interactions (e.g., Le Fèvre et al. 2000; Patton et al. 2002; Conselice et al. 2003).

A full and fair exploration of any or all of these goals is unfortunately beyond the scope of this work. Here, I choose to focus on one particular key issue: the importance of galaxy mergers in driving galaxy evolution in the epoch since $z \sim 3$, and especially at $z \lesssim 1$. Unlike the evolution of, for example, the stellar mass function or SFR function, to which both quiescent evolution and galaxy accretion can contribute, galaxy mergers (especially those at $z \lesssim 2$) are an unmistakable hallmark of the hierarchical assembly of galaxies. Therefore, exploration of galaxy mergers directly probes one of the key features of our current cosmological model.

I will focus here on exploring the number of major galaxy mergers (traditionally defined as those with mass ratios of 3:1 or less) over the last 10 Gyr. There are three complementary approaches to exploring merger rate, all of which suffer from important systematic uncertainties: the evolution of the fraction of galaxies in close pairs, the evolving fraction of galaxies with gross morphological irregularities, and investigation of the evolution of plausible merger remnants. In this work, I will briefly discuss all three methods, highlighting areas of particular uncertainty to encourage future development in this exciting and important field.

3.1. *The evolution of close galaxy pairs*

One of the most promising measures of galaxy merger rate is the evolution in the population of close galaxy pairs. Most close (often defined as being separated by $< 20\,\mathrm{kpc}$), bound galaxy pairs with roughly equal masses will merge within $\lesssim 1$ Gyr owing to strong dynamical friction (e.g., Patton et al. 2000). Thus, if it can be measured, the fraction of galaxies in physical close pairs is an excellent proxy for merger rate.

Accordingly, there have been a large number of studies which have attempted to measure close pair fraction evolution (a few examples are Zepf & Koo 1989; Carlberg, Pritchet, & Infante 1994; Le Fèvre et al. 2000; see Patton et al. 2000 for an extensive discussion of the background to this subject). In Fig. 3, I show the fraction of $M_B - 5\log_{10} h \lesssim -19.5$ galaxies in close $|\Delta_{\mathrm{proj}}| < 20 h^{-1}\,\mathrm{kpc}$ pairs from Patton et al. (2000); Le Fèvre et al. (2000); and Patton et al. (2002). Patton et al. (2000) and Patton et al. (2002) explore the fraction of galaxies in close pairs with velocity differences $< 500\,\mathrm{km s}^{-1}$, whereas Le Fèvre et al. (2000) study projected galaxy pairs, and correct them for projection in a statistical way. Both studies, and indeed most other studies, paint a *broadly* consistent picture that the frequency of galaxy interactions was much higher in the past than at the present, and that the average $\sim L^*$ galaxy has suffered 0.1–1 major interaction between $z \sim 1$ and the present day. Small number statistics, coupled with differences in assumptions about how to transform pair fractions into merger rate, lead to a wide dispersion in the importance of major merging since $z \sim 1$.

Yet, there are significant obstacles to the interpretation of these, and indeed future, insights into galaxy major-merger-rate evolution. Technical issues, such as contamination from foreground or background galaxies, redshift incompleteness, and the construction of equivalent samples across the whole redshift range of interest, must be thought about carefully (see, e.g., Patton et al. 2000, 2002). An further concern is that most close-pair measurements are based on galaxies with similar magnitudes in the rest-frame optical. Bursts of tidally-triggered star formation may enhance the optical brightness of both the galaxies of a pair, and the selection of galaxies with nearly equal optical brightness may not select a sample of galaxies with nearly equal stellar masses. Bundy et al. (2004) used K-band images of a sample of 190 galaxies with redshifts to explore this source of bias,

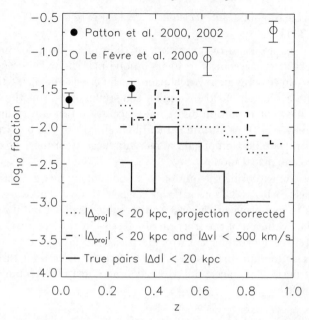

FIGURE 3. The evolution of the close pair fraction. Data points show measurements of the fraction of galaxies with one or more neighbors within a projected distance of $20h^{-1}$ kpc, corrected for projection (open points; Le Fèvre et al. 2000), and within a projected distance of $20h^{-1}$ kpc and a velocity separation of < 500 kms^{-1} (solid points; Patton et al. 2000, 2002). The lines show the evolution of approximately equivalent measurements from the van Kampen et al. (in prep.) mock COMBO-17 catalogs: projection-corrected fraction (dotted line), the fraction of galaxies with $\geqslant 1$ neighbor within $20h^{-1}$ kpc and with velocity difference < 300 kms^{-1} (dashed line), and the real fraction of galaxies with $\geqslant 1$ neighbor within $20h^{-1}$ kpc in real space.

finding a substantially lower fraction of galaxies in nearly equal-mass close pairs than derived from optical data.

Yet, it is also vitally important to build one's intuition as to how the measured quantities relate to the true quantities of interest. As a crude example of this, I carry out the simple thought experiment where we compare the pair fraction derived from projection-corrected projected pair statistics (dotted line in Fig. 3), the pair fraction derived if one used spectroscopy to keep only those galaxies with $|\Delta v| < 300$ km s^{-1} (dashed line), and the true fraction of galaxies with real space physical separation of $< 20h^{-1}$ kpc (solid line). To explore this issue, I use a mock COMBO-17 catalog that is being developed by van Kampen et al. (in prep.; see van Kampen, Jimenez, & Peacock 1999 for a description of this semi-numerical technique); this model is used with their kind permission. It is important to note that this simulation is still under development. The match between the observations and models at $z \lesssim 0.5$ is mildly encouraging, although the generally flat evolution of pair fraction may not be a robust prediction of this model. Nonetheless, this can be used to great effect to gain some insight into sources of bias and uncertainty.

The catalog is tailored to match the broad characteristics of the COMBO-17 survey to the largest extent possible (see Wolf et al. 2003 for a description of COMBO-17): it has $3 \times 1/4$ square degree fields limited to $m_R < 24$ and photometric redshift accuracy mimicking COMBO-17's as closely as possible. Primary galaxies with $m_R < 23$, $0.2 \leqslant z \leqslant 1.0$, and $M_V \leqslant -19$ are chosen; satellite galaxies are constrained only to have $m_R < 24$ and $|\Delta m_R| < 1$ mag (i.e., to have a small luminosity difference). The dotted and dashed lines show the pair fraction recovered by approximately reproducing the methodologies of Le Fèvre et al. (2000; dotted line) and Patton et al. (2000, 2002; dashed line). It is clear that statistical field subtraction, adopted by Le Fèvre et al. (2000) and others, may be rather robust in terms of recovering the trends that one sees with the spectroscopic+imaging data. The solid line shows the fraction of galaxies actually separated by $20h^{-1}$ kpc. It is clear that, at least in this simulation, many galaxies with projected close separation and small velocity difference are members of the primary galaxy's group which happen to lie close to the line of sight to the primary galaxy, but are $\gtrsim 20$ kpc from the primary. In Le Fèvre et al. (2000), Patton et al. (2002), and other studies, the correction was often made by multiplying the observed fraction by $0.5(1 + z)$ following the analysis of Yee & Ellingson (1995); the data shown in Fig. 3 was corrected in this way. Our analysis of these preliminary COMBO-17 mock catalogs suggest that these corrections are uncertain, and that our current understanding of galaxy merger rate from close pairs may be biased. It is important to remember that the simulations discussed here are preliminary; the relationship between 'observed' and true close-pair fractions may well be different from the trends predicted by the model. Yet, it is nonetheless clear that further modeling work is required before one can state with confidence that one understands the implications of close-pair measurements.

3.2. *The evolution of morphologically-disturbed galaxies*

With the advent of relatively wide-area, high-spatial resolution and S/N imaging from *HST*, it has become feasible to search for galaxy interactions by identifying galaxies which are morphologically disturbed. There are a number of ways in which one can evaluate morphological disturbance: a few examples are by visual inspection (e.g., Arp 1966), residual structures in unsharp-masked or model-subtracted images (e.g., Schweizer & Seitzer 1992), automated measures of galaxy asymmetry (e.g., Conselice, Bershady, & Jangren 2000), or by the distribution of pixel brightnesses and second-order moment of the light distribution (e.g., Lotz, Primack, & Madau 2004).

The important hallmarks that these different methodologies probe—multiple nuclei and extended tidal tails—are the non-equilibrium signatures of tidal interaction. These features are common in simulations of interacting galaxies, and cannot result from quiescent secular evolution (e.g., Toomre & Toomre 1972; Barnes & Hernquist 1996). Visual inspection is a subjective way to pick out these structures, even to relatively low surface-brightness limits; in contrast, automated measures of morphology typically quantify the amount of light in bright asymmetric structures, so they pick out multiple bright patches or bright tidal tails rather effectively.

Recently, Conselice et al. (2003) have explored the fraction of galaxies with strong asymmetries in the rest-frame B-band out to redshift ~ 3 in the Hubble Deep Field North. Their results, for a similar absolute magnitude range as explored by Patton et al. (2002) and Le Fèvre et al. (2000), are shown in Fig. 4. While the small number statistics (there are less than 500 galaxies contributing to the fractions)—coupled with asymmetry uncertainties—are clearly an issue, a broadly consistent picture is painted whereby galaxy mergers were substantially more frequent at $z \gtrsim 1$, and have decreased in frequency until the present day. It is interesting to note that the observations are not well reproduced by either the Benson et al. or van Kampen et al. models, which both predict substantially lower merger rates. Indeed, perhaps for the first time, observers are invoking the need for many more mergers than theorists are!

While these first tentative steps are encouraging, there are a number of systematic uncertainties that should be considered carefully. At some level, contamination from projected close-galaxy pairs is bound to be a problem. Samples of highly asymmetric galaxies will, correctly, also include some very close physically-associated galaxy pairs.

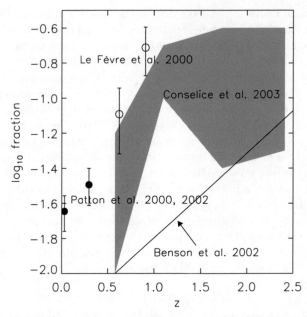

FIGURE 4. The evolution of merger rate as inferred from the fraction of galaxies with gross asymmetries (Conselice et al. 2003; gray shaded region). Inferred merger rates from close pairs are reproduced from Fig. 3 for comparison. A rough fit to predictions of major merger rate (visible for 1 Gyr) from the semi-analytic galaxy formation model of Benson et al. (2002), reproduced from Conselice et al. (2003), is shown by the solid line for reference.

Unfortunately, they will also contain a number of physically unassociated close-galaxy pairs: Le Fèvre et al. (2000) find that only ~30% of close-galaxy pairs are likely to be physically associated (in the sense of being in the same galaxy group; a still smaller fraction are genuine close pairs), and the mock COMBO-17 catalog analysis presented earlier suggests a fraction closer to 10%. These contaminants may be weeded out by source extraction software, as the pair of galaxy images may be parsed into two separate galaxies, each of which has a low asymmetry. Yet, if this image parsing is too enthusiastic, genuine major interactions may be parsed into multiple, individually rather symmetric, sub-units.

Furthermore, there are differences between different classification methodologies as to what constitutes a 'merger.' An example of this is given in Fig. 18 of Conselice et al. (2005), who explore the distribution of morphologically early-type galaxies (those with dominant-bulge components), late-type galaxies (those with dominant disks), and peculiar galaxies (all other types, which includes merging and irregular galaxies) in the concentration (C) and asymmetry (A) plane. C is a measure of the light concentration of a galaxy profile, having high values for centrally concentrated light profiles, and A is a measure of asymmetry, with high values denoting asymmetric galaxies. Inspecting their Fig. 18, one sees scattered trends between visual and automated classifications. Early-type galaxies tend to have higher concentrations and lower asymmetries, later types have lower concentrations and lower asymmetries, and peculiar systems have low concentrations and a wide range of asymmetries, with a tail to very high asymmetries. Yet, there are a number of early-types with low concentrations, and importantly, a large number of peculiar galaxies have low asymmetries, and cannot be distinguished from morphologically normal galaxies in this scheme. Preliminary comparisons between by-eye morphologies and C and A values for F606W images of galaxies in the GEMS (Galaxy Evolution from

Morphology and SEDs; Rix et al. 2004) *HST* survey supports this conclusion, indicating that such automated schemes find it hard to differentiate between irregular galaxies (whose morphologies, to the human eye, indicate sporadic star formation triggered by internal processes) and clearly interacting galaxies with morphological features indicative of tidal interactions, such as multiple bright nuclei or tidal tails. A further complication is that the human eye is more sensitive to faint tidal tails in early- or late-stage mergers than automated schemes (i.e., the timescale probed by visual classifications may be longer than for automated schemes).

This discussion is by no means meant to detract from the value of the intriguing and ground-breaking work on merger rates from morphology; rather, it is to emphasize that the measurement of merger rate is a subtle and difficult endeavor, fraught with systematic uncertainties which will likely have to be modeled explicitly using future generations of *N*-body, semi-analytic and SPH simulations, coupled with the artificial redshifting experiments already commonly used in this type of work.

3.3. *The evolution of plausible merger remnants*

From the earliest days of galaxy morphological classification, a population of galaxies whose light distribution is dominated by a smoothly distributed, spheroidal, centrally-concentrated light distribution was noticed. These *early-type galaxies* are largely supported by the random motions of their stars (Davies et al. 1983). These properties are very naturally interpreted as being the result of violent relaxation in a rapidly changing potential well (Eggen, Lynden-Bell, & Sandage 1962). Therefore, in our present cosmological context, these galaxies are readily identified with the remnants of major galaxy mergers (e.g., Toomre & Toomre 1972; Barnes & Hernquist 1996). Detailed comparisons of simulated major merger remnants broadly supports this notion, although some interesting discrepancies with observations remain, and are perhaps telling us about difficult-to-model gas-dynamical dissipative processes (e.g., Bendo & Barnes 2000; Meza et al. 2003; Naab & Brukert 2003). There is strong observational support for this notion—the correlation between fine morphological structure and residuals from the color-magnitude correlation (Schweizer & Seitzer 1992), the existence of kinematically decoupled cores (e.g., Bender 1988), and the similarity between the stellar dynamical properties of late-stage IR-luminous galaxy mergers and elliptical galaxies (e.g., Genzel et al. 2001). Therefore, study of the evolution of spheroid-dominated, early-type galaxies may be able to give insight into the importance of galaxy merging through cosmic history. There are, as always, a variety of important complications and limitations to this approach. For example, a fraction of the galaxies becoming early-type during galaxy mergers will later re-accrete a gas disk, which will gradually transform into stars, making the galaxy into a later-type galaxy with a substantial bulge component (e.g., Baugh, Cole, & Frenk 1996).

Furthermore, not all galaxy mergers will result in a spheroid-dominated galaxy; some lower mass-ratio interactions will result in a disk-dominated galaxy (e.g., Naab & Burkert 2003). In addition, it would be foolish to *a priori* ignore the possibility that an important fraction of early-type galaxies may be formed rapidly in mergers of very gas-rich progenitors at early epochs—a scenario reminiscent of the classical monolithic collapse picture (Larson 1974; Arimoto & Yoshii 1987). Yet, these interpretive complications, much as in all the cases discussed above, do not lessen the value of placing observational constraints on the phenomenology, with the confidence that our interpretation and understanding of the phenomenology will improve with time.

The rate of progress in this field has largely been determined by the availability of wide-format high-resolution imaging from *HST*. Ground-based resolution is insufficient to

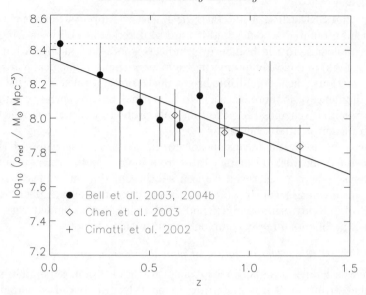

FIGURE 5. The evolution of the stellar mass density in red-sequence galaxies. Stellar masses assume a Kroupa (2001) IMF and $H_0 = 70\,\mathrm{km\,s^{-1}\,Mpc^{-1}}$. The local data point is taken from Bell (2003), and data for $0.2 < z < 1.3$ are taken from Bell et al. (2004b), Chen et al. (2003), and Cimatti et al. (2002). The solid line shows a rough fit to the total stellar mass density in red-sequence galaxies in the Cole et al. (2000) semi-analytic galaxy-formation model. The dotted line shows the expected result if red-sequence galaxies were formed at $z \gg 1.5$ and simply aged to the present day.

robustly distinguish disk-dominated and spheroid-dominated galaxies at cosmologically-interesting redshifts. In the local Universe, the vast majority of morphologically early-type galaxies occupy a relatively tight locus in color-magnitude space—the color-magnitude relation (e.g., Sandage & Visvanathan 1978; Bower, Lucey, & Ellis 1992; Schweizer & Seitzer 1992; Strateva et al. 2001). Therefore, many workers have focused on the evolution of the red galaxy population as an accessible alternative. The efficacy of this approach is only recently being tested. Accordingly, I explore the evolution of the red galaxy population first, turning subsequently to the evolution of the early-type population later.

3.3.1. *The evolution of the total stellar mass in red-sequence galaxies*

It has become apparent only in the last five years that the color distribution of galaxies is bimodal in both the local Universe (e.g., Strateva et al. 2001; Baldry et al. 2004) and out to $z \sim 1$ (Bell et al. 2004b). This permits a model-independent definition of red galaxies—those that lie on the color-magnitude relation. A slight complication is that the color of the red sequence evolves with time as the stars in red-sequence galaxies age, necessitating an evolving cut between the red sequence and the 'blue cloud.' The evolution of this cut means that some workers who explore the evolution of red galaxies using rather stringent color criteria—for example, galaxies the color of local E/S0 galaxies—find much faster evolution in the red galaxy population than those who adopt less stringent or evolving cuts (e.g., Wolf et al. 2003).

Bearing this in mind, I show the evolution in the total stellar mass in red galaxies in Fig. 5. Although many surveys could, in principle, address this question (e.g., of the surveys discussed in §2.2), very few have split their stellar-mass evolution by color, and to date most surveys have only published evolution of luminosity densities in red galaxies. The $z = 0$ stellar-mass density in red galaxies is from SDSS and 2MASS (Bell et al.

2003). The COMBO-17 data points for j_B evolution (Bell et al. 2004b) are converted to stellar mass using color-dependent stellar M/Ls as defined by Bell & de Jong (2001); extrapolation to total stellar-mass density adopts a faint-end slope $\alpha = -0.6$, as found by Bell et al. (2003; 2004b) for red-sequence galaxies at $z \lesssim 1$. Error bars account for stellar-mass uncertainties and cosmic variance, defined by the field-to-field scatter in stellar-mass densities from the three COMBO-17 fields. Stellar masses for the LCIRS sample of red galaxies were estimated using the rest-frame R-band luminosity density presented by Chen et al. (2003), accounting for the mildest possible passive-luminosity evolution so as to minimize any stellar-mass evolution, and using a stellar M/L for early-type galaxies from Bell & de Jong (2001) using a Kroupa (2001) IMF, and adopting a color of $B - R = 1.5$ for early-type galaxies as a $z = 0$ baseline. Error bars include stellar mass estimation uncertainties and estimated cosmic variance, following Somerville et al. (2004). The K20 data point at $z \sim 1.1$, derived from ERO 'old' galaxy space densities from Cimatti et al. (2002), is very roughly calculated using a number of assumptions. A (rather large) stellar mass of $\sim 10^{11}~M_\odot$ is attached to each galaxy, and the densities are multiplied by two to account for the star-forming EROs (the split was roughly 50:50). This stellar-mass density was multiplied by two again to account for fainter, undetected galaxies. Error bars of ± 0.2 dex and ± 0.3 dex, combined in quadrature, account for cosmic variance following Somerville et al. (2004) plus our poor modeling assumptions. Owing to their use of discordant cosmologies, I do not show the inferred stellar mass evolution of the CFRS red galaxies in Fig. 5; however, like Bell et al. (2004b), they infer no evolution in the rest-frame B-band luminosity density to within their errors (Lilly et al. 1995), meaning that their stellar mass evolution would fall into excellent agreement with those of Bell et al. (2004b) or Chen et al. (2003).

The results are shown in Fig. 5. To first order, the luminosity density in the optical in red galaxies is constant out to $z \sim 1$; this is confirmed by a number of surveys (e.g., Lilly et al. 1995; Chen et al. 2003; or Bell et al. 2004b). Coupled with the passive aging of the stellar populations of these red galaxies (as indicated by their steady reddening with cosmic time, and confirmed by study of dynamically derived M/Ls and absorption line ratios; e.g., Wuyts et al. 2004; Kelson et al. 2001), this implies a steadily increasing stellar mass density in red galaxies since $z \sim 1.2$. *To date, to the best of our knowledge, there are no published determinations of red-galaxy stellar mass or luminosity density which contradict this picture.*

The implications of this result are rather far reaching. Bearing in mind that at $z \gtrsim 1$ the red galaxy population may be significantly contaminated and/or dominated by dusty star-forming galaxies (e.g., Yan & Thompson 2003; Moustakas et al. 2004), this evolution may well represent a strong upper limit to the stellar mass in early-type galaxies since $z \sim 1.2$, unless large populations of very bright blue morphologically early-type galaxies are found. I address this question next, by exploring the stellar-mass evolution in morphologically early-type galaxies since $z \sim 1$.

3.3.2. *The evolution of the total stellar mass in early-type galaxies*

Owing to the lack of extensive wide-area *HST*-resolution imaging data, there are even weaker constraints on the evolution of stellar mass in morphologically early-type galaxies. We show published results from Im et al. (2002), Conselice et al. (2005), and Cross et al. (2004), supplementing them with a preliminary analysis of galaxies from GEMS and COMBO-17, which is presented with the permission of the GEMS and COMBO-17 teams.

Im et al. (2002) present a thorough study of the luminosity-density evolution of E/S0 galaxies from the DEEP Groth Strip Survey, defined using bulge-disk decompositions and placing a limit on residual substructure in the model-subtracted images, supplemented with *HST* V- and I-band imaging for 118 square arcminutes. A final sample of 145 E/S0 galaxies with $0.05 < z < 1.2$ are isolated. The fit of average early-type galaxy stellar M/L as a function of redshift from COMBO-17/GEMS, $\log_{10} M/L_B = 0.34 - 0.27z$, was used to transform the published B-band luminosity densities into stellar-mass densities for the purposes of Fig. 6.† Cross et al. (2004) present rest-frame B-band luminosity functions for visually classified E/S0 galaxies in five ACS fields: the B-band luminosity densities were transformed into stellar mass in exactly the same way as for Im et al. (2002). Conselice et al. (2005) presented stellar mass densities directly, and these are reproduced on Fig. 6, after accounting for our use of a Kroupa (2001) IMF.

The preliminary GEMS/COMBO-17 early-type galaxy data points depict the evolution of total stellar mass in galaxies with Sérsic indices $n > 2.5$. In GEMS, Bell et al. (2004a) found that galaxies with $n > 2.5$ included $\sim 80\%$ of the visually classified E/S0/Sa galaxy population at $z \sim 0.7$, with $\sim 20\%$ contamination from later galaxy types (recall, high Sérsic indices indicate concentrated light profiles). Here, in this very preliminary exploration of the issue, a $n = 2.5$ cut is adopted irrespective of redshift, ignoring the important issue of morphological k-correction. To recap, stellar masses are calculated using color-dependent stellar M/Ls from Bell & de Jong (2001), again extrapolating to total using a faint-end slope $\alpha = -0.6$.

The resulting $n > 2.5$ stellar mass evolution for the $30' \times 30'$ GEMS field is shown in the left panel of Fig. 6. One can clearly see strong variation in the stellar-mass density of early-type galaxies, resulting from large-scale structure along the line of sight. From such data, it is clearly not possible to place any but the most rudimentary and uninteresting limits on the evolution of early-type galaxies over the last 9 Gyr. Yet, noting that the bulk of early-type galaxies are in the red sequence at $z \sim 0$ (e.g., Strateva et al. 2001) and at $z \sim 0.7$ (e.g., Bell et al. 2004a), and that the stellar mass density of red-sequence galaxies undergo very similar fluctuations, it becomes interesting to ask if the ratio of stellar mass in red-sequence galaxies to early types varies more smoothly with redshift. This is plotted in the lower left panel of Fig. 6. There is a weak trend in early-type galaxy to red-sequence galaxy-mass density, caused by an increasingly important population of blue galaxies with $n > 2.5$ towards higher and higher redshift (see also Cross et al. 2004, who discuss this issue in much more depth). Importantly, however, there is only a $\sim 15\%$ scatter around this trend despite the nearly order of magnitude variation in galaxy density, arguing against strong environmental dependence in the early-type to red galaxy ratio.

This relatively slow variation in early-type to red-sequence ratio (modeled using a simple linear fit for the purposes of this paper) is used to convert COMBO-17's stellar mass evolution in red galaxies from Fig. 5, which is much less sensitive to large-scale structure, into the evolution of stellar-mass density in morphologically early-type galaxies. This is shown in the right panel of Fig. 6. It is clear that there is a strongly increasing stellar-mass density in morphologically early-type galaxies since $z \sim 1.2$.‡ While there are recent indications that lower-luminosity early-type galaxies are largely absent at $z \gtrsim 0.8$ (e.g., Kodama et al. 2004; de Lucia et al. 2004), the total stellar-mass density is strongly dominated by $\sim L^*$ galaxies, and there is no room to avoid the conclusion that there has

---

† Again, a Kroupa (2001) IMF was used.

‡ Im et al. (2002) report a low stellar-mass density in early-type galaxies at $0.05 < z < 0.6$. We would attribute this deficiency in stellar mass to incompleteness at bright apparent magnitudes, leading to a deficit of nearby E/S0s with large stellar masses (as argued by Im et al. themselves), and perhaps to small number, statistics, and cosmic variance (as the volume probed by this study at $z \lesssim 0.5$ is rather small). Further work is required to explore further this discrepancy.

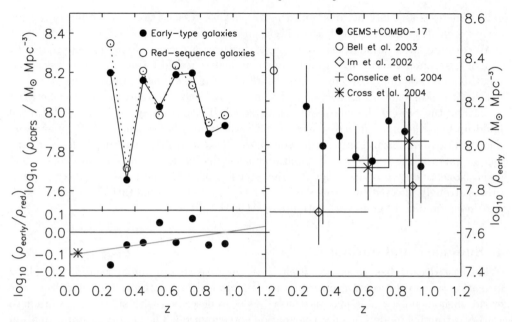

FIGURE 6. The evolution of the stellar-mass density in early-type galaxies. Stellar masses assume a Kroupa (2001) IMF and $H_0 = 70\,\mathrm{km\,s^{-1}\,Mpc^{-1}}$. The left-hand panel shows the stellar-mass densities from GEMS only; solid circles denote early-type galaxies and open circles red-sequence galaxies. The ratio of stellar mass in early-type galaxies to the red sequence is shown in the lower left panel; the asterisk is the local value taken from Bell (2003), and the gray line shows a linear fit to the GEMS data only (RMS $\sim 15\%$). The right-hand panel shows the resulting corrected early-type galaxy stellar-mass density evolution, where again the local data are taken from Bell (2003), the solid circles denote the GEMS+COMBO-17 result, the diamonds show results from Im et al. (2002), the naked error bars show results from Conselice et al. (2005), and the asterisks show results from Cross et al. (2004).

been a substantial build-up in the total number of $\sim L^*$ early-type galaxies since $z \sim 1.2$. Like the results presented in Sections 3.1 and 3.2, these results suggest an important role for $z \lesssim 1$ galaxy mergers in shaping the present-day galaxy population.¶

3.4. *The importance of galaxy mergers since $z \sim 1$*

Three independent methods have been brought to bear on this problem—the evolution of close galaxy pairs, the evolution of morphologically-disturbed galaxies, and the evolution of early-type galaxies as a proxy for plausible merger remnants. All three methods suffer from important systematic and interpretive difficulties. Close galaxy pairs, even supplemented with spectroscopy to isolate galaxies within $\lesssim 500\,\mathrm{km s^{-1}}$ of each other, remain contaminated by galaxies in the same local structures. Some measures of morphological disturbance are susceptible to this source of error to a lesser extent, and a clear consensus on the meaning of the different automated and visual measures of morphological disturbance is yet to emerge. Early-type galaxies will be the result of only a subset of

¶ Disk re-accretion and fading work in opposite directions in this context; disk accretion following a major merger takes galaxies away from the early-type class, whereas fading of disks which formed at earlier times will increase the relative prominence of spheroidal-bulge components and add galaxies to the early-type class. A thorough investigation of these issues, involving bulge-disk decompositions of rest-frame B-band images of the GEMS sample and drawing on galaxy-evolution models to build intuition of the importance and effects of different physical processes, can be found in Häußler et al. (in preparation).

galaxy mergers and interactions, and the role of disk re-accretion and fading in driving early-type galaxy evolution is frustratingly unclear.

Yet despite these difficulties, some broad features are clear. Mergers between $\sim L^*$ galaxies are almost certainly more frequent at higher redshift than at the present day, but this does not imply by any means that galaxy mergers are unimportant at $z \sim 1$: indeed, it is possible that an average $\sim L^*$ galaxy undergoes roughly one merger between $z = 1$ and the present day (Le Fèvre et al. 2000). This is supported by the substantial build-up in stellar mass in red-sequence and early-type galaxies since $z \sim 1$. Substantial uncertainties remain and important questions, like the stellar-mass dependence of the merger rate, or the fraction of dissipationless vs. dissipational mergers, are completely open. Nonetheless, these first, encouraging steps imply that the massive galaxy population is strongly affected by late-galaxy mergers, in excellent qualitative agreement with our understanding of galaxy evolution in a ΛCDM Universe.

4. Summary and outlook

The last decade has witnessed impressive progress in our understanding of galaxy formation and evolution. Despite important technical and interpretive difficulties, the broad phenomenology of galaxy assembly has been described with sufficient accuracy to start constraining models of galaxy formation and evolution. The history of star formation has been attacked from two complementary and largely independent angles—through the evolution of the cosmic-averaged star-formation rate, and through the build-up of stellar mass with cosmic time. Prior to $z \sim 1$, stars formed rapidly, and $\sim 2/3$ of the total present-day stellar mass was formed in this short ~ 4 Gyr interval. The epoch subsequent to $z \sim 1$ has seen a dramatic decline in cosmic-averaged SFR by a factor of roughly 10; however, $\sim 1/3$ of the present-day stellar mass was formed in this 9 Gyr interval. These two diagnostics of cosmic star-formation history paint a largely consistent picture, giving confidence in its basic features.

The assembly of present-day galaxies from their progenitors through the process of galaxy mergers was studied using three largely independent diagnostics—the evolution of galaxy close pairs, the evolution of morphologically disturbed galaxies, and the evolution of early-type galaxies as a plausible major merger remnant. The interpretive difficulties plaguing all three diagnostics are acute; accordingly our understanding of the importance of major mergers in shaping the present-day galaxy population is incomplete. Yet, all three diagnostics seem to indicate that an important fraction (dare I suggest $\gtrsim 1/2$?) of $\sim L^* \sim 3 \times 10^{10} \ M_\odot$ galaxies are affected by galaxy mergers since $z \sim 1$. This contrasts sharply with the form of the cosmic star-formation history, where all of the action is essentially over by $z \sim 1$. It is not by any means indefensible to argue that late mergers shape the properties of the massive-galaxy population in a way which is qualitatively consistent with our understanding of galaxy formation and evolution in a ΛCDM Universe.

Yet, it is clear that much work is to be done to fully characterize the physical processes driving galaxy evolution in the epoch since reionization $z \lesssim 6$. The 'shopping list' is too extensive to discuss properly, so I will focus on three aspects which I feel are particularly important.

4.1. *Our increasing understanding of the local Universe*

Wide area, uniform photometric and spectroscopic surveys, such as the Sloan Digital Sky Survey (SDSS; York et al. 2000), the Two Micron All Sky Survey (2MASS; Skrutskie et al. 1997), and the Two degree Field Galaxy Redshift Survey (2dFGRS; Colless et al. 2001), are revolutionizing our understanding of the local galaxy population. These surveys have allowed one to tie down the $z = 0$ data point for many evolutionary studies, greatly increasing the redshift and time-baseline leverage. Yet, this increased leverage is often difficult to fully apply, because local studies suffer from very different systematics than the lookback studies, and experimental details such as imaging depth, resolution, waveband, etc., are often imperfectly matched. At this stage, few groups have grasped the nettle of repackaging these local surveys in a format which can readily be artificially redshifted, etc., to allow for uniform analysis of the distant and local control samples. This will be an important feature of the most robust of the future works in galaxy evolution, and will greatly increase the scope and discriminatory power of studies of galaxy photometric, dynamical, and morphological evolution.

4.2. *Using models as interpretive tools*

Large and uniform multi-wavelength and/or spectroscopic datasets are becoming increasingly common. An important consequence of this change in the nature of the data used to study galaxy evolution is that often the error budget is completely dominated by systematic and interpretive uncertainties. In this context, models of galaxy formation and evolution can help us to understand and limit these systematic uncertainties through exhaustive analysis of realistic mock catalogs. A crude example of this kind of analysis is given in §3.1; a very nice example is the 2dFGRS group catalog of Eke et al. (2004). Mock simulations of this kind of quality must become a more prominent part of our toolkit in order for the kind of interpretive difficulties faced in §3.1 or §3.3 to be successfully navigated.

4.3. *The importance of multiple large HST fields*

Large-scale structure is an important and frequently neglected source of systematic uncertainty (Somerville et al. 2004).† Comparison of the left-hand and right-hand panels of Fig. 6 illustrates this point powerfully; the broad features of the evolution of early-type galaxy stellar-mass density are discernable using the $30' \times 30'$ Chandra Deep Field South alone, but correction of this result for cosmic variance using the other two COMBO-17 fields yields a significantly more convincing picture. Yet, while this kind of correction for cosmic variance may work when there is significant overlap between populations of interest (although it is debatable how far one should push such an idea), it is not *a priori* clear that rarer and/or optically-obscured phases of galaxy evolution such as AGN or IR-luminous mergers will be well modeled with such techniques. Furthermore, short-timescale astronomical phenomena, such as AGN or galaxy mergers, have lower number density and are potentially very strongly clustered leading to large uncertainties from number statistics and cosmic variance. Yet, these phases of galaxy evolution, where galaxies undergo important and potentially permanent transformations in the cosmic blink of an eye, *require HST*-resolution data in order to explore their physical drivers.

When *HST* could be viewed as an essentially endless resource, a piecemeal approach was perfectly optimal: more and/or larger *HST* fields could be justified on a case-by-case basis, depending on the science goals of interest. Yet, faced with an unclear future for *HST*, it is not obvious that this approach is optimal. *HST* perhaps should be thought of as a fixed-lifetime experiment, where the primary goal could become the creation of an archival dataset which will support 10–20 years of top-class research. In the creation of such an archival dataset, important questions will need to be addressed: availability of

† It is interesting to note that a ×10 increase in area in a single contiguous field gives only a ×2 reduction in cosmic variance, because the various parts of the single field are correlated with each other.

resources will need to be balanced against number statistics and cosmic variance, arguably at least two *HST* passbands will be required to allow some attempt at morphological *k*-correction, and deep multi-wavelength data will be required for study of black hole accretion and obscured star formation, naturally driving the fields into one of a small number of low H I and cirrus holes (see Papovich, these proceedings).

I wish to warmly thank the GEMS and COMBO-17 collaborations—Marco Barden, Steven Beckwith, Andrea Borch, John Caldwell, Simon Dye, Boris Häußler, Catherine Heymans, Knud Jahnke, Martina Kleinheinrich, Shardha Jogee, Daniel McIntosh, Klaus Meisenheimer, Chien Peng, Hans-Walter Rix, Sebastian Sanchez, Rachel Somerville, Lutz Wisotzki, and last but by no means least, Christian Wolf—for their permission to present some GEMS and COMBO-17 results before their publication, for useful discussions, and for their friendship and collaboration. It is a joy to be part of these teams. I wish also to thank Eelco van Kampen and his collaborators for their efforts to construct mock COMBO-17 catalogs, for their permission to share results from these catalogs in this article, and for useful comments. Chris Conselice, Emmanuele Daddi, Sadegh Kochfar, and Casey Papovich are thanked for useful and thought-provoking discussions on some of the topics discussed in this review, and Chris Conselice and Shardha Jogee are thanked for constructive comments on the first version of this review. This work is supported by the European Community's Human Potential Program under contract HPRN-CT-2002-00316, SISCO.

REFERENCES

ARIMOTO, N. & YOSHII, Y. 1987 *A&A* **173**, 23.

ARP, H. 1966 *ApJS* **14**, 1.

BALDRY, I. K., GLAZEBROOK, K., BRINKMANN, J., IVEZIC, Z., LUPTON, R. H., NICHOL, R. C., & SZALAY, A. S. 2004 *ApJ* **600**, 681.

BARNES, J. E. & HERNQUIST, L. 1996 *ApJ* **471**, 115.

BAUGH, C. M., COLE, S., & FRENK, C. S. 1996 *MNRAS* **283**, 1361.

BELL, E. F. 2002 *ApJ* **577**, 150.

BELL, E. F. 2003 *ApJ* **586**, 794.

BELL, E. F. & DE JONG, R. S. 2001 *ApJ* **550**, 212.

BELL, E. F., GORDON, K. D., KENNICUTT, JR., R. C., & ZARITSKY, D. 2002 *ApJ* **565**, 994.

BELL, E. F., MCINTOSH, D. H., KATZ, N., & WEINBERG, M. D. 2003 *ApJS* **149**, 289.

BELL, E. F., ET AL. 2004a *ApJ* **600**, L11.

BELL, E. F., ET AL. 2004b *ApJ* **608**, 752.

BENDER, R. 1988 *A&A* **202**, L5.

BENDO, G. J. & BARNES, J. E. 2000 *MNRAS* **316**, 315.

BENSON, A. J., LACEY, C. G., BAUGH, C. M., COLE, S., & FRENK, C. S. 2002 *MNRAS* **333**, 156.

BOWER, R. G. & BALOGH, M. L. 2004. In *Clusters of Galaxies: Probes of Cosmological Structure and Galaxy Evolution* (eds. J. S. Mulchaey, A. Dressler & A. Oemler). p. 326. Cambridge University Press.

BOWER, R. G., LUCEY, J. R., & ELLIS, R. S. 1992 *MNRAS* **254**, 601.

BRINCHMANN, J. & ELLIS, R. S. 2000 *ApJ* **536**, L77.

BRESSAN, A., SILVA, L., & GRANATO, G. L. 2002 *A&A* **392**, 377.

BRUZUAL, G., & CHARLOT, S. 2003 *MNRAS* **344**, 1000.

BUNDY, K., FUKUGITA, M., ELLIS, R. S., KODAMA, T., & CONSELICE, C. J. 2004 *ApJ* **601**, L123.

CALZETTI, D. 2001 *PASP* **113**, 1449.

CALZETTI, D., KINNEY, A. L., & STORCHI-BERGMANN, T. 1994 *ApJ* **429**, 582–601.

CARLBERG, R. G., PRITCHET, C. J., & INFANTE, L. 1994 *ApJ* **435**, 540.

CHEN, H.-W., ET AL. 2003 *ApJ* **586**, 745.

CIMATTI, A., ET AL. 2002 *A&A* **381**, L68.

COLE, S., LACEY, C. G., BAUGH, C. M., & FRENK, C. S. 2000 *MNRAS* **319**, 168.

COLE, S., ET AL. 2001 *MNRAS* **326**, 255.

COLLESS, M., ET AL. 2001 *MNRAS* **328**, 1039.

CONDON, J. J. 1992 *ARA&A* **30**, 575.

CONSELICE, C. J., BERSHADY, M. A., DICKINSON, M., & PAPOVICH, C. 2003 *AJ* **126**, 1183.

CONSELICE, C. J., BERSHADY, M. A., & JANGREN, A. 2000 *ApJ* **529**, 886.

CONSELICE, C. J., BLACKBURNE, J. A., & PAPOVICH, C. 2005 *ApJ* **620**, 564.

CROSS, N. J. C., ET AL. 2004 *AJ* **128**, 1990.

DAVIES, R. L., EFSTATHIOU, G., FALL, S. M., ILLINGWORTH, G., SCHECHTER, P. L. 1983 *ApJ* **266**, 41.

DE LUCIA, G., ET AL. 2004 *ApJ* **610**, L77.

DICKINSON, M. E., PAPOVICH, C., FERGUSON, H. C., & BUDAVÁRI, T. 2003 *ApJ* **587**, 25.

DRORY, N., BENDER, R., FEULNER, G., HOPP, U., MARASTON, C. SNIGULA, J., & HILL, G. J. 2004 *ApJ* **608**, 742.

EGGEN, O. J., LYNDEN-BELL, D., & SANDAGE, A. R. 1962 *ApJ* **136**, 748.

EKE, V. R., ET AL. 2004 *MNRAS* **348**, 866.

FERRARESE, L., POGGE, R. W., PETERSON, B. M., MERRITT, D., WANDEL, A., & JOSEPH, C. L. 2001 *ApJ* **555**, L79.

FIOC, M. & ROCCA-VOLMERANGE, B. 1997 *A&A* **326**, 950.

FLORES, H., ET AL. 1999 *ApJ* **517**, 148.

FONTANA, A., ET AL. 2003 *ApJ* **594**, L9.

FREEDMAN, W. L., ET AL. 2001 *ApJ* **553**, 47.

GENZEL, R., TACCONI, L. J., RIGOPOULOU, D., LUTZ, D., & TECZA, M. 2001 *ApJ* **563**, 527.

GLAZEBROOK, K., ET AL. 2004 *Nature* **430**, 181.

GOLDADER, J. D., ET AL. 2002 *ApJ* **568**, 651.

GORDON, K. D., ET AL. 2004 *ApJS* **154**, 215.

HAARSMA, D. B., PARTRIDGE, R. B., WINDHORST, R. A., & RICHARDS, E. A. 2000 *ApJ* **544**, 641.

HAEHNELT, M. G. 2004. In *Coevolution of Black Holes and Galaxies* (ed. L. C. Ho). p. 406. Cambridge University Press.

HOGG, D. W. 2001 *PASP*, submitted; astro-ph/0105280.

HOPKINS, A. M. 2004 *ApJ* **615**, 209.

IM, M., ET AL. 2002 *ApJ* **571**, 136.

KAUFFMANN, G., ET AL. 2003 *MNRAS* **341**, 33.

KELSON, D. D., ILLINGWORTH, G. D., FRANX, M., & VAN DOKKUM, P. G. 2001 *ApJ* **522**, L17.

KENNICUTT, JR., R. C. 1983 *ApJ* **272**, 54.

KENNICUTT, JR., R. C. 1998 *ARA&A* **36**, 189.

KODAMA, T., ET AL. 2004 *MNRAS* **350**, 1005.

KONG, X., CHARLOT, S., BRINCHMANN, J., & FALL, S. M. 2004 *MNRAS* **349**, 769.

KROUPA, P. 2001 *MNRAS* **322**, 231.

LARSON, R. B. 1974 *MNRAS* **166**, 585.

LE FÈVRE, O., ET AL. 2000 *MNRAS* **311**, 565.

LILLY, S. J., LE FÈVRE, O., HAMMER, F., & CRAMPTON, D. 1996 *ApJ* **460**, L1.

LILLY, S. J., TRESSE, L., HAMMER, F., CRAMPTON, D., & LE FÈVRE, O. 1995 *ApJ* **455**, 108.

LISENFELD, U., VÖLK, H. J., & XU, C. 1996 *A&A* **314**, 745.

LOTZ, J. M., PRIMACK, J., & MADAU, P. 2004 *AJ* **128**, 163.

MADAU, P., FERGUSON, H. C., DICKINSON, M. E., GIAVALISCO, M., STEIDEL, C. C., & FRUCHTER, A. 1996 *MNRAS* **283**, 1388.

MEZA, A., NAVARRO, J. F., STEINMETZ, M., & EKE, V. R. *ApJ* **590**, 619.

MOUSTAKAS, L. A., ET AL. 2004 *ApJ* **600**, L131.

NAAB, T. & BURKERT, A. 2003 *ApJ* **597**, 893.

NIKLAS, S. & BECK, R. 1997 *A&A* **320**, 54.

PAPOVICH, C., DICKINSON, M. E., & FERGUSON, H. C. 2001 *ApJ* **559**, 620.

PATTON, D. R., ET AL. 2000 *ApJ* **536**, 153.

PATTON, D. R., ET AL. 2002 *ApJ* **565**, 208.

PEEBLES, P. J. E. 1980 *The large-scale structure of the universe.* Princeton, NJ: Princeton University Press.

RIESS, A. G., ET AL. 2004 *ApJ* **607**, 665.

RIX, H.-W., ET AL. 2004 *ApJS* **152**, 163.

RUDNICK, G., ET AL. 2003 *ApJ* **599**, 847.

SANDAGE, A. & VISVANATHAN, N. 1978 *ApJ* **223**, 707.

SCHWEIZER, F. & SEITZER, P. 1992 *AJ* **104**, 1039.

SELJAK, U., ET AL. 2005 *Phys. Rev. D* **71**, 103515.

SÉRSIC, J. L. 1968 *Atlas de Galaxias Australes.* Cordoba: Observatorio Astronomico.

SKRUTSKIE, M. F., ET AL. 1997. In *The Impact of Large Scale Near-IR Sky Surveys* (eds. F. Garzon, et al.). p. 25. Kluwer Academic Publishing Company.

SOMERVILLE, R. S., LEE, K., FERGUSON, H. C., GARDNER, J. P., MOUSTAKAS, L. A., & GIAVALISCO, M. 2004 *ApJ* **600**, L171.

SPERGEL, D. N., ET AL. 2003 *ApJS* **148**, 175.

STANWAY, E. R., BUNKER, A. J., McMAHON, R. G., ELLIS, R. S., TREU, T., & McCARTHY, P. J. 2004 *ApJ* **607**, 704.

STEIDEL, C. C., ADELBERGER, K. L., GIAVALISCO, M., DICKINSON, M., & PETTINI, M. 1999 *ApJ* **519**, 1.

STRATEVA, I., ET AL. 2001 *AJ* **122**, 1861.

TOOMRE, A. & TOOMRE, J. 1972 *ApJ* **178**, 623.

TULLY, R. B., PIERCE, M. J., HUANG, J. S., SAUNDERS, W., VERHEIJEN, M. A. W., & WITCHALLS, P. L. 1998 *AJ* **115**, 2264.

VAN KAMPEN, E., JIMENEZ, R., & PEACOCK, J. A. 1999 *MNRAS* **310**, 43.

WHITE, S. D. M. & FRENK, C. S. 1991 *ApJ* **379**, 52.

WHITE, S. D. M. & REES, M. J. 1978 *MNRAS* **183**, 341.

WITT, A. N. & GORDON, K. D. 2000 *ApJ* **528**, 799.

WOLF, C., MEISENHEIMER, K., RIX, H.-W., BORCH, A., DYE, S., & KLEINHEINRICH, M. 2003 *A&A* **401**, 73.

WORTHEY, G. 1994 *ApJS* **95**, 107.

WYUTS, S., VAN DOKKUM, P. G., KELSON, D. D., FRANX, M., ILLINGWORTH, G. D. 2004 *ApJ* **605**, 677.

YAN, L. & THOMPSON, D. 2003 *ApJ* **586**, 765.

YEE, H. K. C. & ELLINGSON, E. 1995 *ApJ* **445**, 37.

YORK, D. G., ET AL. 2000 *AJ* **120**, 1579.

ZEPF, S. E. & KOO, D. C. 1989 *ApJ* **337**, 34.

Probing the reionization history of the Universe

By Z. HAIMAN

Department of Astronomy, Columbia University, New York, NY 10027, USA

We discuss currently available observational constraints on the reionization history of the inter-galactic medium (IGM), and the extent to which accreting black holes (BHs) and stars can help account for these observations. We argue, based on the combined statistics of Lyman α and β absorption in quasar spectra, that the IGM contains a significant amount of neutral hydrogen with $n_{H\,I}/n_H \gtrsim 0.1$. On the other hand, we argue, based on the lack of a strong evolution in the observed abundance of Lyman α emitting galaxies beyond $z \sim 5.5$, that the mean neutral hydrogen fraction cannot exceed $n_{H\,I}/n_H \approx 0.3$ at the same redshift. We conclude that the IGM is experiencing rapid ionization at redshift $z \sim 6$.

We find that quasar BHs, including faint ones that are individually below the detection thresholds of existing optical and X-ray surveys, are unlikely to drive the evolution of the neutral fraction around this epoch, because they would over-produce the present-day soft X-ray background. On the other hand, the seeds of the $z \sim 6$ quasar BHs likely appeared at much earlier epochs ($z \sim 20$), and produced hard ionizing radiation by accretion. These early BHs are promising candidates to account for the high redshift ($z \sim 15$) ionization implied by the recent cosmic microwave anisotropy data from *WMAP*.

Using a model for the growth of BHs by accretion and mergers in a hierarchical cosmology, we suggest that the early growth of quasars must include a super-Eddington growth phase, and that, although not yet optically identified, the FIRST radio survey may have already detected several thousand $> 10^8\ M_\odot$ BHs at $z > 6$.

1. Black holes and reionization

The recent discovery of the Gunn-Peterson (GP) troughs in the spectra of $z > 6$ quasars in the Sloan Digital Sky Survey (SDSS; see Becker et al. 2001; Fan et al. 2003; White et al. 2003), has suggested that the end of the reionization process occurs at a redshift near $z \sim 6$. On the other hand, the high electron scattering optical depth, $\tau_e = 0.17 \pm 0.04$, measured recently by the *Wilkinson Microwave Anisotropy Probe* (*WMAP*) experiment (Spergel et al. 2003) suggests that ionizing sources were abundant at a much higher redshift, $z \sim 15$. These data imply that the reionization process is extended and complex, and is probably driven by more than one population of ionizing sources (see, e.g., Haiman 2003 for a recent review).

The exact nature of the ionizing sources remains unknown. Natural candidates to account for the onset of reionization at $z \sim 15$ are massive, metal-free stars that form in the shallow potential wells of the first collapsed dark matter halos (see Wyithe & Loeb 2003a; Cen 2003; Haiman & Holder 2003). The completion of reionization at $z \sim 6$ could then be accounted for by a normal population of less massive stars that form from the metal-enriched gas in more massive dark-matter halos present at $z \sim 6$.

The most natural alternative cause for reionization is the ionizing radiation produced by gas accretion onto an early population of black holes ("miniquasars"; Haiman & Loeb 1998). The ionizing emissivity of the known population of quasars diminishes rapidly beyond $z \gtrsim 3$, and bright quasars are unlikely to contribute significantly to the ionizing background at $z \gtrsim 5$ (see Shapiro, Giroux, & Babul 1994; Haiman, Abel, & Madau 2001). However, if low-luminosity, yet undetected miniquasars are present in large numbers, they could dominate the total ionizing background at $z \sim 6$ (Haiman & Loeb 1998). Recent work, motivated by the *WMAP* results, has emphasized the potential significant contribution to the ionizing background at the earliest epochs ($z \sim 15$) from accretion onto the seeds of would-be supermassive black holes (e.g., Madau et al. 2004; Ricotti & Ostriker 2004). The soft X-rays emitted by these sources can partially ionize the IGM early on (e.g., Oh 2001; Venkatesan, Giroux, & Shell 2001).

In this contribution, we address the following issues: (1) What is the fraction of neutral hydrogen in the IGM at $z \sim 6$? (2) Can quasar black holes contribute to reionization either at $z \sim 6$ or at $z \sim 15$? (3) How did BHs grow in a cosmological context, starting from early, stellar-mass seeds at $z \sim 20$? (4) Can we detect massive early ($z > 6$) black holes directly? Numerical statements throughout this paper assume a background cosmology with parameters $\Omega_m = 0.27$, $\Omega_\Lambda = 0.73$, $\Omega_b = 0.044$, and $h = 0.71$, consistent with the recent measurements by *WMAP* (Spergel et al. 2003).

2. What is the neutral fraction of hydrogen at $z \sim 6$?

The ionization state of the IGM at redshift $6 \lesssim z \lesssim 7$ has been a subject of intense study over the past few years. While the *WMAP* results imply that the IGM is significantly ionized out to $z \sim 15$, several pieces of evidence suggest that it has a high neutral fraction at $z \sim 6$–7. Here we describe a *lower* limit, from the analysis of the spectra of the $z \sim 6$ quasars, suggesting $x_{\mathrm{H\,I}} \gtrsim 0.1$, as well as an *upper* limit, from the (lack of) evolution of the abundance of Lyman α emitting galaxies, which suggests $x_{\mathrm{H\,I}} \lesssim 0.3$. The combination of these two methods brackets the mean neutral fraction to an accuracy of a factor of 2–3.

2.1. *A lower limit from quasar spectra*

2.1.1. *Previous arguments for a significant neutral fraction*

One argument against a simple model, in which the IGM is ionized at $z \sim 15$, and stays ionized thereafter, comes from the thermal history of the IGM (see Hui & Haiman 2003; Theuns et al. 2002). The temperature of the IGM, measured from the Lyα forest, is quite high at $z \sim 4$, with various groups finding values around $T \sim 20,000$ K (e.g., Zaldarriaga, Hui, & Tegmark 2001; McDonald et al. 2001; Schaye et al. 2000). As long as the universe is reionized before $z = 10$ *and remains highly ionized thereafter*, the IGM reaches an asymptotic thermal state, determined by a competition between photoionization heating and adiabatic cooling (the latter being due to the expansion of the universe). Under reasonable assumptions about the ionizing spectrum, the IGM then becomes too cold at $z = 4$ compared to observations. Therefore, as argued by Hui & Haiman (2003), there must have been significant (order unity) changes in fractions of neutral hydrogen and/or helium at $6 < z < 10$, and/or singly ionized helium at $4 < z < 10$. An important caveat to this argument is the possible existence of an additional heating mechanism that could raise the IGM temperature at $z \sim 4$. Galactic outflows could heat the IGM, in principle, but observations of close pairs of lines of sight in lens systems suggest that the IGM is not turbulent on small scales, arguing against significant stir-up of the IGM by winds (Rauch et al. 2001). The known quasar population is likely driving the reionization of helium, He II→He III, at $z \sim 3$ (e.g., Heap et al. 2000). If this process starts sufficiently early, i.e., at $z \gtrsim 4$, then He II photoionization heating could explain the high IGM temperature. It would be interesting to extend the search for He II patches that do not correlate with H I absorption to $z \sim 4$ to test this hypothesis.

A second argument for a large neutral fraction comes from the rapid redshift-evolution of the transmission near the redshifted Lyα wavelength in the spectra of distant quasars.

The Sloan Digital Sky Survey (SDSS) has detected large regions with no observable flux in the spectra of several $z \sim 6$ quasars (Becker et al. 2001; Fan et al. 2003; White et al. 2003). The presence of these Gunn-Peterson (GP) troughs by itself only sets a lower limit on the volume weighted hydrogen neutral fraction of $x_{HI} \gtrsim 10^{-3}$ (Fan et al. 2002). However, this strong limit implies a rapid evolution in the ionizing background, by nearly an order of magnitude in amplitude, from $z = 5.5$ to $z \sim 6$ (see Cen & McDonald 2002; Fan et al. 2002; Lidz, Hui, & Zaldarriaga 2002; we note that the Lyman β region of the spectra, which is needed for this conclusion, is dismissed in another recent study by Songaila 2004, which therefore reaches different conclusions). Known ionizing populations (quasars and Lyman break galaxies) do not evolve this rapidly; comparisons with numerical simulations of cosmological reionization (e.g., Gnedin 2004) suggests that we are, instead, witnessing the end of the reionization epoch, with the IGM becoming close to fully neutral at $z \sim 7$. At this epoch, when discrete H II bubbles percolate, the mean-free-path of ionizing photons can evolve very rapidly (e.g., Haiman & Loeb 1999a) and could explain the steep evolution of the background flux.

2.1.2. *Inferring the neutral fraction from the H II region in the quasar spectrum*

However, perhaps the strongest argument for a large neutral fraction comes from the presence of the cosmic Strömgren spheres surrounding high-z quasars. If indeed the intergalactic hydrogen is largely neutral at $z \sim 6$, then quasars at this redshift should be surrounded by large ionized (H II) regions, which will strongly modify their absorption spectra (Madau & Rees 2000; Cen & Haiman 2000). Recent work by Mesinger, Haiman, & Cen (2004) has shown that the damping wing of absorption by neutral hydrogen outside the H II region imprints a feature that is statistically measurable in a sample of ~ 10 bright quasars without any additional assumptions. A single quasar spectrum suffices if the size of the Strömgren sphere is constrained independently (see Mesinger, Haiman, & Cen 2004; Mesinger & Haiman 2004 for details). In addition, with a modest restriction (lower limit) on the age of the source, the size of the H II region itself can be used to place stringent limits on the neutral fraction of the ambient IGM (Wyithe & Loeb 2004a).

To elaborate on these last arguments, based on the quasar's H II region, in Figure 1 we illustrate a model for the optical depth to Lyman α absorption as a function of wavelength towards a $z_Q = 6.28$ quasar, embedded in a neutral medium ($x_{HI} = 1$), but surrounded by a Strömgren sphere with a comoving radius of $R_S = 44$ Mpc. Around bright quasars, such as those recently discovered (Fan et al. 2001; 2003) at $z \sim 6$, the proper radius of such Strömgren spheres is expected to be

$$R_S \approx 7.7\, x_{HI}^{-1/3}\, (\dot{N}_Q/6.5 \times 10^{57}\ \text{s}^{-1})^{1/3}\, (t_Q/2 \times 10^7\ \text{yr})^{1/3}\, [(1 + z_Q)/7.28]^{-1}\ \text{Mpc}$$

(Madau & Rees 2000; Cen & Haiman 2000). Here x_{HI} is the volume averaged neutral fraction of hydrogen outside the Strömgren sphere, and $\dot{N}_Q$, t_Q, and z_Q are the quasar's production rate of ionizing photons, age, and redshift. The fiducial values are those estimated for the $z = 6.28$ quasar J1030+0524 (Haiman & Cen 2002; Wyithe & Loeb 2004a). The mock spectrum shown in Figure 1 was created by computing the Lyman α opacity from a hydrodynamical simulation (kindly provided by R. Cen; the analysis procedure is described in Mesinger & Haiman 2004). The optical depth at a given observed wavelength, λ_{obs}, can be written as the sum of contributions from inside (τ_R) and outside (τ_D) the Strömgren sphere, $\tau_{Ly\alpha} = \tau_R + \tau_D$. The residual neutral hydrogen inside the Strömgren sphere at redshift $z < z_Q$ resonantly attenuates the quasar's flux at wavelengths around $\lambda_\alpha (1 + z)$, where $\lambda_\alpha = 1215.67$ Å is the rest-frame wavelength of the Lyman α line center. As a result, τ_R is a fluctuating function of wavelength (solid curve), reflecting the density fluctuations in the surrounding gas. In contrast, the damping wing

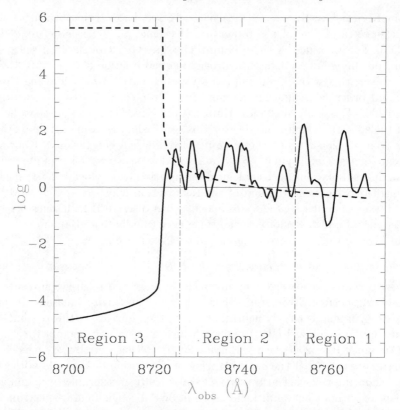

FIGURE 1. Model from a hydrodynamical simulation for the optical depth contributions from within (τ_R) and from outside (τ_D) the local ionized region for a typical line of sight towards a $z_Q = 6.28$ quasar embedded in a fully neutral, smooth IGM, but surrounded by a local H II region of (comoving) radius $R_S = 44$ Mpc. The *dashed curve* corresponds to τ_D, and the *solid curve* corresponds to τ_R. The total Lyman α optical depth is the sum of these two contributions, $\tau_{\mathrm{Ly}\alpha} = \tau_R + \tau_D$. The *dashed-dotted lines* demarcate the three wavelength regions used for our analysis described in the text. For reference, the redshifted Lyman α wavelength is at 8852 Å, far to the right off the plot.

of the absorption, τ_D, is a smooth function (dashed curve), because its value is averaged over many density fluctuations. As the figure shows, the damping wing of the absorption from the neutral universe extends into wavelengths $\lambda_{\mathrm{obs}} \gtrsim 8720$ Å, and can add significantly to the total optical depth in this region.

The sharp rise in τ_D at wavelengths $\lambda_{\mathrm{obs}} \lesssim 8720$ Å is a unique feature of the boundary of the H II region, and corresponds to absorption of photons redshifting into resonance outside of the Strömgren sphere. The detection of this feature has been regarded as challenging: since the quasar's flux is attenuated by a factor of $\exp(-\tau_{\mathrm{Ly}\alpha})$, an exceedingly large dynamical range is required in the corresponding flux measurements. However, Mesinger & Haiman (2004) showed that simultaneously considering the measured absorption in two or more hydrogen Lyman lines can provide the dynamical range required to detect this feature.

In particular, we modeled broad features of the Lyman α and Lyman β regions of the absorption spectrum of the $z = 6.28$ quasar SDSS J1030+0524. The observational input to our analysis is the deepest available absorption spectrum of SDSS J1030+0524, adapted from White et al. (2003), shown in Figure 2. The spectrum exhibits a strong Lyman α Gunn-Peterson (GP) trough, with no detectable flux between wavelengths

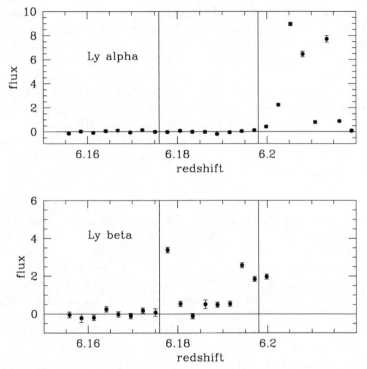

FIGURE 2. The Keck ESI spectrum of the $z = 6.28$ quasar SDSS J1030+0524, adapted from White et al. (2003), in units of 10^{-18} erg s^{-1} cm^{-2} Å^{-1}, and including uncorrelated 1σ errors. The upper and lower panels show the regions of the spectrum corresponding to Lyman α and β absorption in the redshift range $6.15 < z < 6.28$. The Lyman β cross section is smaller than Lyman α, and a significantly higher column of neutral hydrogen is required to block the flux in the Lyman β region than in the Lyman α region. This spectrum therefore requires a steep increase in the opacity over the narrow range from $z = 6.20$ to $z = 6.16$, explainable only by a GP damping wing, as shown by the dashed curve in Figure 1.

corresponding to redshifts $5.97 < z < 6.20$, as well as a somewhat narrower Lyman β trough between $5.97 < z < 6.18$, as shown in Figure 2. The flux detection threshold in the Lyman α and Lyman β regions of this spectrum correspond to Lyman α optical depths of $\tau_{\lim(\text{Ly}\alpha)} \approx 6.3$ and $\tau_{\lim(\text{Ly}\beta)} \approx 22.8$ respectively.†

To summarize these constraints, we have divided the spectrum into three regions, shown in Figure 1. In Region 1, with $\lambda_{\text{obs}} \geqslant 8752.5$ Å, the detection of flux corresponds to the *upper* limit on the optical depth $\tau_{\text{Ly}\alpha} < 6.3$. Region 2, extending from 8725.8 Å $\leqslant \lambda_{\text{obs}} < 8752.5$ Å, is inside the Lyman α trough, but outside the Lyman β trough. Throughout this region, the data requires $6.3 \lesssim \tau_{\text{Ly}\alpha} \lesssim 22.8$. Region 3, with $\lambda_{\text{obs}} < 8725.8$ Å, has a *lower* limit $\tau_{\text{Ly}\alpha} \geqslant 22.8$. As defined, each of these three regions contains approximately eight pixels.

Mesinger & Haiman (2004) modeled the absorption spectrum, attempting to match these gross observed features. We utilized a hydrodynamical simulation that describes

† For our purposes, these optical depths can be taken as rough estimates. Their precise values are difficult to calculate, with $\tau_{\lim(\text{Ly}\beta)}$ especially uncertain (see, for example, Songaila 2004; Lidz, Hui, & Zaldarriaga 2002; Cen & McDonald 2002; Fan et al. 2002). However, we have verified that our conclusions below remain unchanged when the threshold opacities are varied well in excess of these uncertainties. In particular, considering ranges as wide as $5.5 < \tau_{\lim(\text{Ly}\alpha)} < 7$ and $10 < \tau_{\lim(\text{Ly}\beta)} < 30$ would lead to constraints similar to those we derive below.

the density distribution surrounding the source quasar at $z = 6.28$. We extracted density and velocity information from 100 randomly chosen lines of sight (LOSs) through the simulation box. Along each line of sight (LOS), we computed the Lyman α absorption as a function of wavelength. The size of the ionized region (R_S) and the fraction of neutral hydrogen outside it ($x_{\mathrm{H\,I}}$), and the quasar's ionizing luminosity, L_{ion}, were free parameters. Note that changing R_S moves the dashed (τ_D) curve in Figure 1 left and right, while changing $x_{\mathrm{H\,I}}$ moves it up and down; changing L_{ion} moves the solid (τ_R) curve up and down. We evaluated τ_R and τ_D for each LOS, and for each point in a three-dimensional parameter space of R_S, $x_{\mathrm{H\,I}}$, and L_{ion}, we computed the fraction of the LOSs that were acceptable descriptions of the spectrum of SDSS J1030+0524, based on the criteria defined above.

2.1.3. *The minimum allowed neutral fraction*

The procedure outlined above turns out to provide tight constrains on all three of our free parameters *simultaneously*. In particular, we find the allowed range for the radius of the Strömgren sphere to be 42 Mpc $\leqslant R_S \leqslant$ 47 Mpc, and a $\sim 1\sigma$ lower limit on the neutral fraction of $x_{\mathrm{H\,I}} \gtrsim 0.17$. These results can be interpreted as follows. As mentioned previously, the presence of flux in Region 2 ($\tau_{\mathrm{Ly}\alpha} < 22.8$) sets an immediate *lower limit* on R_S. Region 3, however, yields an *upper limit* on R_S, from the requirement that $\tau_{\mathrm{Ly}\alpha} > 22.8$ in that region. This high optical depth cannot be maintained by τ_R alone, without violating the constraint in Region 1 of $\tau_{\mathrm{Ly}\alpha} < 6.3$. We note that tight constraints on the neutral fraction ($x_{\mathrm{H\,I}} \gtrsim 0.1$) can be obtained from the size of the Strömgren sphere, together with an assumed lower limit on its lifetime (Wyithe & Loeb 2004). Our direct determination of the Strömgren sphere size is only slightly larger than the value assumed in Wyithe & Loeb (2004), lending further credibility to this conclusion.

Our direct constraint on the neutral hydrogen fraction, $x_{\mathrm{H\,I}}$, on the other hand, comes from the presence of flux in the Lyman β region of the spectrum corresponding to Region 2. Because of fluctuations in the density field (and hence in τ_R), a strong damping wing is needed to raise $\tau_{\mathrm{Ly}\alpha}$ above 6.3 throughout Region 2, *while still preserving* $\tau_{\mathrm{Ly}\alpha} < 6.3$ in Region 1. This result is derived from the observed sharpness of the boundary of the H II region alone, and relies only on the gross density fluctuation statistics from the numerical simulation. *In particular, it does not rely on any assumption about the mechanism for the growth of the H II region.*

Using a large sample of quasars (and/or a sample of gamma-ray burst afterglows with near-IR spectra) at $z > 6$, it will be possible to use the method presented here to locate sharp features in the absorption spectrum from intervening H II regions, not associated with the target source itself. The Universe must have gone through a transition epoch when H II regions, driven into the IGM by quasars and galaxies, partially percolated and filled a significant fraction of the volume. The detection of the associated sharp features in future quasar absorption spectra will provide a direct probe of the 3D topology of ionized regions during this crucial transition epoch (in particular, it should enhance any constraint available from either the Lyman α or β region alone; Furlanetto, Hernquist, & Zaldarriaga 2004).

2.2. *An upper limit from the abundance of Lyman α emitters*

Lyα emission lines from high-redshift sources can serve as another probe of the ionization state of the IGM. The damping wing of the GP absorption from the IGM can cause a characteristic absorption feature (Miralda-Escudé 1998). In a significantly neutral IGM, the absorption can produce conspicuous effects, i.e., attenuating the emission line, making it asymmetric, and shifting its apparent peak to longer wavelengths (Haiman 2002; Santos

2004). These effects generally increase with $\langle x_{HI} \rangle$, and it has been suggested that a strong drop in the abundance of Lyman α emitters beyond the reionization redshift can serve as a diagnostic of a nearly neutral IGM (Haiman & Spaans 1999). Recently, Malhotra & Rhoads (2004; hereafter MR04) and Stern et al. (2004) carried out the first application of this technique, by comparing the luminosity functions (LFs) of Lyman α emitters at $z = 5.7$ and $z = 6.5$. The LF shows no evolution in this range, and this has been interpreted as evidence against percolation taking place near $z \sim 6$.

A potential caveat for this interpretation is that $z \sim 6$ galaxies will be surrounded by their own local cosmological H II regions, which can significantly reduce the attenuation of the Lyman α line flux (Cen & Haiman 2000; Madau & Rees 2000). It has been shown that this can render Lyman α lines detectable even if the galaxies are embedded in a fully neutral IGM (Haiman 2002; Santos 2004; Cen et al. 2004), especially when the increased transmission due to the clustering of ionizing sources is taken into account (Furlanetto et al. 2004; Gnedin & Prada 2004; Wyithe & Loeb 2004b), and/or if the Lyman α emission has a significant recession velocity with respect to the absorbing gas.

In a recent paper (Haiman & Cen 2004), we studied the impact of IGM absorption on the Lyman α LF. This analysis is similar to preceding work by MR04; the improvement is that we model the attenuation of Lyα lines including cosmological H II regions. While very little attenuation (by a factor of $\lesssim 2$) can be tolerated by the lack of evolution of the LF over the range $5.7 \lesssim z \lesssim 6.5$, we find that the presence of local H II regions allows a neutral fraction as high as $\langle x_{HI} \rangle = 0.25$. On the other hand, $\langle x_{HI} \rangle \sim 1$ is allowed only if the ionizing sources are unusually strongly clustered. In either case, we find that the present Lyman α LFs are consistent with reionization occurring near $z \sim 6$.

Following Malhotra & Rhoads (2004; hereafter MR04) we take as basic inputs to our analysis the Lyman α LFs at $z = 5.7$ and $z = 6.5$. We adopt the forms of the LF at these two redshifts from MR04. Note that at both redshifts, the LFs are inferred by culling several independent datasets (Hu et al. 2002; Rhoads et al. 2003; Kodaira et al. 2004). At $z = 5.7$, the LF is comparatively well determined and well fit by a Schechter function with relatively small errors in normalization ($\Phi_\star$) and characteristic luminosity ($L_\star$), while the faint-end slope (α) is less well determined. Here we adopt the best-fit values of $(\Phi_\star, L_\star, \alpha) = (10^{-4} \text{ Mpc}^{-3}, 10^{43} \text{ erg s}^{-1}, -1.5)$ (displayed by the short-dashed curve in Fig. 4 below). Within the uncertainties quoted in MR04, the values of these parameters do not significantly change our conclusions below.

At $z = 6.5$, the LF is less well determined, as shown by four data-points with errors in Figure 4 below. Of these four data-points, the most constraining is from the deep Subaru field, which revealed the presence of a handful of emitters at $z \sim 6.5$ (Kodaira et al. 2003), and has the smallest error bar. The single, faint-lensed galaxy at this redshift found by Hu et al. (2002; represented by the lowest-luminosity point in Fig. 4 below) implies a surprisingly high abundance of the faintest emitters, but with a single source, it has a large uncertainty. Finally, we follow MR04 and conservatively assume that the intrinsic LF (i.e., prior to processing of the line through the IGM) is the same at $z = 6.5$ as at $z = 5.7$. More realistic models, based on the hierarchical growth of structures, combined with Lyman α line processing by gas and dust interior to the galaxies, predict that the number of emitters were smaller at earlier epochs (Haiman & Spaans 1999; Le Delliou et al. 2004), which would strengthen the limits on $\langle x_{HI} \rangle$ below.

2.2.1. *The attenuation of Lyman α emission lines*

We follow Cen & Haiman (2000), and use a simple model to find the attenuation of the Lyα line as a function of wavelength. The model is straightforward, and we only briefly recap the main features. We start with a Gaussian emission line originating at

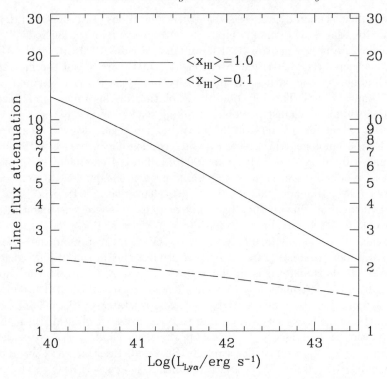

FIGURE 3. The suppression of the total Lyα line flux, as a function of the observed line luminosity, for an IGM with $\langle x_{\rm H\,I} \rangle = 0.1$ (or $\Gamma_{12} = 0.0015$; long-dashed curve), and for and IGM with $\langle x_{\rm H\,I} \rangle = 1.0$ (or $\Gamma_{12} = 0$; solid curve).

$z_s = 6.5$, with a line-width of $\Delta v = 300$ km s^{-1}, typical of normal galaxies (narrower lines would be attenuated more severely; see discussion below). We assume that the source is surrounded by a spherical Strömgren sphere that propagated into the IGM with a neutral fraction $\langle x_{\rm H\,I} \rangle$. The IGM is assumed to have a clumping factor $C \equiv \langle \rho^2 \rangle / \langle \rho \rangle^2 = 10$ with a log-normal distribution. We assume the presence of a uniform ionizing background, characterized by an ionizing rate Γ_{12} (ionizations per 10^{12} s per atom), yielding the given value of $\langle x_{\rm H\,I} \rangle$ in ionization equilibrium for the IGM.

We assume a Salpeter IMF, and the relation (Kennicutt 1983) between star-formation rate and Lyman α luminosity (SFR $= 1$ $M_\odot$/yr) $\rightarrow$ (L(Lyα) $= 10^{42}$ erg s^{-1}). We conservatively assume $f_{\rm esc} = 1$ (escape fraction of ionizing radiation) in computing R_s, and a lifetime $t_s = 10^8$ yrs for each source. With all other parameters in the model fixed, we vary Γ_{12} (or equivalently $\langle x_{\rm H\,I} \rangle$), and compute the factor by which the total line flux is suppressed relative to the unabsorbed Gaussian line. This factor is shown in Figure 3 as a function of the intrinsic luminosity of the source. Faint sources are more severely attenuated: by a factor of up to ~ 15 for $\log L$(Lyman α/erg s^{-1}) $= 40$ [or up to ~ 6 for $\log L$(Lyman α/erg s^{-1}) $= 42$, the present lower limit for the detection of a Lyman α emitter at $z = 6.5$] in a neutral universe. Bright ($\log L$(Lyman α/erg s^{-1}) $= 43$) sources, however, are only attenuated by a factor of ~ 1.5 even if $\langle x_{\rm H\,I} \rangle$ is as high as 0.1.

The attenuation factors in Figure 3 are computed by assuming that the H II regions are being created by a single ionizing source (the Lyman α emitter itself). These H II regions (with typical sizes of a few Mpc in the observed range of source luminosities) may contain many other undetected galaxies, which make the H II regions larger (Furlanetto et al. 2004; Gnedin & Prada 2004; Wyithe & Loeb 2004b) and also more highly ionized.

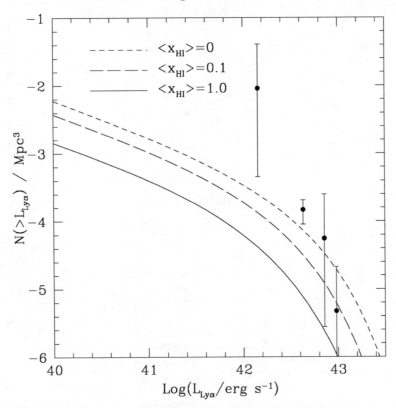

FIGURE 4. The Lyα luminosity functions for a fully ionized IGM (no suppression, short-dashed curve), for an IGM with $\langle x_{HI} \rangle = 0.1$ (or $\Gamma_{12} = 0.0015$; long-dashed curve), and for and IGM with $\langle x_{HI} \rangle = 1.0$ (or $\Gamma_{12} = 0$; solid curve).

Therefore, the results shown in Figure 3 should be viewed as conservative upper limits. These results show that the H II regions of individual galaxies can significantly reduce the attenuation, especially for relatively bright sources.

2.2.2. *The Impact on the Lyα Luminosity Function*

We next use the attenuation factors computed above, and obtain the reduction in the cumulative Lyman α LF. The resulting LF is shown in Figure 4, for three values of $\langle x_{HI} \rangle$: 0, 0.1, and 1 (top to bottom). The plot also shows the data for the $z = 6.5$ LF with error bars (from MR04). The short–dashed curve is a reproduction of the $z = 5.7$ LF (a Schechter function), which, by assumption, would be identical to the $z = 6.5$ LF if the IGM contained no neutral H I.

Note that the suppression of the line flux depends on the luminosity of the source, but the reduction of the LF in Figure 4 is uniform, by a nearly constant factor. The steepening of the LFs at the bright end compensates for the reduced effect of line attenuation; the two conspire to give a nearly constant suppression of the cumulative LF.

2.2.3. *The maximum allowed neutral fraction*

We can now assign a likelihood to each value of $\langle x_{HI} \rangle$, by comparing the model LFs and the data shown in Figure 4. For each value of $\langle x_{HI} \rangle$, we obtain the mean expected number of Lyman α emitters above the flux thresholds shown by the data points (here we use the effective volume probed by each survey, listed in Table 2 in MR04). We assume Poisson fluctuations to compute the likelihood in each bin. We find that the lowest-luminosity

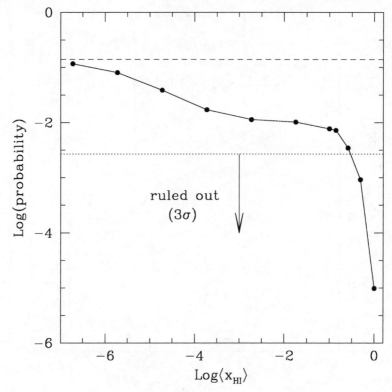

FIGURE 5. The probability of each mean neutral fraction in the IGM, given the observational constraints from the $z = 6.5$ Lyα luminosity function. The model assumes that the intrinsic LF is the same as at $z = 5.7$ (see text for discussion).

data-point has an *a priori* Poisson likelihood that is exceedingly low, even in a fully ionized universe. As Figure 4 shows, $\lesssim 0.03$ objects are expected in the small survey volume in which the single lensed Lyman α emitter, represented by this data-point, was discovered. However, given the difficulties in obtaining an accurate effective volume from the poorly constraint lensing configuration for this object, we follow MR04 and omit this source from our analysis. The result, i.e., the product of the Poisson probabilities for the three higher luminosity bins, are shown in Figure 5 as a function of $\langle x_{HI} \rangle$.

We find that $\langle x_{HI} \rangle = 1$ is ruled out at high significance, in agreement with the conclusions reached by MR04 and Stern et al. (2004, but see discussion below). On the other hand, at the 99.7% confidence level, we find only the relatively weak constraint $\langle x_{HI} \rangle < 0.25$. The upper dashed line corresponds to the limit of no H I absorption and shows that the data is consistent (at the $\sim 10\%$ confidence level) with the intrinsic $z = 5.7$ and $z = 6.5$ LFs being identical (but note that the IGM would have to be ionized to an unrealistically high level to actually approach this limit).

As mentioned above, H II regions can be enlarged due to the clustering of ionizing sources, and also due to their being elongated along our line of sight (e.g., Gnedin & Prada 2004). These effects reduce the Lyman α attenuation, and weaken the upper limit on $\langle x_{HI} \rangle$. It is interesting to ask whether *any* constraint can be placed on the neutral fraction, in the presence of clustering. To address this question, we re-computed the attenuation factors shown in Figure 3, and the implied LFs shown in Figure 4, fixing $\langle x_{HI} \rangle = 1$, but boosting the ionizing emissivity of each source by a constant factor of B. We find that the boosting factor required to render this model acceptable at the 99.7%

confidence level is approximately $B = 60$ (note that the increased luminosity is required to reduce both the GP damping wing, and the residual H I inside the H II region).

In order to assess whether the ionizing emissivity can be boosted by this factor, we computed the total mass in dark matter halos in the mass range M_{min} to M_s and within a sphere of radius R around a halo of mass M_s. We included the spatial correlations between halos with a prescription for linear bias (see eq. 15 in Haiman et al. 2001). For the likely halo mass range of $10^{11} M_\odot \lesssim M \lesssim 10^{12} M_\odot$ (Rhoads et al. 2003) and H II sphere radii ($2 \lesssim R \lesssim 6$ Mpc; e.g., Haiman 2002) of Lyman α emitters at $z = 6.5$, we find that the total collapsed halo mass, down to $2 \times 10^8 M_\odot$ (corresponding to a virial temperature of 10^4 K at $z = 6.5$) is increased by a factor of 2–18 (the satellite halos have linear bias parameters of $2 \lesssim b \lesssim 10$). Unless the star formation efficiency is preferentially higher in the low-mass "satellite" galaxies, it is unlikely that the ionizing emissivity is boosted by the required factor of 60. A recent numerical simulation by Gnedin & Prada (2004, see their Fig. 3) however, suggests that at least 10% of galaxies may transmit their Lyman α lines without any absorption at $z = 6.5$.

Another effect that can further weaken the upper limits on the neutral fraction is a systematic redshift of the Lyman α emission lines relative to the absorbing gas. Lyman break galaxies at lower redshift reveal such systematic shift by several hundred km/s, attributed to galactic winds (Shapley et al. 2003). If velocity offsets as high as this are common for Lyman α emitters at $z \sim 6$, this would render Lyman α line attenuation from the residual H I within the H II region effectively negligible, and will also significantly reduce the GP damping wing absorption. By repeating our analysis above to obtain a likelihood for a neutral universe, we find that a redshift by 600 km s^{-1} is required to for $\langle x_{HI} \rangle = 1$ to yield an acceptable fit for the LF. Shapley et al. (2003) find a decreasing velocity offset with increasing Lyman α equivalent width, with $\langle \Delta v \rangle < 500$ km s^{-1} for $W_\alpha > 50$ Å (see their Fig. 11), suggesting that high-z Lyman α emitters may not have the requisite 600 km s^{-1} offsets.

Finally, we propose a different diagnostic of the neutral fraction that could be available from a future, larger sample of $z \gtrsim 6$ Lyman α emitters. For faint Lyman α emitters, the line attenuation is dominated by the residual H I inside their cosmic H II regions, whereas for bright emitters, the GP damping wing is dominant (Haiman 2002). In Figure 6, we show the predicted full width at half maximum (FWHM) of the transmitted line, as a function of Lyman α luminosity (for two different intrinsic line-widths). The figure reveals a clear imprint of this transition at $\sim 10^{41-42}$ erg s^{-1}. This feature occurs only if the IGM contains a significant amount of neutral hydrogen (solid curves), and we checked that it is present for different assumed intrinsic line-shapes (e.g., a top-hat or a Lorentzian, which have less/more extended wings than a Gaussian). This suggests that a measurement of observed Lyman α line width as a function of the Lyman α luminosity may therefore serve as a more robust diagnostic of the neutral fraction.

3. Did accreting black holes contribute to reionization?

The two most natural types of UV sources that could have reionized the IGM are stars or accreting black holes. Deciding which of these two sources dominated the ionization has been studied for over 30 years (e.g., Arons & Wingert 1972). It has become increasingly clear over the past decade that the ionizing emissivity of the known population of bright quasars diminishes rapidly beyond $z \gtrsim 3$, and they are unlikely to contribute significantly to the ionizing background at $z \gtrsim 5$ (e.g., Shapiro, Giroux, & Babul 1994; Haiman, Abel, & Madau 2001). This, however, leaves open two possibilities. First, if low-luminosity, yet undetected miniquasars are present in large numbers, they could still dominate the total

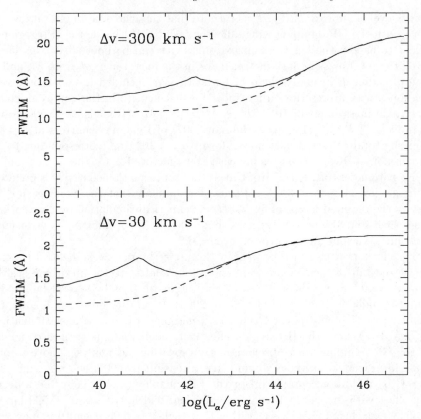

FIGURE 6. The FWHM of the transmitted Lyman α line as a function of the luminosity of the source. The upper (lower) panel assumes an intrinsic Gaussian line shape with a width of 300 (30) km s^{-1}. On both panels, the dashed curve assumes an ionizing background of $\Gamma_{12} = 0.1$ (or $\langle x_{\mathrm{H\,I}} \rangle \sim 10^{-3}$), and the solid curve assumes $\Gamma_{12} = 0$ ($\langle x_{\mathrm{H\,I}} \rangle = 1$). The feature at $\sim 10^{41-42}$ erg s^{-1} is unique to a significantly neutral IGM.

ionizing background at $z \sim 6$ (Haiman & Loeb 1998). Second, the supermassive black holes at $z \sim 6$ must be assembled from lower-mass seeds which accrete and merge during the hierarchical growth of structure. The population of accreting seed BHs can contribute to the ionization of the IGM at $z \sim 20$ (e.g., Madau et al. 2004; Ricotti & Ostriker 2004).

The above two possibilities can both involve faint BHs that are not individually detectable. However, a population of accreting BHs at $z \gtrsim 6$ would be accompanied by the presence of an early X-ray background. Since the IGM is optically thick to photons with energies E below

$$E_{\mathrm{max}} = 1.8[(1+z)/15)]^{0.5} x_{\mathrm{H\,I}}^{1/3} \text{ keV} \quad ,$$

the soft X-rays with $E \lesssim E_{\mathrm{max}}$ would be consumed by neutral hydrogen atoms and contribute to reionization. However, the background of harder X-rays would redshift without absorption and would be observed as a present-day soft X-ray background (SXB). Under the hypothesis that accreting BHs are the main producers of reionizing photons in the high-redshift universe, it is a relatively straightforward exercise to calculate their contribution to the present-day SXB.

Dijkstra et al. (2004) assumed for simplicity that the accreting BHs form in a sudden burst at redshift $z = z_Q$. The total number of BHs was expressed by a normalization constant η, defined as the ratio of the total number of ionizing photons emitted per unit volume produced by the BH population to the number density of hydrogen atoms. Full ionization of the IGM at $z \sim 6$, where recombinations are significant, likely requires $\eta \gtrsim 10$ (Haiman, Abel, & Madau 2001). We find that in order to account for the electron scattering optical depth $\tau_e \sim 0.17$ by partially (pre-)ionizing the IGM at $z \sim 15$, a somewhat smaller η is sufficient (see below).

The spectrum of the ionizing background is a crucial ingredient of the modeling, and depends on the type of accreting BH that is considered. For luminous quasars powered by supermassive black holes, we adopted a composite spectrum from Sazonov, Ostriker, & Sunyaev (2004), based on the populations of lower redshifts ($0 < z < 5$) QSOs. The spectra of lower-mass miniquasars, with BHs whose masses are in the range $M_{\rm bh} \approx 10^{2-4}\ M_\odot$, are likely to be harder. For these sources, we followed Madau et al. (2004) and adopted a two-component template that includes a multi-temperature accretion disk, and a simple power-law emission to mimic a combination of Bremsstrahlung, synchrotron, and inverse Compton emission by a non-thermal population of electrons.

Finally, we took the unresolved soft X-ray band in the energy range 0.5–2.0 keV to be in the range $(0.35$–$1.23) \times 10^{-12}$ erg s^{-1} cm^{-2} deg^{-2}. This range was obtained by Dijkstra et al. (2004) from a census of resolved X-ray sources, which we subtracted from the total SXB. We included the uncertainties in both the measurement of the total SXB (which was dominant) and in the number of resolved point sources. We examined a range of X-ray energies, and found the strongest constraints in the 0.5–2 keV band. We note that a recent study by Bauer et al. (2004) of faint X-ray sources found a significant population of star-forming galaxies among these sources, with a steeply increasing fractional abundance (over AGNs) toward low X-ray luminosities. If this trend continues to a flux limit that is only modestly below the current point-source detection threshold in the deepest *Chandra* fields, then the SXB would be saturated, strengthening our constraints (leaving less room for any additional, high-z quasars).

We find that models in which $z > 6$ accreting BHs alone fully reionize the universe saturate the unresolved X-ray background at the $\geqslant 2\sigma$ level. Pre-ionization by miniquasars requires fewer ionizing photons, because only a fraction of the hydrogen atoms need to be ionized, the hard X-rays can produce multiple secondary ionizations, and the clumping factor is expected to be significantly smaller than in the UV-ionization case (Haiman, Abel, & Madau 2001; Oh 2001). We find that models in which X-rays are assumed to partially ionize the IGM up to $x_e \sim 0.5$ at $6 \lesssim z \lesssim 20$ are still allowed, but could be constrained by improved future determinations of the unresolved component of the SXB.

As emphasized above, the spectral shape of the putative typical high-z accreting BH is uncertain; the existing templates, motivated by lower-redshift sources, can be considered merely as guides. Figure 7 shows which combinations of α (the logarithmic slope of the ionizing spectrum) and η are allowed by the unaccounted flux in the SXB (the solid and dotted curves cover our inferred range of the unresolved SXB). This figure shows that for $\eta = 10$, a power-law shallower than $\alpha \approx 1.2$–1.4 will saturate the unaccounted flux. For comparison, α is in the range $-1.5 \lesssim \alpha \lesssim -0.5$ for $z \lesssim 0.3$, and $-1.2 \lesssim \alpha \lesssim -0.6$ for $1 \lesssim z \lesssim 6$ for optically selected radio quiet quasars (Vignali et al. 2003).

Our constraints derive from the total number of ionizing photons that the population as a whole needs to produce to either fully or partially reionize the universe. Therefore, our conclusions depend mostly on the assumed spectral shape and the required number of ionizing photons per hydrogen atom η. They are independent of the details of the population, such as the luminosity function and its evolution with redshift. Future improvements in resolving the SXB, improving the limits on the unresolved component by a

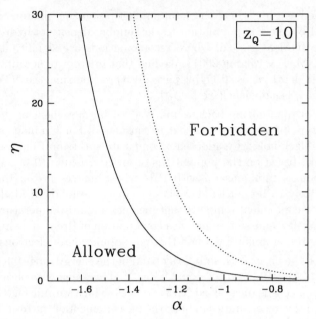

FIGURE 7. Constraints on the number of ionizing photons per H atom, η, and the power-law index for the slope of the ionizing background, α, based on the intensity of the present-day SXB. The quasars are assumed to form at $z_Q = 10$ and have a power-law spectrum, $F_E \propto E^{-\alpha}$ for $E > 13.6$ eV. The curves bracket the allowed parameter space for the mean (solid line) or maximum (dotted) unaccounted flux in the SXB at ~1 keV. The curves shift by +0.1 in α if $z_Q = 15$ is assumed.

factor of a few, would place stringent constraints on the contribution of $z \sim 15$ accreting BHs to the scattering optical depth measured by *WMAP*.

4. How did black holes grow by accretion and mergers?

Since the seeds of early BHs may have played a role in reionization, it is all the more interesting to ask how the earliest BH population was formed, and how it evolved. The remnants of metal-free population III stars that form in the first collapsed dark halos can serve as the initial ~100 $M_\odot$ seeds that later accrete and merge together, to give rise to the supermassive BHs making up the quasar population at $0 < z < 6$, and the remnant BHs found at the centers of local galaxies. Several recent studies, including those by Kauffmann & Haehnelt (2000), Menou, Haiman, & Narayanan (2001), Volonteri, Haardt, & Madau (2003), Islam, Taylor, & Silk (2003), have addressed various aspects of the evolution of the BH populations, using the underlying merger trees of dark matter halos.

Recent work by Favata, Hughes, & Holz (2004) and Merritt et al. (2004) on the generation of gravitational waves during the coalescence of a binary black hole has suggested that the binary experiences a typical gravitational recoil velocity that may be as large as $\gtrsim 100$ km s^{-1}. These velocities exceed the escape velocity $v_{\rm esc}$ from typical dark matter halos at high-redshift ($z \gtrsim 6$), and can therefore disrupt the early stages of growth of BHs by ejecting the earliest seeds from their host galaxies. BHs can then start growing effectively only once the typical dark matter potential wells are sufficiently deep to retain the recoiling BHs.

Relatively little time is available for the growth of few $\times 10^9$ $M_\odot$ SMBHs prior to $z \sim 6$, and their seed BHs must be present as early as $z \sim 10$ (Haiman & Loeb 2001). A model in which stellar seed BHs appear in small progenitor DM halos is consistent with the presence of a $\sim 4 \times 10^9$ $M_\odot$ SMBH at $z \sim 10$, provided that each seed BH can grow at least at the Eddington-limited exponential rate, and that the progenitor halos can form seed BHs sufficiently early on (Haiman & Loeb 2001), in halos with velocity dispersions of $\sigma \sim 30$ km s^{-1}.

We quantified the effect of gravitational recoil on the growth of a few $\times 10^9$ $M_\odot$ black hole (Haiman 2004), by assuming that progenitor holes are ejected from DM halos with velocity dispersions $\sigma < v_{\rm kick}/2$, and do not contribute to the final BH mass. Each halo more massive than this threshold was assumed to host a seed BH that accretes at the Eddington rate, and the BHs were assumed to coalesce when their parents halos merged. We took, as an example, the SMBH powering the most distant SDSS quasar, SDSS 1054+1024 at redshift $z = 6.43$, with an inferred BH mass of $\sim 4 \times 10^9$ $M_\odot$. We find that recoil velocities with $v_{\rm kick} \gtrsim 65$ km s^{-1} must occur infrequently, or else this SMBH must have had a phase during which it gained mass significantly more rapidly than an Eddington-limited exponential growth rate (with a radiative efficiency of $\sim 10\%$) would imply. Yoo & Miralda-Escudé (2004) find that the super-Eddington growth phase can be avoided if seed BHs can form and grow by mergers in dark halos with a velocity dispersion as small as $\sigma \approx 5$ km s^{-1}, the halos have steep density profiles (increasing the escape velocity from the central regions by a factor of $\gtrsim 5$ relative to the naive formula $v_{\rm esc} = 2\sigma$), and the BHs that are retained during the mergers of their halos can grow uninterrupted at the Eddington rate between their birth and $z \approx 6$.

5. Can we detect massive $z > 6$ BHs directly?

A natural question to ask is whether massive BHs at $z > 6$ can be directly detected. While the SDSS has detected a handful of exceptionally bright, few $\times 10^9$ $M_\odot$ black holes (the BH mass is inferred assuming these sources shine at the Eddington luminosity), they are likely a "tip of the iceberg," corresponding to the rare massive tail of the BH mass function. In X-ray bands, the deepest *Chandra* fields have reached the sensitivity to detect nearly ~ 100 times smaller holes (Haiman & Loeb 1999b, provided they radiate the Eddington luminosity, with a few percent of their emission in the X-ray bands). However, due to the small size of these fields, they have revealed only a handful of plausible candidates (Alexander et al. 2001; Barger et al. 2003).

Detections, however, seem promising in the large radio survey FIRST. We used a physically motivated semi-analytic model, based on the mass function of dark matter halos, to predict the number of radio-loud quasars as a function of redshift and luminosity (Haiman, Quataert, & Bower 2004). Simple models, in which the central BH mass scales with the velocity dispersion of its host halo as $M_{\rm bh} \propto \sigma_{\rm halo}^5$, have been previously found to be consistent with a number of observations, including the optical and X-ray quasar luminosity functions (Haiman & Loeb 1998; Wyithe & Loeb 2003b). We find that similar models, when augmented with an empirical prescription for the radio loudness distribution, overpredict the number of faint ($\sim 10 \mu$Jy) radio sources by 1–2 orders of magnitude. This translates into a more stringent constraint on the low-mass end of the quasar black hole mass function than is available from the Hubble and Chandra Deep Fields. We interpret this discrepancy as evidence that black holes with masses $\lesssim 10^7$ $M_\odot$ are either rare or are not as radio-loud as their more massive counterparts. Models that exclude BHs with masses below 10^7 $M_\odot$ are in agreement with the deepest existing radio observations, but still produce a significant tail of high-redshift objects. In the 1–10 GHz bands, at the sensitivity of $\sim 10\mu$Jy, we find surface densities of ~ 100, ~ 10, and ~ 0.3 deg^{-2} for sources located at $z > 6$, 10, and 15, respectively. The discovery of these sources with instruments such as the Allen Telescope Array (ATA), Expanded Very Large Array (EVLA), and the Square Kilometer Array (SKA) would open a new window for the study of supermassive BHs at high redshift. We also find surface densities of ~ 0.1 deg^{-2} at $z > 6$ for mJy sources that can be used to study 21 cm absorption from the epoch of reionization. These models suggest that, although not yet optically identified, the FIRST survey may have already detected several thousand such $> 10^8$ $M_\odot$ BHs at $z > 6$.

I thank Mario Livio and the organizers of the conference for their kind invitation, and for their patience for these proceedings. I also thank my students, Mark Dijkstra and Andrei Mesinger, and my recent collaborators, Geoffrey Bower, Renyue Cen, Lam Hui, Avi Loeb and Eliot Quataert for many fruitful discussions, and for their permission to draw on joint work. Finally, I thank Rick White for an electronic version of the spectrum of SDSS J1030+0524. This work was supported by NSF through grants AST-0307200 and AST-0307291 and by NASA through grants NAG5-26029 and HST-GO-09793.18.

REFERENCES

ALEXANDER, D. M., ET AL. 2001 *AJ* **122**, 2156.

ARONS, J. & WINGERT, D. W. 1972 *ApJ* **177**, 1.

BARGER, A. J., ET AL. 2003 *ApJ* **584**, L61.

BAUER, F. E., ET AL. 2004 *AJ* **128**, 2048.

BECKER, R. H., ET AL. 2001 *AJ* **122**, 2850.

CEN, R. 2003 *ApJ* **591**, 12.

CEN, R. & HAIMAN, Z. 2000 *ApJ* **542**, L75.

CEN, R., HAIMAN, Z., & MESINGER, A. 2004 *ApJ* **613**, 23.

CEN, R. & McDONALD, P. 2002 *ApJ* **570**, 457.

DIJKSTRA, M., HAIMAN, Z., & LOEB, A. 2002 *ApJ* **613**, 646.

FAN, X., ET AL. 2001 *AJ* **122**, 2833.

FAN, X., ET AL. 2002 *AJ* **123**, 1247.

FAN, X., ET AL. 2003 *AJ* **125**, 1649.

FAVATA, M., HUGHES, S. A., & HOLZ, D. E. 2004 *ApJ* **607**, L5.

FURLANETTO, S., HERNQUIST, L., & ZALDARRIAGA, M. 2004 *MNRAS* **354**, 695.

GNEDIN, N. Y. 2004 *ApJ* **610**, 9.

GNEDIN, N. Y. & PRADA, F. 2004 *ApJ* **608**, L77.

GUNN, J. E. & PETERSON, B. A. 1965 *ApJ* **142**, 1633.

HAIMAN, Z. 2004. In *Carnegie Observatories Astrophysics Series: Coevolution of Black Holes and Galaxies* (ed. L. C. Ho). Vol. 1, pp. 67. Cambridge University Press.

HAIMAN, Z. 2002 *ApJ* **576**, L1.

HAIMAN, Z. 2004 *ApJ* **613**, L36.

HAIMAN, Z., ABEL, T., & MADAU, P. 2001 *ApJ* **551**, 599.

HAIMAN, Z. & CEN, R. 2002 *ApJ* **578**, 702.

HAIMAN, Z. & HOLDER, G. P. 2003 *ApJ* **595**, 1.

HAIMAN, Z. & LOEB, A. 1998 *ApJ* **503**, 505.

HAIMAN, Z. & LOEB, A. 1999a *ApJ* **519**, 479.

HAIMAN, Z. & LOEB, A. 1999b *ApJ* **521**, 9.

HAIMAN, Z. & LOEB, A. 2001 *ApJ* **552**, 459.

HAIMAN, Z., QUATAERT, E., & BOWER, G. C. 2004 *ApJ* **612**, 698.

HAIMAN, Z. & SPAANS, M. 1999 *ApJ* **518**, 138.

HEAP, S. R., ET AL. 2000 *ApJ* **534**, 69.

HU, E. M., ET AL. 2002 *ApJ* **568**, L75. [Erratum: 2002 *ApJ* **576**, 99]

HUI, L. & HAIMAN, Z. 2003 *ApJ* **596**, 9.

ISLAM, R. R., TAYLOR, J. E., & SILK, J. 2003 *MNRAS* **340**, 647.

KAUFFMANN, G. & HAEHNELT, M. 2000 *MNRAS* **311**, 576.

KENNICUTT, R. C., JR. 1983 *ApJ* **272**, 54.

KODAIRA, K., ET AL. 2003 *PASJ* **55**, L17.

LE DELLIOU, M., ET AL. 2004 *MNRAS* **357**, L11.

LIDZ, A., HUI, L., ZALDARRIAGA, M., & SCOCCIMARRO, R. 2002 *ApJ* **579**, 491.

MADAU, P. & REES, M. J. 2000 *ApJ* **542**, L69.

MADAU, P., REES, M. J., VOLONTERI, M., HAARDT, F., & OH, S. P. 2004 *ApJ* **604**, 484.

MADAU, P. & QUATAERT, E. 2004 *ApJ* **606**, L17.

MALHOTRA, S. & RHOADS, J. 2004 *ApJ* **617**, L5.

MCDONALD, P., ET AL. 2001 *ApJ* **562**, 52.

MENOU, K., HAIMAN, Z., & NARAYANAN, V. K. 2001 *ApJ* **558**, 535.

MERRITT, D., ET AL. 2004 *ApJ* **607**, L9.

MESINGER, A. & HAIMAN, Z. 2004 *ApJ* **611**, L69.

MESINGER, A., HAIMAN, Z., & CEN, R. 2004 *ApJ* **613**, 23.

MIRALDA-ESCUDÉ, J. 1998 *ApJ* **501**, 15.

OH, S. P. 2001 *ApJ* **553**, 25.

RAUCH, M., ET AL. 2001 *ApJ* **562**, 76.

RHOADS, J., ET AL. 2003 *AJ* **125**, 1006.

RICOTTI, M. & OSTRIKER, J. P. 2004 *MNRAS* **352** 547.

SANTOS, M. R. 2004 *MNRAS* **349**, 1137.

SAZONOV, S. Y., OSTRIKER, J. P., & SUNYAEV, R. A. 2004 *MNRAS* **347**, 144.

SCHAYE, J., ET AL. 2000 *MNRAS* **318**, 817.

SHAPIRO, P. R., GIROUX, M. L., & BABUL, A. 1994 *ApJ* **427**, 25.

SHAPLEY, A. E., STEIDEL, C. C., PETTINI, M., & ADELBERGER, K. L. 2003 *ApJ* **588**, 65.

SONGAILA, A. 2004 *AJ* **127**, 2598.

SPERGEL, D. N., ET AL. 2003 *ApJS* **148**, 175.

STERN, D., ET AL. 2005 *ApJ* **619**, 12.

THEUNS, T., ET AL. 2002 *ApJ* **567**, L103.

VENKATESAN, A., GIROUX, M. L., & SHULL, J. M. 2001 *ApJ* **563**, 1.

VIGNALI, C., ET AL. 2003 *AJ*, **125**, 2876.

VOLONTERI, M., HAARDT, F., & MADAU, P. 2003 *ApJ* **582**, 559.

WHITE, R. L., BECKER, R. H., FAN, X., & STRAUSS, M. A. 2003 *AJ* **126**, 1.

WYITHE, J. S. B. & LOEB, A. 2003a *ApJ* **586**, 693.

WYITHE, J. S. B. & LOEB, A. 2003b *ApJ* **595**, 614.

WYITHE, J. S. B. & LOEB, A. 2004 *Nature* **427**, 815.

WYITHE, J. S. B. & LOEB, A. 2005 *ApJ* **625**, 1.

YOO, J. & MIRALDA-ESCUDÉ, J. 2004 *ApJ* **614**, L25.

ZALDARRIAGA, M., HUI, L., & TEGMARK, M. 2001 *ApJ* **557**, 519.

Studying distant infrared-luminous galaxies with *Spitzer* and *Hubble*

By CASEY PAPOVICH, EIICHI EGAMI,
EMERIC LE FLOC'H, PABLO PÉREZ-GONZÁLEZ,
GEORGE RIEKE, JANE RIGBY, HERVÉ DOLE,
AND MARCIA RIEKE

Steward Observatory, University of Arizona, 933 N. Cherry Avenue, Tucson, AZ 85741, USA

New surveys with the *Spitzer Space Telescope* identify distant star-forming and active galaxies by their strong emission at far-infrared wavelengths, which provides strong constraints on these galaxies' bolometric energy. Using early results from *Spitzer* surveys at 24 μm, we argue that the faint sources correspond to the existence of a population of infrared-luminous galaxies at $z \gtrsim 1$ that are not expected from predictions based on previous observations from *ISO* and *IRAS*. Combining *Spitzer* images with deep ground-based optical and *Hubble Space Telescope* imaging, we discuss the properties of galaxies selected at 24 μm in the region of the Chandra Deep Field South, including redshift and morphological distributions. Galaxies with $z \lesssim 1$ constitute roughly half of the faint 24 μm sources. Infrared-luminous galaxies at these redshifts span a wide variety of normal to strongly interacting/merging morphologies, which suggests that a range of mechanisms produce infrared activity. Large-area, joint surveys between *Spitzer* and *HST* are needed to understand the complex relation between galaxy morphology, structure, environment, and activity level, and how this evolves with cosmic time. We briefly discuss strategies for constructing surveys to maximize the legacy of these missions.

1. Introduction

Infrared (IR) luminous galaxies represent highly active stages in galaxy evolution that are not generally inferred in optically selected galaxy surveys (e.g., Rieke & Low 1972; Soifer, Neugebauer, & Houck 1987). High IR emission is typically generated in heavily enshrouded starbursts associated with morphologically disturbed or merging galaxies (Sanders et al. 1988); in comparison, optical studies probe less obscured star-forming regions often located in galaxy disks (e.g., Kennicutt 1998). At the present day, most of the light emitted from galaxies comes at optical wavelengths, with only one-third of the bolometric luminosity density coming in the IR (Soifer & Neugebauer 1991). However, the cosmic background implies that the far-IR emission from galaxies in the early Universe is as important energetically as the emission in the optical and UV combined (Hauser et al. 1998), and IR number counts from *ISO* indicate that these sources evolved faster than those inferred directly from UV/optical observations. The interpretation of these counts, combined with models of the cosmic IR background (Elbaz et al. 2002; Dole et al. 2003), argues that IR-luminous stages of galaxy evolution were frequently more common at high redshift. As a result, IR-luminous galaxies may be responsible for a substantial fraction of the global star-formation and metal-production rate (e.g., Franceschini et al. 2001).

Studying the mechanisms for this apparent rapidly evolving IR-luminous galaxy population has been problematic, primarily due to low-number statistics of sources at high redshifts and difficulty in measuring their multi-wavelength properties and internal structure. The improvements in IR sensitivity and survey efficiency now possible with *Spitzer* allow major advances in the study of the IR-luminous stages of galaxy evolution, particularly in true panchromatic datasets. Measurements at 24 μm with the *Spitzer*/MIPS

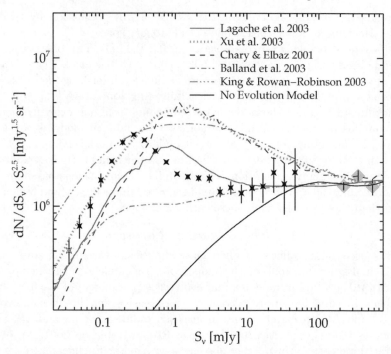

FIGURE 1. Differential *Spitzer* 24 μm number counts (from Papovich et al. 2004), normalized to a Euclidean slope, $dN/dS_\nu \sim S_\nu^{-2.5}$. The solid stars show the average counts from all the *Spitzer* fields. Each flux bin is $\Delta(\log S_\nu) = 0.15$ dex. The shaded diamonds correspond to *IRAS* 25 μm number counts from Hacking & Soifer (1991). The curves show the predictions from various contemporary models from the literature (see figure inset; and adjusted slightly to match the observed *IRAS* counts), and a model based on the local *ISO* 15 μm luminosity function and assuming non-evolving galaxy SEDs.

instrument are key because of their sensitivity to IR emission and angular resolution (see Rieke et al. 2004). Deep, large-area surveys with *Spitzer* are currently underway, and early results demonstrate that *Spitzer*-selected IR-luminous galaxies are not only common at high-redshifts ($z \gtrsim 1$; e.g., Le Floc'h et al. 2004), but may bridge the gap between the far-IR/sub-mm galaxy populations at $z \sim 3$ and galaxies at more moderate redshifts—placing them in the context of the cosmic star-formation-rate density (e.g., Egami et al. 2004).

In these proceedings, we present early results from IR surveys with *Spitzer* at 24 μm from time allocated to the Guaranteed Time Observers (GTOs), and we consider implications for IR stages of galaxy evolution. We then describe the properties of 24 μm-selected galaxies using deep ground-based and *HST* imaging, and we discuss questions that arise when considering IR-luminous galaxies within contemporary theories of galaxy evolution. Wherever appropriate, we assume a cosmology with $H_0 = 70$ km s^{-1} Mpc^{-1}, $\Omega_m = 0.3$, and $\Lambda = 0.7$.

2. *Spitzer* Observations of Distant IR-luminous galaxies

Spitzer provides efficient, deep observations of large areas of sky containing a high source surface density. Figure 1 shows the 24 μm differential number counts that have been derived using roughly 50,000 galaxies from five fields spanning approximately 10.5 square degrees (Papovich et al. 2004). At bright flux densities, $S_\nu \gtrsim 5$ mJy, the differential 24 μm source counts increase at approximately the Euclidean rate, $dN/dS_\nu \sim S^{-2.5}$, which extends the trends observed by the *IRAS* 25 μm population by two orders of magnitude (Hacking & Soifer 1991; Shupe et al. 1998). For $S_\nu \simeq 0.4$–4 mJy, the counts increase at a super-Euclidean rate, and peak near 0.2–0.4 mJy. This observation is similar to the trend observed in the *ISO* 15 μm source counts (Elbaz et al. 1999), but the peak in the 24 μm differential source counts occurs at fluxes fainter by a factor of ≈ 2.0. The counts converge rapidly at $\lesssim 0.2$ mJy, with a faint-end slope of $dN/dS_\nu \sim S_\nu^{-1.5\pm0.1}$.

The thick line in Figure 1 shows the expected counts from non-evolving models of the local IR-luminous population. While the non-evolving fiducial model is consistent with the observed 24 μm counts for $S_\nu \gtrsim 20$ mJy, it underpredicts the counts at $S_\nu \lesssim 0.4$ mJy by more than a factor of 10. The *Spitzer* 24 μm number counts require strong evolution in the IR-luminous galaxy population. This is similar to conclusions based on data from *IRAS* and *ISO*, but the *Spitzer* counts extend them to fainter fluxes (and higher redshifts, see below) than those probed from these earlier missions.

2.1. *Interpretation of 24 μm sources*

The 24 μm source counts differ strongly from predictions of various contemporary models (as labeled in Figure 1). Four of the models are phenomenological in approach (so-called 'backwards-evolution' models), and evolve the parameters of the local luminosity function back in time (generally accounting for density and luminosity evolution by changing ϕ^* and L^*) to match counts from *ISO*, radio, sub-mm, and other datasets. Several models (Chary & Elbaz 2001; King & Rowan-Robinson 2001; Xu et al. 2003) predict a rapid increase in the number of sources with super-Euclidean rates at relatively bright flux densities ($S_\nu \gtrsim 10$ mJy). They predict 24 μm counts that peak near 1 mJy, and overpredict the measured counts at this flux density by factors of 2–3. These models expected there to be more luminous IR galaxies (LIRGs) and ultra-luminous IR galaxies (ULIRGs) selected by *Spitzer* 24 μm near $z \sim 1$, based largely on the redshift distribution of the *ISO* 15 μm sources. Lagache et al. (2003) predicted roughly Euclidean counts for $S_\nu > 10$ mJy. The shape of the counts in this model is similar to the observed distribution, but it peaks at $S_\nu \sim 1$ mJy, at higher flux densities than the observed counts. This model has a redshift distribution that peaks near $z \sim 1$, but tapers more slowly, with a significant population 24 μm sources out to $z \gtrsim 2$ (Dole et al. 2003).

The peak in the 24 μm differential number counts occurs at fainter flux densities than predicted from the models based on the ISO results. Because the number counts are essentially just the integral of the galaxy luminosity function over redshift and flux down to the survey flux limit, they are likely dominated by galaxies with 'L^*' luminosities (modulo variations in the faint-end slope of the luminosity function). Models that reproduce the IR background require far-IR luminosity functions with $L^*(\mathrm{IR}) \gtrsim 10^{11}\, L_\odot$ (see Hauser & Dwek 2001). Elbaz et al. (2002) observed that the redshift distribution of objects with these luminosities in deep *ISO* 15 μm surveys spans $z \simeq 0.8$–1.2, and that these objects constitute a large fraction of the total cosmic IR background. Assuming the 24 μm number counts at 0.1–0.4 mJy correspond to L^* galaxies, their redshifts must lie at $z > 1$.

2.2. *Challenges to galaxy evolution theories*

Recently, Lagache et al. (2004) have updated their phenomenological model in order to reproduce the measured *Spitzer* number counts. To do this, they required a minor modification of the redshift distribution of 24 μm sources, such that galaxies with $z \gtrsim 1$ contribute more than half of the counts at faint fluxes ($\sim$0.2 mJy). They also required an adjustment to the flux density in the mid-IR region of galaxy SEDs (3–30 μm) of up to a factor of two. The implications are that: 1) stochastically heated emission features at mid-IR wavelengths (UIBs and PAHs) likely persist at high redshifts ($z \gtrsim 2$); and 2) the relative strength of the various mid-IR features may evolve with redshift. The second implication is not wholly unexpected as higher redshift galaxies may have very different metallicity and chemistry, and the cosmic UV radiation field is more intense (the latter contributes to the heating of the grains responsible for the mid-IR emission features, e.g., Désert et al. 1990). The intriguing prospect is that the mid-IR SEDs of IR-luminous galaxies may evolve with redshift, which complicates modeling efforts. Forthcoming spectroscopy of high-redshift galaxies at mid-IR wavelengths with the *Spitzer* Infrared Spectrograph will measure the strength of these features and will help to constrain any evolution observationally.

Although backwards-evolution models provide a useful framework for parameterizing the strong evolution of IR-luminous galaxies, they are unable to explain the physics responsible for this evolution. Models of galaxy formation and evolution that start from first principles (so called 'forward-evolution' models) currently lack the means of producing the strong evolution observed either in the IR number counts or in the cosmic IR background (see, e.g., Hauser & Dwek 2001). For example, the dot-dashed line in Figure 1 shows the model of Balland, Devriendt, & Silk (2003), which is based on semi-analytical hierarchical models within the Press-Schecter formalism. In that model, galaxies identified as 'interacting' are assigned IR-luminous galaxy SEDs. This model includes additional physics in that the evolution of galaxies depends on their local environment and merger/interaction histories. Although this model predicts a near-Euclidean increase in the counts for $S_\nu \gtrsim 10$ mJy, the counts shift to sub-Euclidean rates at relatively bright flux densities. Semi-analytic models by R. Somerville, J. Primack, et al. (in preparation), which broadly reproduce optical–near-IR properties of galaxies from $z \sim 0\text{–}3$, predict an IR background intensity that is too faint by a factor of several. These examples are typical of the general status of forward-evolution modeling efforts. *We are faced with a lack of understanding why such rapid evolution occurs in the IR-luminous galaxy population.*

3. Ground-based observations of distant *Spitzer* galaxies

The *Spitzer* GTO extragalactic survey fields were selected to have low zodiacal and Galactic backgrounds (see §5; Table 1), and to have the highest-quality ancillary data available at other wavelengths. The GTOs used *Spitzer* to observe a 1×0.5 sq. degree region of the Chandra Deep Field South (CDF-S) in early February 2004. The CDF-S has exceptional ancillary data from X-ray to radio wavelengths. For the remainder of this contribution we will discuss only a portion of these data—focusing on the optical imaging and redshift distribution of the *Spitzer*-selected galaxies. Studies of the *Spitzer* sources in this field using other ancillary data have been carried out, or are in progress. For example, Rigby et al. (2004) study the properties of X-ray–selected *Spitzer* 24 μm sources in this field.

The region around the CDF-S has been the target of several ground-based imaging surveys. Of these, the COMBO-17 survey (Classifying Objects by Medium-Band Observations in 17 filters; Wolf et al. 2003) has observed a 30×30 sq. arcmin region around the CDF-S field with imaging from 0.3–1 μm. Using a suite of medium-band filters, they provide highly reliable photometric redshifts for galaxies with $R \leqslant 24$ to $z \lesssim 1.3$, and for AGN out to substantially higher redshift (Wolf et al. 2004). Nearly the entire COMBO-17 field overlaps with the *Spitzer* field. Most *Spitzer* 24 μm sources brighter than 60 μJy (the estimated 50% completeness limit) are readily identified in the COMBO-17 images: 3850 of the 4720 24 μm sources in this region have optical counterparts to $R \leqslant 25$ within $2''$ (the *Spitzer* 24 μm PSF is roughly $6''$ FWHM). Of these, roughly 2970 have good

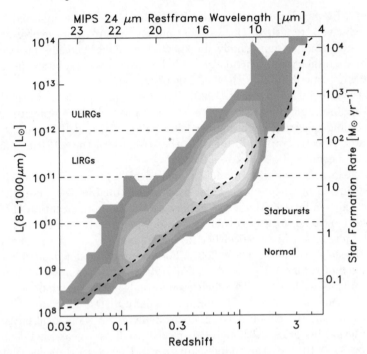

FIGURE 2. Redshift and luminosity distribution of optically selected *Spitzer*/MIPS 24 μm sources. Redshifts correspond to values published for the COMBO-17 survey (Wolf et al. 2004), with additional spectroscopic redshifts from VLT/FORS2 (Vanzella et al. 2004) and VIRMOS (Le Fèvre et al. 2004). Contours indicate regions containing 1, 2, 4, 8, 16, 32, and 64 galaxies. The heavy, dashed line shows the estimated 80% completeness limit of the 24 μm imaging (see Papovich et al. 2004). The right-hand axis shows the SFR corresponding to $L(8$–1000 μm) for the assumption that all the IR luminosity results from star formation, and using the relation established by Kennicutt (1998). The top axis shows the rest-frame wavelength observed at 24 μm. Regions separated by dashed lines show fiducial IR-galaxy classes.

photometric redshift estimates. Several large spectroscopic surveys from the VLT (Le Fèvre et al. 2004; Vanzetta et al. 2004) provide an additional 290 *Spitzer* 24 μm sources.

3.1. *The redshift distribution of* Spitzer *24 μm sources*

Figure 2 shows the redshift and luminosity distribution of *Spitzer*-selected sources with counterparts in the photometric- and spectroscopic-redshift catalogs from the CDF-S. The total IR luminosity, $L(8$–1000 μm), is calculated by converting the measured 24 μm flux density to a luminosity using the reported redshift, then extrapolating to the total IR luminosity using the semi-empirical SEDs of Dale et al. (2001). It is important to note that there is some scatter between far-IR colors and total IR luminosity which is not included in the figure (see Chapman et al. 2003). Much of this scatter can be reduced by including *Spitzer* 70 μm data to constrain the mid- to far-IR 'color' (see, e.g., Papovich & Bell 2002).

IR-luminous galaxies are readily identified out to $z \sim 1.3$. Galaxies at higher redshifts generally lie beyond the COMBO-17 limits. At low redshifts ($z \lesssim 0.2$) most of the 24 μm-selected sources correspond to relatively normal star-forming galaxies with some starbursts. This reflects the limited volume probed by the survey for these redshifts ($\sim$14000 Mpc^3), in which few IR-luminous galaxies are expected. The majority of 24 μm sources with $z \sim 0.4$–1 correspond to LIRGs, and these sources likely dominate the IR luminosity density at these redshifts. ULIRGs are generally not common in this field

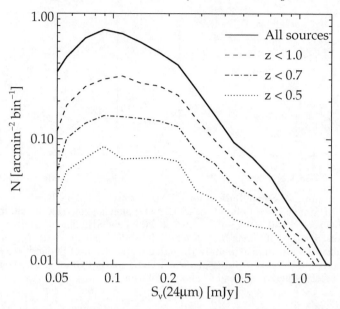

FIGURE 3. Differential *Spitzer* 24 μm number counts in the region of the CDF-S field covered by COMBO-17, which covers roughly 900 sq. arcmin. The solid curve shows the counts from all 24 μm sources in this area. The broken lines show the contribution to the counts from optically selected galaxies with redshifts below 0.5 (dotted line), 0.7 (dot-dashed line), and 1.0 (dashed line). Each bin is $\Delta(\log S_\nu) = 0.1$ dex. Optically selected IR galaxies with redshifts $z < 1.0$ contribute roughly half of the 24 μm source counts at the faint end (0.1–0.4 mJy).

until $z \gtrsim 0.8$, and LIRGs generally seem to dominate the IR emission at these redshifts as well. ULIRGs appear scarce even at these high redshifts, although we are certainly missing optically faint ULIRGs with $R \gtrsim 24$ (see, e.g., Egami et al. 2004). To study their properties will require both large survey areas and deep optical data.

Figure 3 shows the 24 μm differential number counts in the *Spitzer*-COMBO-17 overlap areas, and the contribution of galaxies as a function of redshift. At the bright end, most of the counts are due to galaxies with $z \lesssim 1$. At fainter 24 μm flux densities, higher-redshift galaxies dominate the counts. Galaxies with $z \lesssim 0.7$ make up only one-quarter to one-third of the total counts at ~ 0.2 mJy (near the peak in Figure 1). Similarly, 24 μm sources at these flux densities with $z \lesssim 1$ contribute only $\sim$50% of the total counts. There is roughly one *Spitzer* 24 μm source per sq. arcmin with no counterpart in the optical images to $R \sim 25$. Elbaz et al. (2002) found a redshift distribution of *ISO* 15 μm sources with a median at $z \sim 0.7$ and a small tail to $z \sim 1$. In contrast, the *Spitzer* 24 μm data is very sensitive to galaxies at $z \gtrsim 1$. *The ISO populations make up only a fraction of the faint 24 μm sources.*

3.2. *Evolution of the IR-luminosity density*

The available redshifts allow a crude estimate for the evolution in the IR luminosity density relative to that in rest-frame UV and visible bands in the CDF-S. From $z \sim 0.2$–1, the luminosity density in the rest-frame U and V bands increases by roughly a factor $\sim$3 (uncorrected for extinction or incompleteness effects), consistent with findings from previous studies (e.g., Lilly et al. 1996). In comparison, the IR luminosity density grows by roughly a factor $\gtrsim 8$, where the inequality symbol denotes the fact that this estimate does not include the contribution from IR-luminous galaxies fainter than the magnitude limits ($R \sim 24$) of the ground-based surveys. This underlines the fact that *the*

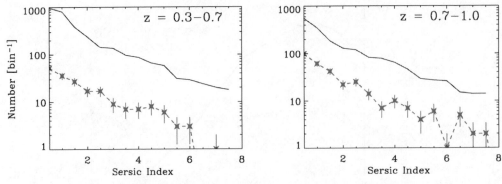

FIGURE 4. Distribution of Sersic indices for GEMS-selected galaxies based on two-dimensional fits to the F850LP images (see Peng et al. 2002). The panels show the distribution for galaxies with redshifts between $z = 0.3$–0.7 and $z = 0.7$–1.0 (as labeled). The stars connected by dashed lines show the distribution of Sersic indices for IR-luminous galaxies with $L_{IR} > 10^{10.5} L_\odot$. The IR-luminous galaxies show approximately the same distribution of Sersic indices as the general galaxy population in both redshift intervals, which suggests that these morphological parameters are not indicative of IR-active stages of galaxy evolution.

IR-luminous galaxy population appears to evolve more rapidly than that directly measured from UV/optical-selected galaxies.

4. *HST* observations of distant *Spitzer* galaxies

The Advanced Camera for Surveys (ACS) has greatly improved the efficiency of imaging with *HST*. The region of the CDF-S has extensive *HST*/ACS imaging from the Galaxies Evolution through Morphologies and SEDs survey (GEMS; Rix et al. 2004), which provides F606W and F850LP imaging over roughly 780 sq. arcmin, i.e., most of the COMBO-17 field.† To date, *HST* is the most efficient means of obtaining kpc-scale resolution of distant galaxies, as terrestrial adaptive-optics techniques are currently only effective over small patches of sky. Combined with the high-quality redshifts, these wide-area ACS data allow us to test whether structural properties and environmental effects correlate with IR-luminous stages of galaxy evolution.

What is the distribution of galaxy morphological types that are in IR-active evolutionary stages? As a first experiment, one can parameterize morphological type from the *HST* images simply in terms of the Sersic index, n_s (also called the generalized de Vaucouleur profile), where $I(R) \sim \exp(-R^{1/n_s})$. Objects with exponential surface brightness profiles have Sersic indices $n_s \sim 1$, which is typical of disk-like galaxies. Objects with more concentrated surface-brightness profiles have higher Sersic indices, as in the case of spheroids and bulges. Classical $r^{1/4}$-law galaxies have $n_s = 4$. Crudely speaking, the Sersic index quantifies the bulge-to-disk ratio of a galaxy's light emission, and it can be used to discriminate between late-type, disk-dominated galaxies ($n_s \leqslant 2.5$), and early-type, bulge-dominated galaxies ($n_s > 2.5$).

Figure 4 shows the distribution of Sersic indices for all galaxies in the GEMS catalogs with redshifts $z = 0.3$–0.7 and 0.7–1.0. The distribution is skewed towards large numbers of galaxies with lower Sersic indices, which illustrates the fact that the majority of galaxies are disk-dominated. *Interestingly, the distribution of Sersic indices for the IR-luminous galaxies, $L_{IR} \geqslant 10^{10.5} L_\odot$, is nearly identical to that of the general galaxy population*

---

† The Great Observatories Origins Deep Survey has deeper ACS imaging in a smaller 160 sq. arcmin area within the GEMS field; see Giavalisco et al. (2004).

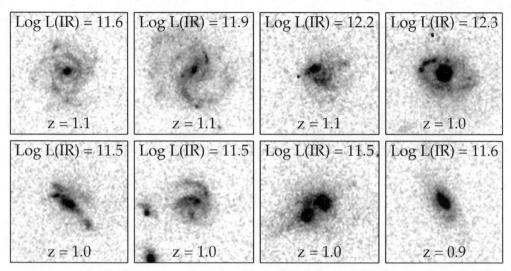

FIGURE 5. ACS F850LP images of luminous IR galaxies at $z \sim 1$ in the CDF-S/GEMS areas. Each image is 5 arcsec per side, which is roughly 40 kpc at these redshifts. Total IR luminosities are given in units of log $L_\odot$. The galaxies have a wide range of morphologies. Clear distortions are evident (the two bottom, leftmost galaxies), including rings (upper rightmost galaxy), and multiple nuclei (second from right, bottom row). However, the two leftmost galaxies on the top row appear to be fairly regular late-type spirals, and the right-most galaxy on the bottom row appears to have a normal spheroidal component.

regardless of redshift. The implication is that these morphological indicators alone are a poor discriminator of IR-activity.

Even the most luminous IR galaxies, $L_{IR} \gtrsim 10^{11.5} L_\odot$, span a range of morphological types. As an illustration, Figure 5 shows the ACS/F850LP images of several fiducial IR-luminous galaxies at $z \sim 1$ from the GEMS data. At these redshifts, the F850LP filter probes roughly the rest-frame B-band of these galaxies. It is clear that while many of these types of galaxies have highly disturbed morphologies or evidence of strong mergers, there are also clear examples of fairly normal galaxy types. Systematic studies using these data at $z \sim 0.7$ (E. Bell et al., in preparation), and as a function of IR-luminosity (C. Papovich et al., in preparation) will help to understand the relation between galaxy morphology, environment, and IR-activity.

5. Selection of deep, extragalactic survey fields

We close with a discussion on how to choose the location of deep fields for studying IR-luminous galaxies. The dominant sources of IR background are zodiacal light and emission from cirrus clouds in the Milky Way. Zodiacal light dominates at mid-IR wavelengths (3–40 μm), decreases rapidly with ecliptic latitude, and shows strong seasonal changes (e.g., Price et al. 2003). Galactic cirrus is the dominant source of background at far-IR wavelengths (40–200 μm). It scales roughly linearly with the Galactic column density, N(HI) (Lockman et al. 1986; Boulanger & Perault 1988), and produces two noise components for IR observations. The first is due simply to the elevated sky brightness, which limits the flux sensitivity of an observation by a factor roughly $\sim \sigma_{cirrus}^{-1}$. Fields near the plane will require exposures several times longer than fields near the poles to achieve comparable depth.

The second component is confusion noise from structure within cirrus clouds (e.g., Low et al. 1984; Helou & Beichman 1990; Kiss et al. 2001, 2003). Helou & Beichman expressed

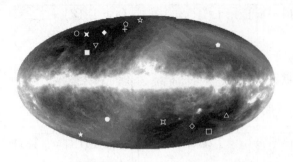 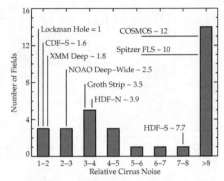

FIGURE 6. *Left*: Location of selected extragalactic survey fields on the Galactic *IRAS* 100 μm image. Galactic north is up, and east to the left. Galactic cirrus dominates the image, although zodiacal emission is evident and runs roughly from the bottom left to the upper right. The symbols denote fiducial fields: HDF-N, *filled star*; HDF-S, *open star*; CDF-S, *open square*; Marano, *open diamond*; XMM-Deep, *open pentagon*; XMM-LSS, *filled pentagram*; NOAO Boötes, *cross*; Groth strip, *filled diamond*; ELAIS-N1, *downward triangle*; ELAIS-S2, *upward triangle*; COSMOS, *pentagon*; Lockman Hole, *open circle*; SSA22, *filled circle*; Subaru Deep, *open pentagram*; *Spitzer* FLS, *filled square*. *Right*: Far-IR cirrus noise of the extragalactic survey fields listed in Table 1 relative to that of the Lockman Hole. Estimates are based on the Galactic N(HI), which correlates linearly with cirrus brightness. High cirrus noise substantially increases the confusion noise at all angular scales and is a strongly limiting factor for far-IR observations.

the cirrus confusion noise as $\sigma_{\mathrm{cirrus}} \sim \lambda^{2.5} D^{-2.5} B_\lambda^{1.5}$, where λ is the emitted wavelength, D is the diameter of the telescope aperture, and B_λ is the mean sky brightness. Because cirrus brightness correlates with Galactic hydrogen column density, an increase in N(HI) by a factor of five corresponds to an increase in the relative confusion noise by a factor of ten. Far-IR observations in fields with high cirrus sky brightness pay a substantial penalty in terms of cirrus-confusion noise, and this imposes a hard limit on the ultimate survey depth in such fields.

Table 1 lists the properties of known extragalactic survey fields (updated and adapted from a compilation by Stiavelli et al. 2003). For each field, the Galactic extinction (parameterized by the color excess, $E(B-V)$) and hydrogen column density are taken from the maps of Schlegel et al. (1998) and Dickey & Lockman (1990), respectively. The left panel of Figure 6 shows the location of several of these fiducial fields superimposed on an *IRAS* 100 μm all-sky image. The right panel of Figure 6 shows the distribution of Galactic cirrus confusion noise relative to that of the Lockman Hole, the sightline with the minimum N(HI).

Fields that are both far from the ecliptic (*low zodiacal light*) *and the Galactic plane* (*low N*(HI) *and cirrus*) *have the lowest backgrounds and confusion noise, and are, in a sense, chosen by nature to be the ideal locations for full multi-wavelength extragalactic surveys.* Future IR missions (e.g., *JWST, Herschel, SAFIR*) will gravitate to these fields, as well as future X-ray telescopes (e.g., *Constellation-X*). To optimally study IR-luminous stages of galaxy evolution will require full panchromatic surveys in these fields, including high angular resolution *HST* imaging. These multi-wavelength data will be crucial for dissecting the mechanisms for galaxy evolution, not only in the *Spitzer* era, but for decades to come.

We are indebted to all our collaborators on this project for their work and for allowing us to present some of this material prior to publication. We wish to acknowledge the fellow members of the *Spitzer* MIPS, IRAC, and IRS GTO teams for stimulating conversations, efficient processing and analysis of the data, and much hard work. We

Name	R.A. (J2000.0)	Decl. (J2000.0)	l (deg)	b (deg)	$E(B-V)$	N(HI) $(10^{20}\,\mathrm{cm}^{-2})$
DEEP-1	0:17:00	16:00:00.0	111.0	−46.1	0.049	4.18
WHT Deep	0:22:33	00:20:57.0	107.6	−61.7	0.025	2.73
ELAIS-S1	0:34:44	−43:28:12.0	313.5	−73.3	0.008	2.52
FORS Deep	1:06:04	−25:45:46.0	191.1	−86.5	0.018	1.88
XMM-LSS	2:21:20	−04:30:00.0	170.3	−58.8	0.027	2.61
CNOC2	2:23:00	00:00:00.0	165.7	−55.1	0.039	2.96
DEEP-2	2:30:00	00:00:00.0	168.1	−54.0	0.022	2.89
CFDF	3:00:00	00:00:00.0	177.0	−48.9	0.096	6.99
Marano	3:15:09	−55:13:57.0	270.2	−51.8	0.016	2.45
CDF-S†	3:32:30	−27:48:47.0	223.6	−54.4	0.008	0.79
ELAIS-S2	5:02:24	−30:35:55.0	232.6	−35.7	0.012	1.43
CNOC2	9:20:00	37:00:00.0	186.6	44.7	0.011	1.47
COSMOS	10:00:29	02:12:21.0	236.8	42.1	0.017	2.90
Lockman Hole†	10:52:43	57:28:48.0	149.3	53.1	0.008	0.57
EIS Deep	11:20:45	−21:42:00.0	276.4	36.5	0.045	4.18
HDF-N†	12:36:49	62:12:58.0	125.9	54.8	0.012	1.41
SSA13	13:12:21	42:41:21.0	109.0	73.9	0.014	1.46
Subaru Deep	13:24:21	27:29:23.0	37.6	82.7	0.019	1.19
XMM Deep†	13:34:37	37:54:44.0	85.6	75.9	0.006	0.83
Groth Strip†	14:16:00	52:10:00.0	96.3	60.4	0.013	1.30
ELAIS-N3	14:29:06	33:06:00.0	54.7	68.1	0.008	1.11
NOAO Boötes†	14:32:06	34:16:47.5	58.2	67.7	0.012	1.04
ELAIS-N1	16:10:01	54:30:36.0	84.3	44.9	0.005	1.38
ELAIS-N2	16:36:58	41:15:43.0	65.3	42.2	0.007	1.07
DEEP-2	16:52:00	34:55:00.0	57.4	38.3	0.016	1.78
Spitzer FLS	17:18:00	59:30:00.0	88.3	34.9	0.023	2.66
CFHT Legacy	22:15:31	−17:44:05.0	39.3	−52.9	0.026	2.39
SSA22	22:17:35	00:15:30.0	63.1	−44.0	0.066	4.64
DEEP-2	23:30:00	00:00:00.0	85.0	−56.7	0.037	4.04
HDF-S	22:32:56	−60:30:02.7	328.3	−49.2	0.027	2.22
EIS Deep	22:50:00	−40:12:59.0	357.5	−61.7	0.011	1.47

† Denotes field included in the *Spitzer* GTO cosmological surveys.

TABLE 1. Compilation of the properties of known extragalactic survey fields

would also like to thank the members of the COMBO-17 and GEMS teams for their assistance and useful conversations, in particular Eric Bell, Dan McIntosh, Hans-Walter Rix, Rachel Somerville, and Christian Wolf. C.P. acknowledges highly interesting conversations with other participants at the Aspen Center for Physics, where much of this work was completed. Finally, we would like to extend our deep appreciation to the symposium organizers for the invitation to present this material, and for planning such an interesting and successful meeting. Support for this work was provided by NASA through contract 960785 issued by JPL/Caltech.

REFERENCES

Balland, C., Devriendt, J. E. G., & Silk, J. 2003 *MNRAS* **343**, 107.
Chapman, S. C., Helou, G., Lewis, G. F., & Dale, D. A. 2003 *ApJ* **588**, 186.
Chary, R. R. & Elbaz, D. 2001 *ApJ* **556**, 562.
Dale, D. A., Helou, G., Contursi, A., Silbermann, N. A., & Kolhatkar, S. 2001 *ApJ* **549**, 215.
Désert, F.-X., Boulanger, F., & Puget, J.-L. 1990 *A&A* **237**, 215.

DICKEY, J. M. & LOCKMAN, F. J. 1990 *ARAA* **28**, 215.

DOLE, H., LAGACHE, G., & PUGET, J.-P. 2003 *ApJ* **585**, 617.

EGAMI, E., ET AL. 2004 *ApJS* **154**, 130.

ELBAZ, D., ET AL. 1999 *A&A* **351**, L37.

ELBAZ, D., ET AL. 2002 *A&A* **384**, 848.

FRANCESCHINI, A., AUSSEL, H., CESARSKY, C. J., ELBAZ, D., & FADDA, D. 2001 *A&A* **378**, 1.

GIAVALISCO, M., ET AL. 2004 *ApJ* **600**, L93.

HACKING, P. & SOIFER, B. T. 1991 *ApJ* **367**, L49.

HAUSER, M. G., ET AL. 1998 *ApJ* **508**, 25.

HAUSER, M. G. & DWEK, E. 2001 *ARAA* **39**, 249.

HELOU, G. & BEICHMAN, C. A. 1990. In *From Ground-Based to Space-Borne Sub-mm Astronomy* (ed. B. Kaldeich). p. 117. ESA.

KENNICUTT, R. C., JR. 1998 *ApJ* **498**, 541.

KING, A. J. & ROWAN-ROBINSON, M. 2003 *MNRAS* **339**, 260.

KISS, CS., ÁBRAHÁM, P., KLAAS, U., JUVELA, M., & LEMKE, D. 2001 *A&A* **379**, 1161.

KISS, CS., ET AL. 2003 *A&A* **399**, 177.

LAGACHE, G., DOLE, H., & PUGET, J.-L. 2003 *MNRAS* **338**, 555.

LAGACHE, G., ET AL. 2004 *ApJS* **154**, 112.

LE FÈVRE, O., ET AL. 2004 *A&AA* **428**, 1043.

LE FLOC'H, E., ET AL. 2004 *ApJS* **154**, 170.

LOW, F. J., ET AL. 1984 *ApJ* **278**, L19.

PAPOVICH, C. & BELL, E. F. 2002 *ApJ* **579**, L1.

PAPOVICH, C., ET AL. 2004 *ApJS* *254*, 70.

PENG, C. Y., HO, L. C., IMPEY, C. D., & RIX, H.-W. 2002 *AJ* **124**, 266.

PRICE, S. D., NOAH, P. V., MIZUNO, D., WALKER, R. G., & JAYARAMAN, S. 2003 *AJ* **125**, 962.

RIEKE, G. & LOW, F. 1972 *ApJ* **176**, 95.

RIEKE, G., ET AL. 2004 *ApJS* **154**, 25.

RIGBY, J., ET AL. 2004 *ApJS* *154*, 160.

RIX, H.-W., ET AL. 2004 *ApJS* **152**, 163.

SANDERS, D. B., ET AL. 1988 *ApJ* **325**, 74.

SCHLEGEL, D. J., FINKBEINER, D. P., & DAVIS, M. 1998 *ApJ* **500**, 525.

SHUPE, D. L., FANG, F., HACKING, P. B., & HUCHRA, J. P. 1998 *ApJ* **501**, 597.

SOIFER, B. T., NEUGEBAUER, G., & HOUCK, J. R. 1987 *ARAA* **25**, 187.

SOIFER, B. T. & NEUGEBAUER, G. 1991 *AJ* **101**, 354.

STIAVELLI, M., PANAGIA, N., & FERGUSON, H. 2003, *Field Selection Criteria for the ACS Ultra Deep Field* (www.stsci.edu/hst/udf/planning_doc_files/field2)

VANZELLA, E., ET AL. 2004 *A&A* **423**, 761.

WOLF, C., MEISENHEIMER, K., RIX, H.-W., BORCH, A., DYE, S., & KLEINHEINRICH, M. 2003 *A&A* 401, 73.

WOLF, C., ET AL. 2004 *A&A* **421**, 913.

XU, C., LONSDALE, C. J., SHUPE, D. L., FRANCESCHINI, A., MARTIN, C., & SCHIMINOVICH, D. 2003 *ApJ* **587**, 90.

Galaxies at $z \approx 6$–i'-drop selection and the GLARE Project

By ELIZABETH R. STANWAY,[1]
KARL GLAZEBROOK,[2] ANDREW J. BUNKER[3]
AND THE GLARE CONSORTIUM†

[1]Institute of Astronomy, Madingley Road, Cambridge, CB3 0HA, UK

[2]Department of Physics & Astronomy, The Johns Hopkins University, 3400 North Charles Street, Baltimore, MD 21218-2686, USA

[3]School of Physics, University of Exeter, Stocker Road, Exeter, EX4 4QL, UK

Within the last few years, a number of public and legacy projects have generated very deep photometric datasets. The *Hubble Space Telescope* (*HST*) leads the way in this field, with the high spatial resolution and ability to detect very faint galaxies essential for this challenging work. The Advanced Camera for Surveys (ACS) on *HST* has now carried out several large deep surveys, including the Great Observatories Origins Deep Survey (GOODS) and the Hubble Ultra Deep Field (HUDF). These have been designed to allow the systematic broadband selection of very high redshift galaxies ($z > 5$) using the SDSS-i' and z' filters. This endeavor to identify faint and distant galaxies has been complemented by advances in spectroscopy. The current generation of spectrographs on 8m-class telescopes and the development of new techniques such as Nod & Shuffle have allowed the spectroscopic limit to be pushed to ever fainter magnitudes. The Gemini Lyman-Alpha at Reionization Era (GLARE) project is a spectroscopic campaign which aims to obtain 100-hour Gemini/GMOS spectra for a large number of $z \approx 6$ galaxy candidates in and around the Ultra Deep Field. We describe the use of the i'-drop photometric technique to identify very high-redshift candidates in the data of the public GOODS and HUDF surveys. We comment on confirmed high-redshift galaxies discovered using this technique. We then discuss the photometric and spectroscopic characteristics of the galaxy sample resulting from the first 7.5 hours of GLARE observations.

---

1. Introduction

In recent years the observational horizon has expanded rapidly and radically for those observing distant galaxies. Large format red-sensitive detectors on wide-field imaging instruments, such as the Advanced Camera for Surveys (ACS) on the refurbished *HST*, have pushed the limits to which we can routinely detect star-forming distant galaxies progressively from redshifts of one to beyond $z = 6$. At the highest redshifts currently accessible, narrow-band emission line searches using the Lyman-α line have moved on from redshifts of 4 (Hu & McMahon 1996; Fujita et al. 2003), to 5.7 (Hu et al. 1999) and now reach to $z \sim 6.5$ (Kodaira et al. 2003; Hu et al. 2002). In recent months increasing numbers of candidate galaxies have been also been found by the use of Lyman break methods at redshifts of $z \approx 6$ (e.g., Bouwens et al. 2004; Dickinson et al. 2004; Stanway et al. 2003).

While large ground-based telescopes have made spectroscopic confirmation of redshifts for galaxies beyond $z \sim 6$ possible (e.g., Hu et al. 2002), this process is expensive in telescope time; until recently, it has only been possible for those objects with strong emission lines, or which are lensed by intervening objects (e.g., Ellis et al. 2001).

Spectroscopic surveys for fainter objects are essential to shape our understanding of the universe at these redshifts and, with modern instrumentation and new techniques for

† See http://mrhanky.pha.jhu.edu/~kgb/GLARE.

faint-object spectroscopy, such surveys are now possible. The GLARE survey, described in Stanway et al. (2004b) and these proceedings, aims to obtain extremely deep spectra of very faint high redshift star-forming galaxies, selected via i–z color (Stanway et al. 2003, 2004a; Bunker et al. 2004) from the HUDF (Beckwith et al. 2003) and GOODS (Giavalisco et al. 2004) data, using the GMOS spectrograph on Gemini South.

In Section 2 we describe the i'-drop Lyman break technique. In Section 3 we go on to describe follow-up spectroscopy of candidate $z \approx 6$ galaxies as part of the GLARE survey, while in Section 4 we discuss candidate selection using the i'-drop method on the deep GOODS and HUDF imaging. Finally, in Section 5 we present spectra obtained as part of this survey.

Throughout we adopt a Λ-dominated, 'concordance' cosmology with $\Omega_\Lambda = 0.7$, $\Omega_M = 0.3$ and $H_0 = 70 \, h_{70} \, \mathrm{km \, s^{-1} \, Mpc^{-1}}$. All magnitudes in this paper are quoted in the AB system (Oke & Gunn 1983) and the Madau (1995) prescription, extended to $z = 7$, is used where necessary to estimate absorption due to the intergalactic medium.

2. The i'-drop technique

Photometric redshifts are now routinely used to process large datasets and identify high redshift candidates. An application of this method, the continuum-based Lyman-break photometric technique pioneered at $z \sim 3$ by Guhathakurta et al. (1990) and Steidel et al. (1995), has been extended progressively to $z \sim 4$ and $z \sim 5$ (e.g., Steidel et al. 1999; Bremer et al. 2004), and using i'-band drop selection further extended to $z \sim 6$ (Stanway et al. 2003, 2004a; Bunker et al. 2004; Bouwens et al. 2003; Dickinson et al. 2004; Yan et al. 2003).

The 'classical' Lyman break technique used by Steidel et al. (1996) uses three filters, one redward of rest frame Lyman-α ($\lambda_{\mathrm{rest}} > 1216$ Å), a second in the spectral region between rest-frame Lyman-α and the rest-frame Lyman limit (912 Å) and a third at $\lambda_{\mathrm{rest}} < 912$ Å. At $z \sim 3$ the technique relies on the step in the spectra of the stellar component of galaxies at 912 Å due to photospheric absorption, supplemented by optically-thick Lyman limit system absorption caused by neutral hydrogen in the galaxy in question or in the intervening IGM.

At higher redshifts, the evolution in the Lyman-α forest absorption—particularly in the spectral region 912–1216 Å—means that the optical depth of the IGM is $\gg 1$ and the effective break migrates redward to the Lyman-α region. As a result, two filters—one to either side of $\lambda_{\mathrm{rest}} = 1216$ Å—are sufficient to identify the large continuum break, and hence candidate galaxies, at $z \approx 6$. As illustrated by Figures 1 and 2, the *HST*/ACS F775W (SDSS-i') and F850LP (SDSS-z') filters neatly bracket the continuum break at redshifts $5.6 < z < 7.0$ and the resulting large i'–z' color is indicative of a high redshift source. Thus, provided one works at a sufficiently high signal-to-noise ratio, candidate $z' \approx 6$ galaxies can be safely identified through detection in the single redder band.

As Figure 2 demonstrates, lower redshift ($z \simeq 1$–2) elliptical galaxies (or Extremely Red Objects, e.g., Cimatti et al. 2002) exhibit large i'–z' colors due the 4000 Å break in the spectrum of old stellar populations. An i'–z' color-cut criterion of $(i'$–$z')_{AB} > 1.5$ should exclude the majority of these, but contamination would still be possible due to scatter in the photometric properties of these objects and some $z \approx 5.6$ galaxies would likely be lost. As a result, we work at a more relaxed color cut of $(i'$–$z')_{AB} > 1.3$ and note the potential contamination. Deep v- and b-band imaging can assist in identifying these galaxies which are likely to have significant flux in the v-band and at lower wavelengths, and objects with b-band or strong v-band detections are eliminated from an i'-drop sample.

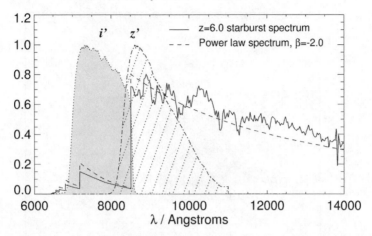

FIGURE 1. An illustration of the i'-drop technique. Normalized transmission profiles of the $F775W(i')$ and $F850LP(z')$ filters on the ACS instrument on *HST* overplotted with the profiles of a starburst galaxy placed at $z \simeq 6$ (solid line, Kinney et al. 1996) and also a power-law profile with $f_\lambda \propto \lambda^{-2.0}$ (expected for the UV spectrum of a starburst galaxy). The Madau (1995) prescription for intergalactic absorption at $z = 6$ has been applied to both spectra.

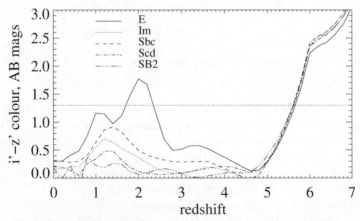

FIGURE 2. Model color-redshift tracks for galaxies with non-evolving stellar populations (from Coleman, Wu, & Weedman 1980 template spectra). The 'hump' in $(i'-z')$ color seen at $z \approx 1$–2 is due to the 4000 Å break redshifting beyond the i'-filter while large colors at high redshift are caused by the Lyman continuum break.

The region of color space occupied by L-, M- and T-class stars is also similar to that occupied by high redshift galaxies (Hawley et al. 2002). These have been shown to be a large contaminant population, accounting for all bright, fully unresolved i'-drops in the GOODS fields (Stanway et al. 2004a; Dickinson et al. 2004). Hence unresolved i'-drops are often excluded from i'-drop samples (Bouwens et al. 2003; Dickinson et al. 2004), or treated with caution (Stanway et al. 2004a).

Despite these potential contaminants, the spectroscopy presented here and elsewhere (e.g., Dickinson et al. 2004; Stanway et al. 2004b; Bunker et al. 2003) demonstrates that the i'-drop technique effectively detects starbursting galaxies at $z \approx 6$.

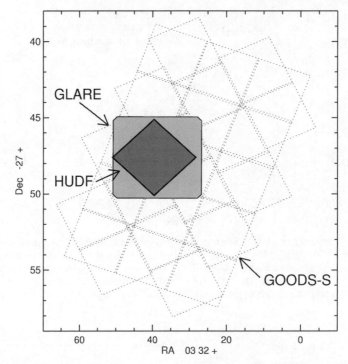

FIGURE 3. The region surveyed by the GLARE project, together with the HUDF and the GOODS-South pointings. The 3.5×3.5 arcmin^2 field of view of the GMOS instrument on Gemini South allows the entire HUDF and some of the surrounding GOODS-S region to be surveyed in one pointing.

3. Gemini/GMOS spectroscopy

The spectra described in this paper were obtained as part of the GLARE project on the 8m Gemini South telescope using the Gemini Multi-Object Spectrograph (GMOS, Hook et al. 2003) targeting sources in the region of the HUDF (Beckwith et al. 2003). This program aims to determine the redshift distribution and measure the luminous properties of faint $5.5 < z < 7$ starburst galaxies.

As Figure 3 illustrates, the large (3.5×3.5 arcmin^2) field of view of the GMOS instrument on Gemini South allows the entire HUDF and some of the surrounding GOODS-S region to be surveyed in one pointing. Hence, the goal of the GLARE program was to obtain a total of 100 hours exposure on a single multi-object slitmask, centered on the HUDF.

Observations were made with the R150 grating at 9000 Å central wavelength and a custom-made 7800 Å longpass filter ('RG780') giving a spectral range from 7800 Å up to the CCD cutoff (about 10,000 Å) and a spectroscopic resolution of 15 Å (4 pixels FWHM at 3.5 Å/pix) with $0\rlap{.}''7$ wide slits, a resolving power of $\lambda / \Delta\lambda_{\mathrm{FWHM}} \approx 550$. As the seeing disk ($< 0\rlap{.}''5$ FWHM) was smaller than the slit width ($0\rlap{.}''7$), the true resolution is somewhat better for a source which does not fill the slit. The data was taken using the 'Nod & Shuffle' (N&S) observing mode (Glazebrook & Bland-Hawthorn 2001; Abraham et al. 2004, hereafter GDDS1) in order to efficiently subtract skylines and to allow easy identification of weak emission lines (see Section 5). Our N&S setup and observational scheme followed that in GDDS1 except that the slits were $2\rlap{.}''5$ long with a $1\rlap{.}''5$ nod.

Data reduction of the slitmask spectra was carried out using the GDDS pipeline, also following that described in GDDS1 (Appendix B and C). Absolute flux calibration was done by normalizing the spectra of the mask alignment stars to their z'-band photometric fluxes, and sky lines were used for wavelength calibration.

A total of 22.5 hours integration time on source were obtained in Semester 2003B and the program will continue, with a revised slitmask, in 2004B.

4. i'-drop galaxies in and around the HUDF

For our program of spectroscopic follow up, candidate high-redshift star-forming galaxies were selected via $i - z$ color from both the HUDF and the GOODS imaging.

The GOODS v1.0 data release† (Giavalisco et al. 2004) comprises publicly available, reduced, co-added imaging from five 'epochs' of observations, reaching 3σ magnitude limits of $F606W(v)_{\rm lim} = 29.44$, $i'_{\rm lim} = 28.83$ and $z'_{\rm lim} = 28.52$ in two fields, one (GOODS-S) in the Chandra Deep Field South and the second in the region of the Hubble Deep Field North. Here we focus on the GOODS-S field, which surrounds and includes the region surveyed by the HUDF.

We constructed source catalogs from this imaging using the SExtractor v2.2.2 software package (Bertin & Arnouts 1996), requiring at least five adjacent pixels above a flux threshold of 2σ per pixel ($0.01\,{\rm counts\,pixel^{-1}\,s^{-1}}$) on the drizzled data for source extraction (with a pixel scale of $0\rlap{.}''03\,{\rm pixel^{-1}}$). Magnitudes were measured in $0\rlap{.}''3$ apertures and corrected to total magnitudes using a fixed aperture correction gauged for compact sources on the images. We selected $z \approx 6$ candidates satisfying $(i'-z')_{\rm AB} > 1.3$ and $z'_{\rm AB} < 27.2\,{\rm mag}$ from this imaging, checking that each candidate was undetected in the v-band. Although this pushes into the noise of the GOODS v1.0 imaging (the faint end reaches the 3σ magnitude limit on the $i'-z'$ color), the GOODS survey is the deepest available dataset in the region surrounding the HUDF and the number density of such faint candidate sources is sufficiently high to populate a slitmask as described in Section 3.

Although we have now performed a full analysis (in Bunker et al. 2004) on the publicly released HUDF data (released in March 2004), this imaging was unavailable when the GLARE slitmask was designed (November 2003). As a result, we utilized a similar i'-drop candidate selection made from the catalog of 100 reddest sources in the first third of HUDF imaging. This catalog was released by the HUDF team‡ at the Space Telescope Science Institute in November 2003 in order to facilitate such ground-based spectroscopy. Each object in the catalog was assigned to a color category ($0.9 < i'-z' < 1.5$, $1.5 < i'-z' < 2.0$ and $i'-z' > 2.0$). We prioritized candidates with $i'-z' > 1.5$, although bluer candidates were targeted where slitlet geometry permitted.

In total, 20 candidates for spectroscopy were drawn from the HUDF catalog in the center of the mask, and a further 10 sources from the GOODS v1.0 selection were placed in the outlying regions. Remaining space on the slit mask was used for a blank-sky survey. Of the objects for which spectra are presented here, GLARE 1042 was identified as a candidate in both catalogs, while GLARE 3001 and GLARE 3011 lie outside the HUDF field and were selected from the GOODS v1.0 data.

5. Spectroscopically confirmed GLARE galaxies

While the GLARE project is ongoing, an initial analysis of the first 7.5 hours of observations (carried out in November 2003) revealed three bright line emitters with single

---

† available from `ftp://archive.stsci.edu/pub/hlsp/goods/v1/`
‡ http://www.stsci.edu/hst/udf/

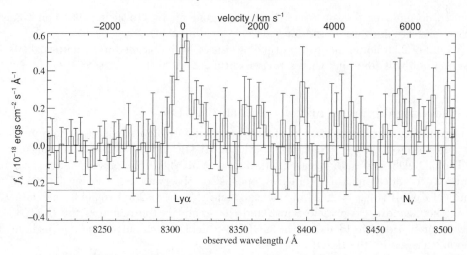

FIGURE 4. The unsmoothed (top) and gaussian-smoothed (middle) 2D spectrum of GLARE 1042. A 1D spectrum extracted from both channels and smoothed over three pixels is shown below for reference. The 2D spectra show the distinctive positive/negative emission-line profile that arises due to the Nod & Shuffle technique.

ID	z	Peak Å	Flux ergs cm^{-2} s^{-1}	FWHM Å	z'_{AB}	$(i'-z')_{AB}$	R_h (z')	EW$_{rest}$ Å
1042	5.83	8309.1	0.97×10^{-17}	16.3	25.48 ± 0.03	1.48 ± 0.09	$0''.09$	20
3001	5.79	8252.7	0.79×10^{-17}	20.5	26.37 ± 0.06	1.66 ± 0.20	$0''.14$	30
3011	5.94	8434.0	1.0×10^{-17}	24.5	27.15 ± 0.12	$> 1.68 \ (3\sigma)$	$0''.13$	100

TABLE 1. The spectroscopic and photometric properties of the three emission line candidates. Photometry was measured on the GOODS v1.0 images. Wavelengths are measured in air. Error on the fluxes due to slit losses and photon noise is ≈20%. Flux is taken between zero power points. The gaussian FWHM of the line is shown in the fifth column (spectral resolution ≈16.5 Å). All objects are undetected at 3σ in v ($v > 29.4$) and also resolved in the z' band (stellar half-light radius $R_h = 0''.05$). Equivalent widths are calculated from broadband magnitudes.

isolated emission lines in the spectroscopically surveyed sample. More recent observations identify further line-emitter candidates in the data, but the analysis of these is as yet incomplete. In this section we focus on the brightest emission-line galaxies, reported in Stanway et al. (2004b).

An advantage of the N&S data-reduction technique is that lines appear in a positive-negative dipole pattern in the two-dimensional images, making them easy to distinguish from residual CCD defects and other non-astrophysical effects. Figure 4 illustrates this: Using the example of GLARE 1042, the dipole emission pattern of this line is easy to distinguish from the sky lines and other noise which surrounds the line. All 2D spectra obtained by the GLARE project were inspected for evidence of lines and, as mentioned above, three lines were found at high signal to noise in the early data analysis.

One-dimensional spectra of these galaxies are shown in Figure 5. The emission line is clearly visible in each galaxy (with signal to noise of between 12 and 15 in each case).

GLARE 1042 (J2000 $3^h32^m40^s.0$–$27°48'15''.0$), is an i-drop already confirmed as a $z = 5.83$ Lyman-α emitter (Stanway et al. 2004a; Dickinson et al. 2004). GLARE 3001 ($3^h32^m46^s.0$–$27°49'29''.7$) and GLARE 3011 ($3^h32^m43^s.2$–$27°45'17''.6$) are new. Their

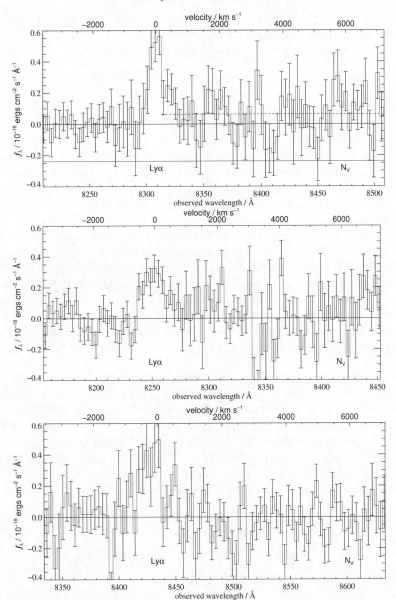

FIGURE 5. Unbinned 7.5 hour extracted GMOS-S spectra of GLARE 1042, GLARE 3001 and GLARE 3011 around the Lyman-α candidate line emission. The continuum level is indicated on 1042.

properties are given in Table 1. All three emission lines are almost unresolved given our dispersion, although #1042 is known to be highly asymmetric from published spectra and shows evidence for this in the GMOS spectra.

The most plausible line identifications given the red i–z selection are Lyman-α at $z = 5.83$, $z = 5.79$ and $z = 5.94$ for #1042, #3001 and #3011 respectively. Hβ $\lambda\,4861.3$ Å or [O III] $\lambda\lambda\,5006.8, 4958.9$ Å at $z \approx 0.7$ or Hα $\lambda\,6562.8$ Å at $z \approx 0.25$ are ruled out be the absence of nearby lines and because galaxies at these redshifts do not have strong i–z continuum breaks. An unresolved [O II] $\lambda\lambda\,3726.1, 3728.9$ Å doublet at redshifts $z \approx 1.2$ is a possibility for #3001 and #3011 (#1042 is already known to be strongly asymmetric ruling this out, Stanway et al. 2003). The best evidence against this is the red i–z color. The reddest possible color from an early-type SED at $z = 1.2$ is i'–$z' = 1.2$ (see Figure 2) and our two new objects are redder than this at 95% confidence.

6. Conclusions

In the preceding sections we have described and discussed the application of i'-drop color selection to identify $z \approx 6$ galaxies. We have presented photometry and spectra for three objects with extreme i'–z' colors and line emission which may be Lyman-α at $z \approx 5.8$, observed as part of the GLARE project. Our main conclusions can be summarized as follows:

(i) i'–z' color selection is a successful and efficient technique by which to identify objects lying at $z \approx 6$.

(ii) The recent, publicly released and extremely deep surveys carried out using the ACS instrument on *HST* are valuable resources, ideal for i'-drop color selection.

(iii) The GLARE project has detected three very high-redshift objects ($z = 5.83$, 5.79, 5.94) in the first 7.5 hours of integration time on Gemini/GMOS-S. To the best of our knowledge, the two new Lyman-α emitters are fainter in z' than any previous Lyman break selected objects with a spectroscopic redshift.

(iv) Spectroscopic confirmation of line emission at the faint end of the galaxy luminosity function at $z = 6$ is now within the reach of 8m telescopes.

This research is based on observations made with the NASA/ESA *Hubble Space Telescope*, obtained at the STScI, which is operated by AURA, under NASA contract NAS 5-26555. These observations are associated with programs #DD9978, #9425, and 9583. Also based on observations obtained at the Gemini Observatory, which is operated by AURA under a cooperative agreement with the NSF on behalf of the Gemini partnership: NSF (U.S.), PPARC (U.K.), NRC (Canada), CONICYT (Chile), ARC (Australia), CNPq (Brazil) and CONICET (Argentina). K. G. & S. S. acknowledge generous funding from the David and Lucille Packard Foundation. E. R. S. thanks the UK PPARC for funding.

REFERENCES

ABRAHAM, R. G., GLAZEBROOK, K., MCCARTHY, P. J., CRAMPTON, D., MUROWINSKI, R., JØRGENSEN, I., ROTH, K., HOOK, I. M., SAVAGLIO, S., CHEN, H., MARZKE, R. O., & CARLBERG, R. G. 2004 *AJ* **127**, 2455 (GDDS1).

BECKWITH, S. V. W., CALDWELL, J., CLAMPIN, M., DE MARCHI, G., DICKINSON, M., FERGUSON, H., FRUCHTER, A., HOOK, R., JOGEE, S., KOEKEMOER, A., LUCAS, R., MALHOTRA, S., GIAVALISCO, M., PANAGIA, N., RHOADS, J., SODERBLOM, D. R., ROYLE, T., STIAVELLI, M., SOMERVILLE, R., CASERTANO, S., MARGON, B., & BLADES, J. C. 2003 The Hubble Ultra Deep Field. *American Astronomical Society Meeting* **202**, #1705.

BERTIN, E. & ARNOUTS, S. 1996 *A&AS* **117**, 393.

BOUWENS, R. J., ILLINGWORTH, G. D., ROSATI, P., LIDMAN, C., BROADHURST, T., FRANX, M., FORD, H. C., MAGEE, D., BENÍTEZ, N., BLAKESLEE, J. P., MEURER, G. R., CLAMPIN, M., HARTIG, G. F., ARDILA, D. R., BARTKO, F., BROWN, R. A., BURROWS, C. J., CHENG, E. S., CROSS, N. J. G., FELDMAN, P. D., GOLIMOWSKI, D. A., GRONWALL, C., INFANTE, L., KIMBLE, R. A., KRIST, J. E., LESSER, M. P., MARTEL, A. R., MENANTEAU, F., MILEY, G. K., POSTMAN, M., SIRIANNI, M., SPARKS, W. B., TRAN, H. D., TSVETANOV, Z. I., WHITE, R. L., & ZHENG, W. 2003 *ApJ* **595**, 589.

BOUWENS, R. J., ILLINGWORTH, G. D., THOMPSON, R. I., BLAKESLEE, J. P., DICKINSON, M. E., BROADHURST, T. J., EISENSTEIN, D. J., FAN, X., FRANX, M., MEURER, G., & VAN DOKKUM, P. 2004 *ApJ* **606**, L25.

BREMER, M. N., LEHNERT, M. D., WADDINGTON, I., HARDCASTLE, M. J., BOYCE, P. J., & PHILLIPPS, S. 2004 *MNRAS* **347**, L7.

BUNKER, A. J., STANWAY, E. R., ELLIS, R. S., & MCMAHON, R. G. 2004 *MNRAS* **355**, 374.

BUNKER, A. J., STANWAY, E. R., ELLIS, R. S., MCMAHON, R. G., & MCCARTHY, P. J. 2003 *MNRAS* **342**, L47.

CIMATTI, A., DADDI, E., MIGNOLI, M., POZZETTI, L., RENZINI, A., ZAMORANI, G., BROAD-HURST, T., FONTANA, A., SARACCO, P., POLI, F., CRISTIANI, S., D'ODORICO, S., GIALLONGO, E., GILMOZZI, R., & MENCI, N. 2002 *A&A* **381**, L68.

DICKINSON, M., STERN, D., GIAVALISCO, M., FERGUSON, H. C., TSVETANOV, Z., CHORNOCK, R., CRISTIANI, S., DAWSON, S., DEY, A., FILIPPENKO, A. V., MOUSTAKAS, L. A., NONINO, M., PAPOVICH, C., RAVINDRANATH, S., RIESS, A., ROSATI, P., SPINRAD, H., & VANZELLA, E. 2004 *ApJ* **600**, L99.

ELLIS, R., SANTOS, M. R., KNEIB, J., & KUIJKEN, K. 2001 *ApJ* **560**, L119.

FUJITA, S. S., AJIKI, M., SHIOYA, Y., NAGAO, T., MURAYAMA, T., TANIGUCHI, Y., OKAMURA, S., OUCHI, M., SHIMASAKU, K., DOI, M., FURUSAWA, H., HAMABE, M., KIMURA, M., KOMIYAMA, Y., MIYAZAKI, M., MIYAZAKI, S., NAKATA, F., SEKIGUCHI, M., YAGI, M., YASUDA, N., MATSUDA, Y., TAMURA, H., HAYASHINO, T., KODAIRA, K., KAROJI, H., YAMADA, T., OHTA, K., & UMEMURA, M. 2003 *AJ* **125**, 13.

GIAVALISCO, M., FERGUSON, H. C., KOEKEMOER, A. M., DICKINSON, M., ALEXANDER, D. M., BAUER, F. E., BERGERON, J., BIAGETTI, C., BRANDT, W. N., CASERTANO, S., CESARSKY, C., CHATZICHRISTOU, E., CONSELICE, C., CRISTIANI, S., DA COSTA, L., DAHLEN, T., DE MELLO, D., EISENHARDT, P., ERBEN, T., FALL, S. M., FASSNACHT, C., FOSBURY, R., FRUCHTER, A., GARDNER, J. P., GROGIN, N., HOOK, R. N., HORN-SCHEMEIER, A. E., IDZI, R., JOGEE, S., KRETCHMER, C., LAIDLER, V., LEE, K. S., LIVIO, M., LUCAS, R., MADAU, P., MOBASHER, B., MOUSTAKAS, L. A., NONINO, M., PADOVANI, P., PAPOVICH, C., PARK, Y., RAVINDRANATH, S., RENZINI, A., RICHARDSON, M., RIESS, A., ROSATI, P., SCHIRMER, M., SCHREIER, E., SOMERVILLE, R. S., SPINRAD, H., STERN, D., STIAVELLI, M., STROLGER, L., URRY, C. M., VANDAME, B., WILLIAMS, R., & WOLF, C. 2004 *ApJ* **600**, L93.

GLAZEBROOK, K. & BLAND-HAWTHORN, J. 2001 *PASP* **113**, 197.

GUHATHAKURTA, P., TYSON, J. A., & MAJEWSKI, S. R. 1990 *ApJ* **357**, L9.

HAWLEY, S. L., COVEY, K. R., KNAPP, G. R., GOLIMOWSKI, D. A., FAN, X., ANDERSON, S. F., GUNN, J. E., HARRIS, H. C., IVEZIĆ, Ž., LONG, G. M., LUPTON, R. H., MCGEHEE, P. M., NARAYANAN, V., PENG, E., SCHLEGEL, D., SCHNEIDER, D. P., SPAHN, E. Y., STRAUSS, M. A., SZKODY, P., TSVETANOV, Z., WALKOWICZ, L. M., BRINKMANN, J., HARVANEK, M., HENNESSY, G. S., KLEINMAN, S. J., KRZESINSKI, J., LONG, D., NEILSEN, E. H., NEWMAN, P. R., NITTA, A., SNEDDEN, S. A., & YORK, D. G. 2002 *AJ* **123**, 3409.

HOOK, I., ALLINGTON-SMITH, J. R., BEARD, S. M., CRAMPTON, D., DAVIES, R. L., DICKSON, C. G., EBBERS, A. W., FLETCHER, J. M., JORGENSEN, I., JEAN, I., JUNEAU, S., MUROWINSKI, R. G., NOLAN, R., LAIDLAW, K., LECKIE, B., MARSHALL, G. E., PURKINS, T., RICHARDSON, I. M., ROBERTS, S. C., SIMONS, D. A., SMITH, M. J., STILBURN, J. R., SZETO, K., TIERNEY, C., WOLFF, R. J., & WOOFF, R. 2003. In *Instrument Design and Performance for Optical/Infrared Ground-based Telescopes* (eds. M. Iye & A. F. M. Moorwood). Proceedings of the SPIE, Volume 4841, p. 1645. SPIE.

HU, E. & MCMAHON, R. G. 1996 *Nature* **382**, 281.

HU, E. M., COWIE, L. L., MCMAHON, R. G., CAPAK, P., IWAMURO, F., KNEIB, J.-P., MAIHARA, T., & MOTOHARA, K. 2002 *ApJ* **568**, L75.

HU, E. M., MCMAHON, R. G., & COWIE, L. L. 1999 *ApJ* **522**, L9.

KINNEY, A. L., CALZETTI, D., BOHLIN, R. C., MCQUADE, K., STORCHI-BERGMANN, T., & SCHMITT, H. R. 1996 *ApJ* **467**, 38.

KODAIRA, K., TANIGUCHI, Y., KASHIKAWA, N., KAIFU, N., ANDO, H. KAROJI, H., ET AL. 2003 *PASJ* **55**, L17.

MADAU, P. 1995 *ApJ* **441**, 18.

OKE, J. B. & GUNN, J. E. 1983 *ApJ* **266**, 713.

STANWAY, E. R., BUNKER, A. J., & MCMAHON, R. G. 2003 *MNRAS* **342**, 439.

STANWAY, E. R., BUNKER, A. J., MCMAHON, R. G., ELLIS, R. S., TREU, T., & MCCARTHY, P. J. 2004a *ApJ* **607**, 704.

STANWAY, E. R., GLAZEBROOK, K., BUNKER, A. J., ABRAHAM, R. G., HOOK, I., RHOADS, J., MCCARTHY, P. J., BOYLE, B., COLLESS, M., CRAMPTON, D., COUCH, W., JØRGENSEN, I., MALHOTRA, S., MUROWINSKI, R., ROTH, K., SAVAGLIO, S., & TSVETANOV, Z. 2004b *ApJ* **604**, L13.

STEIDEL, C. C., ADELBERGER, K. L., GIAVALISCO, M., DICKINSON, M., & PETTINI, M. 1999 *ApJ* **519**, 1.

STEIDEL, C. C., GIAVALISCO, M., PETTINI, M., DICKINSON, M., & ADELBERGER, K. L. 1996 *ApJ* **462**, L17.

STEIDEL, C. C., PETTINI, M., & HAMILTON, D. 1995 *AJ* **110**, 2519.

YAN, H., WINDHORST, R. A., & COHEN, S. H. 2003 *ApJ* **585**, L93.

The Hubble Ultra Deep Field with NICMOS

By RODGER I. THOMPSON,[1]
RYCHARD J. BOUWENS,[2]
AND GARTH ILLINGWORTH[2]

[1]Steward Observatory, University of Arizona, Tucson, Arizona 85721, USA

[2]Astronomy Department, University of California, Santa Cruz, California 95064, USA

The Advanced Camera for Surveys (ACS) observations of the Hubble Ultra Deep Field (HUDF) provide the highest sensitivity optical observations of galaxies and stars ever achieved. The Near Infrared Camera and Multi-Object Spectrometer (NICMOS) observations in the central portion of the field extend the wavelength coverage by a factor of two to beyond 1.6 microns. Although not as sensitive as the ACS images due to a much smaller field and less observing time, the NICMOS observations extend the redshift range of the HUDF to redshifts as high as 13. Even though the observations are sensitive to redshift 13 objects, we confine our investigation to objects between redshifts of 7 and 9 where there is flux in both the F110W and F160W bands. Candidate sources in this redshift region are identified by requiring a non-detection in the ACS bands and a detection in both the F110W and F160W bands. All of the candidates have an almost flat or blue color in the F110W and F160W bands. The extremely high sensitivity of the ACS observations make this a very stringent criterion. We identify five candidates for objects in this redshift range and discuss tests of the reality of these sources. Although the sources are selected to have flux in both NICMOS bands and none in the ACS bands, we also present the results of a photometric redshift analysis of the candidates. This shows them to be very blue galaxies with redshifts between 7.3 and 7.9. One source yielded an anomalous redshift and spectral type due to flux from an adjacent galaxy falling in the photometric aperture.

1. Introduction

The spectacular success of the northern and southern Hubble Deep Fields (NHDF and SHDF) provided the motivation for the HUDF observations. The HUDF, located in the Chandra Deep Field South (CDFS), is also a prime target for mid-infrared observations with the *Spitzer Infrared Observatory*. The optical observations, done with Director's Discretionary Time orbits of the *Hubble Space Telescope* (*HST*), utilized the wide field and high sensitivity capabilities of the ACS, with observations in four broad wavelength bands centered on 0.435, 0.606, 0.775 and 0.850 microns. An additional set of near-infrared observations with the NICMOS were awarded in an HST Treasury Program allocation. These observations are in two bands centered on 1.1 and 1.6 microns. Only the central portion of the ACS field was covered in the NICMOS program due to the much smaller field of view of the NICMOS camera and the significantly smaller number of orbits allocated to the Treasury Program. Although the NICMOS observations have less sensitivity than the ACS observations, they extend the redshift coverage of the observations to redshifts on the order of 13 as is shown in Figure 1.

One of the interesting aspects of Figure 1 is the very shallow slope of the flux versus redshift curve for the NICMOS F160W band between redshifts of 4 and 11. This is due to cosmological surface-brightness dimming being offset by the transition of the high intensity blue and UV light into the band. There is also an interesting region between redshifts of about 7 and 9 where there is flux in both NICMOS bands, but no flux in any of the ACS bands. This paper concentrates on possible candidates in this redshift region. We do not consider objects at possibly even higher redshifts that have flux in only the F160W band. It is very difficult to discriminate between extremely red faint galaxies

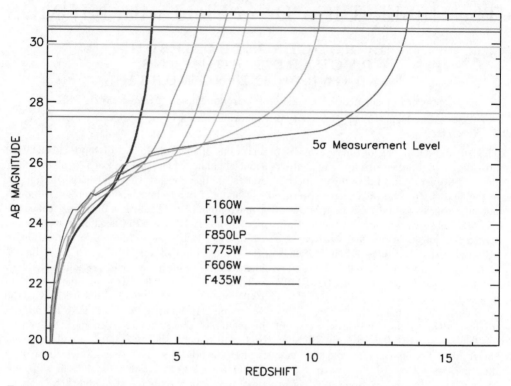

FIGURE 1. The predicted AB magnitudes in the 6 HUDF bands for a L* late-type galaxy are plotted as a function of redshift. The approximate 5σ sensitivity limits in the bands are also plotted on the diagram. Note the very shallow slope of the flux versus redshift curve for the F160W NICMOS band between redshifts of 4 and 11.

and very high redshift galaxies with flux in only one band. The eventual availability of *Spitzer* observations in this region may enable such a discrimination.

2. Observations

The NICMOS observations in the UDF are centered on the position 3hr 32min 39.0sec, −27deg 47min 29.1sec (J2000) at approximately the center of the ACS UDF observations, but near the eastern edge of the Great Observatories Origins Deep Survey (GOODS) image. The images lie in a 3 × 3 grid with centers separated by 45 arcseconds. The grid centers are dithered by five arcseconds in a 4 × 4 square pattern tilted at 22.5 degrees to the x axis of the images to reduce the effect of intra-pixel sensitivity variations.

The NICMOS UDF program consists of 144 orbits broken into two epochs, August 30, 2003 to September 14, 2003 and November 2, 2003 to November 27, 2003. The scheduled start of observations in mid-August 2003 was delayed by a safing of the NICMOS Cooling System (NCS) just prior to the beginning of observations. The epochs are separated partially to enable the detection of type Ia supernova candidates and to provide orbits free of SAA crossings for high sensitivity. The images in the two epochs are rotated by 90 degrees to accommodate the roll restrictions of the spacecraft. The first epoch had the y axis of the NICMOS camera 3 oriented 40.925 degrees east of north, with the second epoch at 130.925 degrees east of north. These orientations are the same as the two ACS orientations in an effort to align the NICMOS and ACS images without the need for a rotation. The drizzled ACS images, however, were produced with the standard north up and east left orientation. The individual NICMOS and ACS images, on the other hand, should have the same orientations.

Each orbit contains a F110W and F160W band SPARS64, NSAMP = 24, 1344 second integration. The F110W integration always preceded the F160W integration. Three orbits comprise a single visit with the location of the image specified by a POSTARG from a reference position. It is possible that the F160W images from the first orbit of a visit may have received a small amount of earthshine in the last readout of the NSAMP = 24 readout sequence. This is due to the guide star acquisition time on the first orbit of the visit, which takes longer than the subsequent guide star reacquisition in the last two orbits of the visit. No correction has been made in the images for this effect. A single three by three grid is completed in three visits. Each visit produces a three-image strip stepped in the detector x direction. The visits were ordered to finish each 3×3 grid before starting another, and always in the same order. This was done to maximize the time coverage for a supernova event, should one occur. After the initial delay due to the NCS safing, all of the visits occurred in their expected order.

Subsequent analysis of the images indicated that the second-half images were not an exact repeat of the first-half images. In the second half, three of the intended three-orbit visits in the center of the grid were actually taken in the outer part of the grid. This reduced the significance of the central part of the image slightly. The weight images produced with the data images accurately reflect this redistribution of the images.

3. Data reduction

The data reduction of the NICMOS images was carried out independently with analysis software developed by RIT and the NICMOS Instrument Definition Team. A detailed account of the data-reduction techniques will be published in a separate document (Thompson et al. 2004). The data-reduction techniques are similar to the techniques utilized in previous NICMOS observations of the NHDF (Thompson et al. 1999), except for some modifications due to the warmer detector temperature with the NCS. The primary modification is a technique to remove the varying dark-current contribution from warm pixels. Warm pixels are pixels that have increased dark current, but not at a level that warrants declaring them bad pixels. At the higher detector temperature there are more warm pixels than during cryogenic operation. The technique is described fully in Thompson et al. (2005).

The mosaic images were assembled utilizing the DRIZZLE software of Fruchter & Hook (2002). Offsets for the DRIZZLE procedure were obtained by matching NICMOS F110W images, drizzled to 0.09 arc second pixels, to the ACS F850LP image that was rebinned to 0.09 arcsecond resolution. The large wavelength overlap between the two images minimized any erroneous offsets due to differences in morphology at different wavelengths. The F160W images were always taken immediately after the F110W images, in the same orbit with no telescope motion. For this reason, we used the same offsets to construct the F160W drizzled mosaic.

4. Source extraction

The source extractions utilize the SExtractor (SE) photometric extraction program of Bertin & Arnouts (1996). Since we are searching for faint, high-redshift sources that are visible in both NICMOS bands, we used the sum of the F110W and F160W images as the detection image with aggressive extraction parameters. Source extraction was performed independently by RIT and RJB. In the RIT source extraction the detection threshold

ID	F435W	F606W	F775W	F850LP	F110W	F160W
825−950	≥ 30.5	≥ 30.8	≥ 30.6	≥ 30.9	27.2	26.9
491−880	≥ 30.5	30.3	30.0	≥ 30.9	26.9	26.7
819−310	32.2	29.5	30.4	29.6	27.2	27.2
983−964	≥ 30.5	30.9	≥ 30.6	30.5	27.1	27.1
818−886	31.8	32.5	≥ 30.6	≥ 30.9	27.2	27.2

TABLE 1. AB magnitudes of the candidate sources in the ACS and NICMOS bands in an 11 pixel (0.99 arcsecond) diameter aperture centered on the source. If the total flux in the band is negative, the magnitude is given as greater than or equal to the 5σ magnitude.

was set at 1.6σ, with the requirement that five contiguous pixels must be above that threshold to be considered a source. This yielded 1,481 sources for consideration.

The independent source extraction by RJB utilized a two-band Szalay procedure (Szalay, Connolly, & Szokoly 1999) on the F110W and F160W images. The source detection threshold was set at 0.21σ, but with a requirement that threshold be met by 20 contiguous pixels. The deblending parameter was set at 0.0003 to reduce merging of high redshift objects with foreground galaxies. The initial 1,704-object catalog was then subjected to the criterion that the six-pixel aperture diameter detections be 3σ in both F110W and F160W. Both extraction procedures detected all of the objects discussed here.

5. High redshift candidate selection

The initial selection criteria required a source to be fainter than an AB magnitude of 29.0 in all ACS bands and to be at least one magnitude brighter in the NICMOS F110W and F160W band. This group of candidates was further culled by visual inspection of the images. Any object with observable flux in any ACS band was eliminated from the group. Since the 5σ flux levels for the ACS bands range between 29.8 and 30.7 for a six-pixel aperture, this procedure produced a candidate list of eight objects with a very large difference in flux between the NICMOS bands and the ACS bands. Three of the candidates were eliminated by the tests described below. The photometric properties of the five remaining candidates are listed in Table 1.

The identification numbers in Table 1 refer to the x and y position in the NICMOS HUDF image. The colors of the candidates place them at redshifts of 7.3 or greater. All of the candidates have F110W AB fluxes at least 2.4 magnitudes greater than the F850LP fluxes, and flat or very slightly red F110W to F160W color. This is the signature of a high-redshift object. All of the objects are extended, and are therefore galaxies—rather than stars with unusual spectra—such as brown dwarfs. The only remaining possibility among objects with known properties is an early galaxy at low redshift with an extreme 4000 Å break. This possibility is discussed in Section 7.1.

Figure 2 shows the five Z-dropout high-redshift candidates. The stretch is linear and the scale for each image is the same in electrons per second. The higher signal to noise of the ACS images is obvious in this image. The right-most image in each row is the sum of the F110W and F160W image. For conciseness F435W, F606W, F775W, F814LP, F110W and F160W are labeled B, V, I, Z, J, and H, respectively, even though they do not exactly match the standard photometric bands. This is particularly true of the F110W band, which is much bluer than the photometric J band.

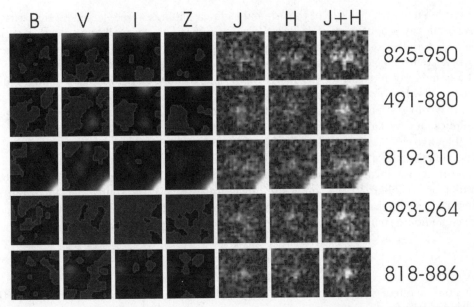

FIGURE 2. Postage-stamp images of the five Z-dropout sources. The images are 25 0.09 arcsecond pixels on a side. The stretch is linear and the scale is identical for all images in electrons per second. The *HST* filters have been labeled by their roughly equivalent standard photometric band names. The source is clearly evident in the $J + H$ image and is at the same location in all subsequent images of the source.

6. Source reality tests

The very high signal to noise of the ACS data makes the possibility that these objects have significant optical flux very unlikely. The reality of the sources relies primarily on the reliability of the NICMOS near-infrared detections. We performed several tests to determine the reality of the sources, which are described in the following sections. Again independent tests were performed by RIT and RJB on the same input data.

6.1. *Negative and subtracted images*

The susceptibility of the detection algorithm to noise can be estimated by using images that are good representations of the noise. The three noise images in this test are the negative image, the first half subtracted from the second half, and its negative. Both RIT and RJB performed their original extraction procedures on the negative images. RIT found one detection in the negative image that only had detectable flux in the F110W image, and therefore would not have made the candidate list. RJB performed extractions on all three images and found 1, 2, and 0 sources in the negative, first half minus second half, and its negative. This suggests that one, or possibly two, of our five sources may be simple noise sources. See also Section 6.5 for a further discussion of noise sources. The next three subsections describe further tests performed by RIT.

6.2. *Inspection of individual images*

Real sources should not have their flux contributed by a single image. In the 3×3 grid used for the observations, about 16 images will include an average source position in both the F110W and F160W bands. Sources that lie near the intersection of images on the grid points may appear in more than 16 images. All individual images that contained a candidate source were examined by eye. None of the final five sources were dominated by an individual image. Three of the original eight candidates were rejected by this criterion. A cosmic ray can hit part of *HST* or NICMOS and produce a small shower, which appears as an extended source on the detector. If this occurs before an integration, there can be a persistence image that does not trigger the cosmic ray removal software—which looks for cosmic ray hits during the integration. The cosmic-ray-persistence residual current decays slowly over a period of an hour. Since the order of observation for all orbits is an F110W integration followed by an F160W integration, the cosmic-ray-persistence signal mimics an extremely blue, high-redshift object.

The five candidate sources were weakly visible in about one third and one half of the individual images, consistent with comparable signal images with optical counterparts. The next test involved removing the two or three individual images where the source was most visible from the composite image. In each case the source persisted, but at a lower signal level. The net result of the inspection of the individual images is that the five candidate sources exhibited the characteristics of real sources.

6.3. *Median images*

A median image is a further test against sources induced by either cosmic-ray persistence or noise anomalies in a single or small number of images. It should be noted here that, although the 3×3 grid pattern was repeated eight times in each of the two orientations, the starting location of each repetition was altered slightly, so that in all images sources appeared at different positions on the detector. This prevented a small group of pixels with elevated dark current from mimicking a source. The only exception to this strategy is that in a given grid point, the F160W image was taken immediately after the F110W image, without any telescope motion to insure accurate registration of the two images. These were taken in one orbit, and the telescope was moved to the next grid point on the next orbit.

All five sources persisted in both the F110W and F160W median images, as was expected from the results in Section 6.2. This procedure was repeated for both the first- and second-half images; again, all five sources persisted in each image.

6.4. *Comparison with an independently processed image*

As part of the continued monitoring of NICMOS performance, a team at STScI performed an independent production of the NICMOS UDF images. The team graciously provided the images for comparison with our images, noting that three of the original eight sources were not present in their images (Bergeron, Stiavelli, & Mobasher 2004). One of these was a source that was excluded after the inspection of the individual images described in Section 6.2. Source 491−880 has weak fluctuations at its position in the STScI image, while source 818−886 has absolutely no corresponding signal in the STScI image. The primary difference between the images is the orientation. The images used here were drizzled onto the original orientation of the first-half images, and with a rotation of 90 degrees for the second half. The STScI images were drizzled to the standard astronomical orientation, with north up. The difference in orientation slightly alters the distribution of input and output pixels. Also, the exact offsets provided to the drizzle procedure were independently determined, and are most likely different by small fractions of a pixel. However, since the sources span many pixels, the small differences in flux distribution should not produce such different output. At this point, the difference between the two images is not understood. The other three sources appear with the same morphology in both images. In particular, source 983−964 appears with even higher contrast in the STScI image, but has the same integrated flux in both images.

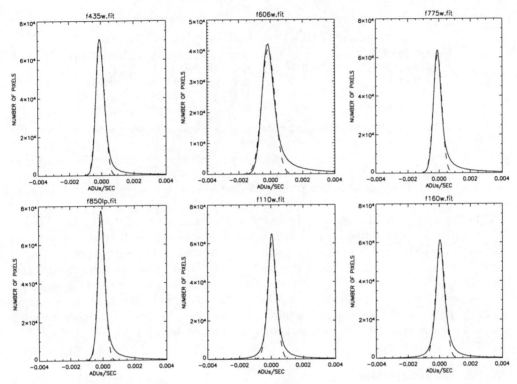

FIGURE 3. The histograms of all of the pixel flux values in adus/sec for the UDF bands. Note that one ADU/sec equals one electron/sec in the ACS bands, but 6.5 electrons/sec for the NICMOS bands. The solid line is the histogram and the dashed line is the Gaussian fit.

6.5. *Noise statistics*

The noise statistics in the NICMOS and ACS images were determined by fitting a histogram of the flux in every pixel with a Gaussian. Since the majority of pixels in the image do not contain sources, this is an excellent determination of the background noise—which is dominated by the zodiacal sky noise. The plots are shown in Figure 3 for all of the bands.

The positive extension of the histogram relative to the Gaussian fit are the pixels that contain sources. The much higher signal to noise of the ACS images is demonstrated by the larger number of pixels above the noise. It is clear that the Gaussian fit gives a good representation of the noise distribution for the ACS images. The NICMOS images, however, have a negative extension and most likely a positive extension hidden by source pixels above the Gaussian fit. This is probably due to the correlation noise introduced by the drizzle process. The formulas given by Fruchter & Hook (2002) equations 9 and 10 indicate a ratio of the true sigma to the observed sigma of 1.8 for the drizzle parameters used in creating the image. Since this error does not have a Poisson distribution, it is not a simple widening of the Gaussian distribution.

If we use a formal error of 1.8 times the Gaussian widths in Figure 3, we find that the five sources have total fluxes that are 3.6 to 2.4σ above the noise in both NICMOS bands over an average of 20 pixels. If we divide the image up into 20 pixel sections there are 3×10^4 sections. For a 3.5σ signal there is a 5×10^{-4} chance that the signal is noise, which says that there should be approximately 15 sources at that level both positive and negative. Since we only consider positive signals that reduces the expected number to

ID	Redshift	$E(B-V)$	Template	RA	DEC
825−950	7.9	0.0	6.0	3:32:39.534	−27:47:17.46
491−880	7.6	0.0	6.0	3:32:40.941	−27:47:41.89
819−310	2.2	0.08	2.4	3:32:36.731	−27:48:01.43
983−964	7.3	0.0	6.0	3:32:38.798	−27:47:07.18
818−886	7.7	0.0	6.0	3:32:39.288	−27:47:22.17

TABLE 2. The redshift, extinction, and SED template returned by the photometric redshift analysis.

seven or eight. This would indicate that the five sources are consistent with expected noise fluctuations.

The factor of 1.8 used above may be too conservative, particularly given the tests with the negative and subtracted images. Inspection of the chi-squared distribution generated in the Szalay source extraction process indicates that the σ value, without multiplication by 1.8, provides a better fit than any higher value. If we use the uncorrected value, the total fluxes are 4.3 to 6.5σ above the noise. The number of expected 5σ sources in the image is zero for these signal to noise values. All of the tests discussed above indicate that the true σ of the observations is closer to the lower value than the higher, therefore we will assume that the noise contamination in our sample is 1 ± 1 source.

7. Source properties

So far, the evidence presented for the sources being high-redshift galaxies is the extended nature of the images and the absence of optical flux, both consistent with the Lyman break at high redshift. Additional source information can be obtained through a chi-squared analysis of the fluxes against Spectral Energy Distribution (SED) templates. The sources were analyzed with the same procedure used on the NHDF sources in Thompson (2003). The procedure is fully documented in that publication and will not be duplicated here. The only change in procedure is an extension of the allowed redshifts to ten, rather than the eight used in the previous work. The procedure returns three values: the redshift, an extinction expressed in $E(B-V)$, and a template SED between one and six in increments of 0.1, with one being an early elliptical SED and six a hot SED of a 50 million-year-old galaxy. Table 2 lists the properties returned by the procedure.

The parameters of the galaxies show several interesting aspects. First, the high-redshift nature of the galaxies is confirmed in all but one source. Second is the very blue nature of the sources, where the best fit is the bluest SED with no extinction. These galaxies would actually be better fit by an even younger galaxy than our bluest 50 million-year-old galaxy SED. The photometric redshifts lie in the region between 7.3 and 7.9, which is expected given Figure 1. The F110W flux is starting to fall fast by redshift eight and our requirement for detection in both bands would be hard to satisfy at higher redshifts.

Perhaps the most surprising set of parameters is the set for source 819−310, which is for a low-redshift early galaxy with modest extinction. The redshift of 2.2 puts the 4000 Å break at 1.28 micron, roughly in the middle of the F110W filter. The possibility of low redshift, early-galaxy interlopers is discussed in the following section.

7.1. *Early galaxies*

The primary reason for the early-galaxy selection is the indication of flux in the F850LP and F606W bands as shown in Table 1. Inspection of the images shows that this is flux from an adjacent faint galaxy that falls in the photometric aperture. Since on the average one third of the galaxies overlap in the NICMOS images, this is not surprising. The overlapping flux gives 819−310 the smallest flux ratio between the F850LP and the F110W bands. It does point out, however, that a faint galaxy with a large 4000 Å break can be mistaken for a high-redshift galaxy if the flux shortward of 4000 Å is below the detection level. In fact, the chi-squared-versus-redshift plot for most high-redshift galaxies shows two minima—one for the Lyman break and a second, smaller minimum for a redshift that puts the 4000 Å break at the wavelength of the Lyman break. This is where the much higher sensitivity of the ACS images provides a great advantage. For true high-redshift galaxies, the limit set on the optical flux relative to the infrared flux is much lower than can be achieved by the 4000 Å break. Only when there is contamination by overlapping galaxies will a mistake be made.

8. Conclusions

We have identified five sources which are properly characterized as Z-band dropouts and are legitimate candidates for redshift 7.3 to 7.9 objects. The reality of the candidates was discussed with the conclusion that approximately one of the sources might be due to noise. The sources were selected strictly as sources with detectable infrared flux, but no optical flux. The high sensitivity of the ACS images makes this a very stringent limit. Analysis with a photometric redshift program established probable redshifts, extinctions and SEDs for the sources. All but one source had the best fit with the bluest, zero-extinction template, and would have been better fit by an even bluer template. The one exception is a galaxy whose photometry was contaminated by another faint galaxy that overlapped the photometric aperture.

REFERENCES

BERGERON, M., STIAVELLI, M., & MOBASHER, B. 2004; *private communication to Rodger Thompson*.

BERTIN, E. & ARNOUTS, S. 1996 *A&A* **311**, 356.

FRUCHTER, A. S. & HOOK, R. N. 2002 *PASP* **114**, 144.

SZALAY, A. S., CONNOLLY, A. J., & SZOKOLY, G. P. 1999 *AJ* **117**, 68.

THOMPSON, R. I., ET AL. 1999 *AJ* **117**, 17.

THOMPSON, R. I. 2003 *ApJ* **596**, 748.

THOMPSON, R. I., ET AL. 2005 *AJ*, **130**, 1.